Water Policy in Minnesota

Water Policy
in Minnesota
Issues, Incentives, and Action

EDITED BY

K. William Easter and Jim Perry

RFF PRESS
RESOURCES FOR THE FUTURE

New York • London

First published in paperback 2024

First published 2011
by RFF Press
605 Third Avenue, New York, NY 10158

Simultaneously published in the UK and ROW
by RFF Press
4 Park Square, Milton Park, Abingdon, Oxon OX14 4RN

RFF Press and Earthscan are imprints of the Taylor & Francis Group, an informa business

Earthscan publishes in association with the International Institute for Environment and Development

Publisher's Note
The publisher has gone to great lengths to ensure the quality of this reprint but points out that some imperfections in the original copies may be apparent.

Library of Congress Cataloging-in-Publication Data

Water policy in Minnesota: issues, incentives, and action / edited By K. William Easter and Jim Perry.
   p. cm.
   Includes bibliographical references and index.
   ISBN 978-1-61726-086-5 (hardback: alk. paper) 1. Water-supply—Government policy—Minnesota. 2. Water resources development—Government policy—Minnesota. 3. Water quality management—Government policy—Minnesota. I. Easter, K. William. II. Perry, Jim.
   HD1694.M6W38 2011
   333.91009776—dc22

                                                                    2010045394

A catalogue record for this book is available from the British Library

ISBN: 978-1-61726-086-5 (hbk)
ISBN: 978-1-03-292172-3 (pbk)
ISBN: 978-1-315-06473-4 (ebk)

DOI: 10.4324/9781315064734

Copyedited by Joyce Bond
Typeset by OKS Press Services
Cover design by Maggie Powell

At Earthscan we strive to minimize our environmental impacts and carbon footprint through reducing waste, recycling and offsetting our CO_2 emissions, including those created through publication of this book.

About Resources for the Future and RFF Press

Resources for the Future (RFF) improves environmental and natural resource policymaking worldwide through independent social science research of the highest caliber. Founded in 1952, RFF pioneered the application of economics as a tool for developing more effective policy about the use and conservation of natural resources. Its scholars continue to employ social science methods to analyze critical issues concerning pollution control, energy policy, land and water use, hazardous waste, climate change, biodiversity, and the environmental challenges of developing countries.

RFF Press supports the mission of RFF by publishing book-length works that present a broad range of approaches to the study of natural resources and the environment. Its authors and editors include RFF staff, researchers from the larger academic and policy communities, and journalists. Audiences for publications by RFF Press include all of the participants in the policymaking process—scholars, the media, advocacy groups, NGOs, professionals in business and government, and the public. RFF Press is an imprint of **Earthscan**, a global publisher of books and journals about the environment and sustainable development.

Resources for the Future

The RFF Press Water Policy Series

Books in the *RFF Press Water Policy Series* are intended to be accessible to a broad range of scholars, practitioners, policymakers, and general readers. Each book focuses on critical issues in water policy with the mission to draw upon and integrate the best scholarly and professional expertise concerning the physical, ecological, economic, institutional, political, legal, and social dimensions of water use. The interdisciplinary approach of the series, along with an emphasis on real world situations and on problems and challenges that recur globally, are intended to enhance our ability to apply the full body of knowledge that we have about water resources—at local, country, regional, and international levels.

We welcome new contributions to the series. For editorial queries about the *RFF Press Water Policy Series*, please write to *waterpolicy@rff.org*.

Contents

Figures, Tables, and Boxes

FIGURES

TABLES

BOXES

Editors and Contributors

EDITORS

K. William Easter is a professor in the Department of Applied Economics at the University of Minnesota and has worked on the economics of water resources internationally and domestically for 50 years. He has coauthored or coedited 12 books and numerous articles on water resource and environmental topics. He worked in India on a Ford Foundation project addressing a range of irrigation and food production issues and served as head of a U.S. Agency for International Development (USAID) project on water management in India, Thailand, Egypt, and Pakistan. Twice Bill was a resident fellow at the East-West Center in Hawaii, where he completed three books. In 1992 to 1993, he worked at the World Bank and was lead author of its *Water Resources Management Policy Paper*, which led to new research on water markets and transaction costs and his 1998 book on water markets. Since completing the book, he has served as the director of the university's Center for Food and Agricultural Policy, was a member of the World Bank's expert panel reviewing Bangladesh's Water Planning Exercise, chaired the third External Program and Management Review of the International Water Management Institute in Sri Lanka, and completed two edited volumes on water resources.

Jim Perry is a Morse-Alumni Professor at the University of Minnesota. He has been a water resource professional for more than 40 years. He studied fish and invertebrates in Colorado streams, served as a Peace Corps ichthyologist in Costa Rica, and was senior water quality specialist for the state of Idaho and area manager for Centrac Associates in Salt Lake City. He was a fellow of Green College at Oxford, a senior fellow of the American Institute for Indian Studies in Tamil Nadu, and a senior fellow of the U.S. National Academy of Sciences in Lodz, Poland. He also served for six years as deputy director of the USAID Environmental Training Project for Central and Eastern Europe. Jim collaborates with the United Nations Environment Programme (UNEP), recently completing a global eLearning platform for integrated environmental assessment, an eLearning implementation of climate change vulnerability assessment, and a global training program for ecosystem management at the watershed scale. He has published more than 100 peer-reviewed papers in water quality and environmental management. His current research focuses on the impacts of climate change on the World Heritage Sites listed by the United Nations Educational, Scientific and Cultural Organization (UNESCO); he has developed a global model to identify the sites most at risk and is developing strategies for climate change adaptation at these sites. He also serves as one of four leaders of a large project working to understand and develop adaptation strategies in response to the impacts of climate change on trout streams in southeast Minnesota's Karst region.

Bill and Jim have worked together for more than 25 years. Their collaboration began with an interdisciplinary watershed-scale project in southeastern Minnesota and has included a large effort on integrated watershed management in the Minnesota River Basin and watershed analyses in Thailand. Their work in the basin led to the concept of the "Scale Incompatibility Dilemma in Water Resources," which suggests that scholars and scientists increasingly recognize that effective water resource management requires an integrated and relatively coarse-scale view (e.g., watersheds, landscapes, ecoregions). In contrast, water policy around the world has progressively devolved to more fine-scale and local influences, as nations pass authority to states and provinces, which in turn pass responsibility to communities. Jim's and Bill's proposals to address this dilemma served as an early cornerstone for framing this book.

CONTRIBUTORS

E. Calvin Alexander, Jr., is a Morse-Alumni Professor in the Geology and Geophysics Department at the University of Minnesota. His research focuses on understanding groundwater flow and quality in karst and fractured aquifers and how best to manage human impacts on these important water resources. He has served on numerous regional, state, and national technical and advisory panels, as well as on the organizing committees of several biennial interdisciplinary sinkhole conferences. He is program cochair and coeditor of the Proceedings of the 12th Conference of the Geological Society of America held in January 2011.

Scott Alexander is an environmental health specialist and manager of the hydrogeochemistry lab in the Geology and Geophysics Department at the University of Minnesota. His research interests have most recently addressed deep saline fluids found across Minnesota, including in the Soudan Mine in the northern part of the state. Scott has joined colleagues in conducting interdisciplinary research into the hydrogeology, geochemistry, mineralogy, and microbiology of crustal fluids. One such project focused on combined CO_2 sequestration and geothermal electricity production in deep saline aquifers across the U.S. Midwest.

Kenneth N. Brooks is a professor in the Department of Forest Resources at the University of Minnesota. He teaches and conducts research in watershed management and forest and wetland hydrology. Ken is certified as a professional hydrologist by the American Institute of Hydrology. He has authored or coauthored more than 100 publications, including three editions of the widely used textbook *Hydrology and the Management of Watersheds*.

Jay S. Coggins is an associate professor in the Department of Applied Economics at the University of Minnesota. He has published widely on issues related to environmental economics and policy. His special research interests include the Clean Air Act, water quality, market-based approaches to environmental protection, and the financial approach to environmental policy. His articles have appeared in *Ecological Economics*, the *Journal of Environmental Economics and Management, Review of Agricultural Economics, International Economic Review,* and *Social Choice and Welfare.*

Charles Dayton is a retired environmental lawyer. He was instrumental in the passage and implementation of many environmental laws in Minnesota, including the Minnesota Environmental Policy Act, and in the protection of the Boundary Waters Canoe Area Wilderness. He received the Distinguished Service Award from the Environment and Natural Resources Section of the Minnesota State Bar Association. He currently works as an advocate and frequent speaker on climate change and energy issues, and he recently served on the Minnesota Climate Change Advisory Group.

James D. Fallon is a hydrologist with the U.S. Geological Survey (USGS) Minnesota Water Science Center. He has worked for the USGS for more than 25 years, first measuring and sampling floods, and later writing about them. He has authored or coauthored several journal articles and USGS reports related to surface water and its quality.

Jerry Fruin is an associate professor in the Department of Applied Economics and a Center for Transportation Faculty Scholar at the University of Minnesota. His areas of research center on national and international transportation, the logistics for agricultural commodities and biofuels, and the economics of transportation on U.S. rivers and lakes. As a member of the National Academy of Sciences' Committee on the St. Lawrence Seaway, Jerry was involved in phase I of the project Options to Eliminate Introduction of Nonindigenous Species into the Great Lakes. He currently serves as an external peer reviewer to the U.S. Army Corps of Engineers for a feasibility study to rebuild locks on the Upper Ohio Navigation System.

Steven Heiskary has been employed with the Minnesota Pollution Control Agency since 1978 and is currently a research scientist in the Environmental Analysis and Outcomes Division. He has done extensive work on lake and stream assessment, with an emphasis on eutrophication impacts. He is currently focusing on the development of stream nutrient criteria and is a member of the National Nutrient Criteria Development Workgroup at the U.S. Environmental Protection Agency (EPA).

John Helland worked as a nonpartisan legislative research analyst for the Minnesota House of Representatives for 36 years. He served as chief research staffer for the House Environment and Natural Resource Policy and Finance

Committees and worked on all major Minnesota environmental legislation between 1970 and 2007. He currently serves on several Minnesota environmental boards and advises new committees that are developing plans for the long-term expenditure of the new legacy funds for natural resource programs.

Frances Homans is a professor in the Department of Applied Economics at the University of Minnesota. Her research focuses on natural resource economics. Her recent research on the economics of invasive forest species has been supported by the U.S. Department of Agriculture's Program on the Economics of Invasive Species Management (PREISM), and she is an active participant in the NSF-IGERT Graduate Training Grant in Risk Analysis for Introduced Species and Genotypes, University of Minnesota. She has had articles published in the *Journal of Environmental Economics and Management, Land Economics, Ecological Economics*, and *Agricultural and Resource Economics Review.*

Rob Johansson received a PhD in agricultural economics from the Department of Applied Economics at the University of Minnesota, where his research compared policy mechanisms for managing phosphorus runoff and water quality in the Minnesota River. He has also evaluated air, water, and conservation policy at the Economic Research Service of the U.S. Department of Agriculture (USDA), the Office of Information and Regulatory Affairs at the Office of Management and Budget, and the Congressional Budget Office. He is currently examining climate policy in the Climate Change Program Office in the Office of the Chief Economist at the USDA.

Bradley C. Karkkainen is a professor at the University of Minnesota Law School, where he holds the Henry J. Fletcher Chair. His recent research focuses on the law and governance of both freshwater and marine aquatic systems. He serves on the water policy team for the Minnesota Water Sustainability Framework and has served as a guest investigator at the Woods Hole Oceanographic Institution Marine Policy Center. His publications have appeared in leading law journals, including the *Columbia, Cornell*, and *Texas Law Reviews* and *Georgetown Law Journal.*

John Kolb is a partner in a Minnesota law firm, where his work focuses on wetlands and natural resources. John speaks frequently on issues related to stormwater and wastewater management and agricultural drainage. He has advised municipalities in the establishment of flood control and stormwater and wastewater management projects throughout the state. John is licensed in state and federal courts and appears regularly for clients in court and before various state and federal regulatory agencies.

Jay A. Leitch is a distinguished professor and emeritus dean in the School of Natural Resources at North Dakota State University. His research focuses on natural resource management policy at all government levels, primarily in the area of water resources. He recently served on an international task force to evaluate flooding and flood resilience in the Upper Midwest, where he contributed to

Living with the Red, an International Joint Commission report. His wetland policy work includes two annotated bibliographies on wetland values and policy issues.

Greg Lindsey is interim dean and a professor at the Humphrey Institute of Public Affairs at the University of Minnesota. He specializes in environmental planning and management and has worked in the areas of water resource management, land use, transportation, and recreation. His work has been published in the *Journal of the American Planning Association, Journal of Environmental Planning and Management, Journal of Recreation and Park Administration, Journal of Water Resources Planning and Management*, and *Landscape and Urban Planning*.

Kent Lokkesmoe is the director of the Minnesota Department of Natural Resources' Division of Waters, where he has served since 1974. His duties initially involved review of public drainage projects and litigation involving such projects. He served as regional hydrologist for the Metropolitan Region from 1978 to 1988 and assistant director from 1988 to 1991. His responsibilities today include policy, fiscal, and legislative issues, as well as administration of programs for permits, shoreland and floodplain management, climatology, and surface water and groundwater technical studies.

David L. Lorenz is a hydrologist with the USGS Minnesota Water Science Center. He has been involved with several studies of floods in Minnesota and has authored or coauthored more than 40 publications on a wide variety of topics related to groundwater, surface water, or water quality. His publications include three regional flood-frequency studies in Minnesota, as well as a regional groundwater recharge study and a study relating precipitation to base flows in streams. He also teaches two classes in environmental statistics for the USGS.

Jason Menard is a geographer with the U.S. Geological Survey at the Minnesota Water Science Center in Mounds View, Minnesota. He has been involved in studies using GIS hydrology in both groundwater and surface water with the USGS, and in spatial analysis, GIS modeling of environmental change, and with GPS at the University of Minnesota. He holds MS degrees from Michigan Technological University and the University of Minnesota.

Raymond M. Newman is a professor of fisheries in the Department of Fisheries, Wildlife, and Conservation Biology at the University of Minnesota. He serves as the director of graduate studies for the Water Resources Science Graduate Program and leads a formal graduate minor and a major training grant on Risk Analysis for Introduced Species and Genotypes. Much of his research focuses on the biology and control of aquatic invasive species. His articles have frequently been published in the *Journal of Aquatic Plant Management* and *Canadian Journal of Fisheries and Aquatic Sciences*.

Don Pereira is the fisheries research and policy manager for the Minnesota Department of Natural Resources' Division of Fish and Wildlife. He also is an adjunct assistant professor in the Department of Fisheries, Wildlife and Conservation Biology at the University of Minnesota. His research interests are diverse, including fish population and community dynamics, as well as large-scale aquatic habitat issues. He considers climate change to potentially be the single biggest driver of ecosystem change and is currently working on plans to advance climate change adaptation for fisheries management in Minnesota.

Gyles W. Randall is a professor at the University of Minnesota's Southern Research and Outreach Center in Waseca and in the university's Soil, Water and Climate Department in St. Paul. His research focuses on soil and crop management systems that improve crop productivity, farm profitability, and environmental quality for midwestern farmers. He has written widely on hydrology and constituent losses in subsurface tile drainage from poorly drained Minnesota soils. His publications have included bulletins on best management practices, journal articles, and book chapters, including coauthoring one of the initial chapters addressing nutrient loads to surface water, groundwater, and the Gulf of Mexico as part of a hypoxia assessment report to the congressional Committee on the Environment and Natural Resources.

Faye Sleeper is the codirector of the Water Resources Center at the University of Minnesota, where she works on issues related to impaired waters and total maximum daily load, water policy, citizen engagement, and professional training. Before this, she worked at the Minnesota Pollution Control Agency (MPCA) in the construction grants and wastewater enforcement programs, and as the project leader for an EPA-funded wastewater project with the Russian Federation. During her last eight years at the MPCA, she managed the Watershed Section, which was responsible for nonpoint sources and for guiding development of Minnesota's impaired waters program.

Louis N. Smith is an attorney in the law firm of Smith Partners PLLP, where he devotes his practice to water resource law and public-private partnerships. He wrote the *Watershed Rulemaking Handbook* for the Minnesota Association of Watershed Districts. His current research projects include an evaluation of Minnesota's drainage laws for the Legislative-Citizen Commission on Minnesota Resources and developing a model of watershed partnerships to guide cross-sector collaboration in conservation corridors. He is a past president of the Rivers Council of Minnesota and past chair of the Minnesota Clean Water Council.

James R. Stark is the director of the USGS Minnesota Water Science Center in Mounds View. He began his USGS career in Michigan and also worked in Utah before coming to Minnesota. Jim also served as a groundwater specialist and directed the National Water Quality Assessment and Studies Programs in Minnesota, and he is a licensed professional geologist in Minnesota and a professional hydrologist with the American Institute of Hydrology. He authored or coauthored more than 60 scientific reports and journal articles focusing on groundwater.

Richard D. Stewart is a professor at the University of Wisconsin–Superior, the director of the Transportation and Logistics Research Center, and a codirector of the Great Lakes Maritime Research Institute. He has commanded oceangoing merchant ships and holds a current unlimited ocean master's license and STCW-95 certification. He is an examiner for the American Society of Transportation and Logistics and is certified in transportation and logistics by the organization. His publications include coauthoring an international logistics textbook, as well as authoring or coauthoring book chapters, journal articles, and numerous papers on transportation management, marine environmental management, port operations, and transportation education.

Steven J. Taff is an associate professor and extension economist with the Department of Applied Economics and an adjunct associate professor with the Department of Forest Resources at the University of Minnesota. He specializes in the economics of agricultural and natural resource policies, with special emphasis on land management decisions in both rural and urban settings. He is widely known for his attempts to bring economic science to bear on practical policymaking. His recent research examines the economics of alternative energy systems, long-term carbon reduction policies for energy producers, landownership change in forested areas, and performance measures in landscape design.

Bob Tipping is a senior scientist with the Minnesota Geological Survey at the University of Minnesota, where his work is a combination of research and outreach, including teaching, serving on technical advisory boards, and responding to public inquiries about anything groundwater-related in Minnesota. His research interests include geology and hydrogeology of fractured and karst terrain, groundwater–surface water interaction, aquifer characterization, groundwater chemistry, and applications of geographic information systems (GIS) to geologic and hydrogeologic research.

John R. Wells is the Minnesota Environmental Quality Board (EQB) staff leader for water and sustainable development, a cochair of the federal-state Minnesota River Basin Integrated Watershed Study, and the leader of the EQB's efforts to design a new state approach to community assistance. He is cochair of the national Sustainable Water Resources Roundtable, vice chair of the policy committee of the American Water Resources Association, and a member of the USGS advisory committee to assist in design of the National Water Census. John played a leadership role in the passage of the Comprehensive Local Water Management Act (1985) and establishment of the Board of Water and Soil Resources (1987). He coordinated the development of the Minnesota Ground Water Protection Act (1989) and was the strategic planning director of a 2007 Environmental Quality Board report titled *Protecting Minnesota's Waters: Priorities for the 2008–2009 Biennium.*

Foreword

W hether traveling or at home, whether in conversations with friends or strangers, as a water scientist I get asked two questions more than any other: Are Minnesota's lakes and rivers getting better or worse? And, with so much water, Minnesota doesn't need to worry about its water resources, does it? My answer to the first question is, "It depends . . . ," and the answer gets complicated. My answer to the second is an emphatic "Yes, we do!" followed by a heartfelt lecture on water supply and demand, population growth, surface water–groundwater interactions, climate change, conservation, and public policy needs.

Minnesota has more water resources than any other state in the Lower 48. The headwaters of the three largest drainage basins in the North American continent can be found here, including the birthplace of the mighty Mississippi; the westernmost drainage of the largest body of fresh water on earth, the Great Lakes; and the humble beginnings of the Red River of the North, which eventually becomes the largest tributary to Hudson Bay. Yes, Minnesota is water wealthy. And all of this water makes us complacent, and in spite of Sig Olson's warning, we often take it for granted. We also sit on a complex system of layered aquifers and assume that this equates to added abundance.

We need to understand that our environment and our water resources are not static or stationary. Our state, our country, our continent, and our world are undergoing a state of rapid change. Minnesota's population will grow by a million people by 2030. Where people live, and their distribution by age and socioeconomic class, will change significantly. The baby boom generation is aging and will draw on social services while contributing less to taxes. These same folks are buying second homes on lakes "up north." This puts pressure on shoreland protection and land use practices and affects the economic resources to address these pressures. Increases in population are driven in part by immigrants, adding to our cultural diversity as well as our use and perceptions of water. The biggest driver of change, beyond those above, is the change in climate, which we currently can see occurring through all sorts of records in temperature, precipitation, ice-in and ice-out, and most believe we will see escalating effects of climate change on water resources in the future.

Our policies should ensure that we have a clean, adequate, and sustainable water supply for all. The Minnesota legislature defines sustainable as "when the use does not harm ecosystems, degrade water quality, or compromise the ability of future generations to meet their own needs." Professor Jerald Schnoor, the National Water Research Institute's 2010 Athalie Richardson Irvine Clarke Prize recipient for outstanding achievement in water science and technology, defined water sustainability in his award speech as "the continual supply of clean water for human uses and for other living things." The concept is clear: not just enough water, but enough *clean* water, for ecosystems, for humans as part of these ecosystems, and for all time.

Minnesota, like all other states as well as the federal government, manages and regulates water in a piecemeal fashion. Water quality is regulated at the federal level; water quantity, in contrast, is regulated by a patchwork of different and often conflicting state-based statutes. In Minnesota, surface water quality is managed by federal and state laws administered by the Minnesota Pollution Control Agency; groundwater quantity (extraction) is managed largely by the Minnesota Department of Natural Resources (DNR); drinking water for human consumption is managed by a different set of regulations administered by the Minnesota Department of Health, even though it may be the same water being managed by DNR. Land use adjacent to water bodies, which has huge impacts on water quality, is managed by local zoning and planning boards. No formal process exists for ensuring that these entities coordinate their efforts, although they often do so informally. (Many of these issues, and thoughtful steps toward their resolution, are discussed in detail in the chapters of this book.)

Thus the state has no overarching water policy, and the federal government has no overarching water policy either. There needs to be a shared vision for our water resources, so that they will be preserved and protected and available into the future. The duck hunter, the jet skier, the farmer, and the urban citizen all need a common understanding of water availability, water usage, drivers of change for water resources, and the implications of their choices. More important, they need to agree to a set of policies that will protect water for all uses, in a fair manner. Such policies must ensure that this shared vision will last for and adapt to all future generations. For this to occur, we need a cultural shift, not just a policy shift. Policy results from societal values. Our shared vision emanates from our culture, which must come first. The vision needs to incorporate conservation as a way of life, acceptance and management of water reuse (which we already do routinely in an unmanaged way), a clear understanding and respect that humans are part of a larger ecosystem with multiple water needs, an intrinsic understanding of the interconnectedness of groundwater and surface water, an acceptance of the economic value of water, and finally, mutual acceptance that water is a common good and that allocations and acceptable quality must be decided and agreed upon. Thus the biggest change that must occur is one of culture.

Minnesotans made a giant step in this direction with the 2008 passage of the Clean Water, Land, and Legacy Amendment to the state constitution. The amendment provides for a three-eighths of 1% increase in the state's sales tax for 25 years, with one-third of these funds dedicated to the Clean Water Fund. More than 60% of our citizens voted to raise the sales tax, even as we faced the biggest economic recession in several generations, creating a robust fund that should last at least until 2033 and will be spent on restoring, protecting, and enhancing the state's surface water and groundwater. As codirector of the University of Minnesota Water Resources Center, I am honored to be leading the effort to provide a road map for the legislature on how to ensure that this investment will result in our water resources being managed in a sustainable manner, and that the fund will not be spent in a piecemeal fashion. This road map, known as the Minnesota Water Sustainability Framework, was completed in January 2011. More than 200 water professionals, researchers, practitioners, stakeholders, and citizens were involved in developing this vision and its implementation plan for the state. The result is the most comprehensive, integrated water management strategy in the country. Minnesota has a rare opportunity where political will, funding, and public desire all have converged.

Given this point in our history, the publication of this book could not be more timely. Its chapters present what has and has not worked in Minnesota water policy thus far, along with clear recommendations for improvement. They cover water policy in detail from several specific fronts, including urban and agricultural nonpoint-source issues, invasive species, water transportation, flood management, economic issues and policies, and perspectives from local water management to international roles and responsibilities. Bill Easter and Jim Perry, internationally respected water professionals in their own right, have assembled a set of talented experts to author these insightful chapters. This book is a must-read for water managers and policymakers, water resource students, and interested citizens. The authors use Minnesota water policy to provide perspectives that can apply to every other U.S. state, as well as many countries.

The challenges that lie in front of us to craft equitable and sustainable water policy may seem daunting. This compelling book will make the task less intimidating and leave its readers better informed, with potential solutions to these challenges and, above all, a sense of hope.

Deborah L. Swackhamer, PhD
Professor and Charles M. Denny Chair in Science, Technology, and Public Policy,
University of Minnesota, and Codirector, Minnesota Water Resources Center

Acronyms and Abbreviations

APHIS	Animal and Plant Health Inspection Service
ASCS	Agricultural Stabilization and Conservation Service
BMP	best management practice
BWCAW	Boundary Waters Canoe Area Wilderness
BWSR	Board of Water and Soil Resources
C	Celsius
CAFO	concentrated animal feeding operation
CBOD	chemical biological oxygen demand
CDF	confined disposal facility
CERCLA	Comprehensive Environmental Response, Compensation and Liability Act
cfs	cubic feet per second
CHF	North Central Hardwood Forests
cm	centimeter(s)
CO	carbon monoxide
COLA	coalition of lake associations
CRP	Conservation Reserve Program
CWA	Clean Water Act
CWC	Clean Water Council
CWLA	Clean Water, Land, and Legacy Amendment
CWP	Clean Water Partnership

DI-P	diatom-inferred phosphorus
DMMP	Dredging Materials Management Plan
DMRP	Dredge Material Research Program
DMT	Drainage Management Team
DNR	Department of Natural Resources
DOER	Dredging Operations and Environmental Research
DWG	Drainage Work Group
DWINS	Drinking Water Infrastructure Needs Survey and Assessment
EPA	U.S. Environmental Protection Agency
EQB	Environmental Quality Board
ESA	Endangered Species Act
ESC	erosion and sediment control
F	Fahrenheit
FEMA	Federal Emergency Management Agency
FIRM	Flood Insurance Rate Maps
FWPCA	Federal Water Pollution Control Act
FWS	U.S. Fish and Wildlife Service
GCM	general circulation model
GEIS	generic environmental impact statement
GIS	geographic information system(s)
GLSLS	Great Lakes–St. Lawrence Seaway
GPS	global positioning system
G16	Group of 16
ha	hectare(s)
HC	hydrocarbon(s)
HGM	Hydrogeomorphic (Approach)
HMT	harbor maintenance tax
hr	hour
HTAC	Harbor Technical Advisory Committee
IJC	International Joint Commission
ILTF	Interagency Levee Task Force
IMO	International Maritime Organization

IPCC	Intergovernmental Panel on Climate Change
IRRB	International Red River Board
IWM	integrated watershed management
JPO	joint powers organization
L	liter(s)
LID	lake improvement district
LIDAR	Light Detection and Ranging
LUG	local unit of government
m	meter(s)
MCC	Minnesota Chamber of Commerce
MCEA	Minnesota Center for Environmental Advocacy
MCES	Metropolitan Council Environmental Services
MCWD	Minnehaha Creek Watershed District
MDA	Minnesota Department of Agriculture
MDH	Minnesota Department of Health
MDOT	Minnesota Department of Transportation
MEI	Minnesota Environmental Initiative
mg	milligram(s)
MISAC	Minnesota Invasive Species Advisory Council
MPCA	Minnesota Pollution Control Agency
MPR	Minnesota Public Radio
MRBI	Mississippi River Basin Initiative
mS	milliSiemen(s)
MSA	metropolitan statistical area
MS4	Municipal Separate Storm Sewer System
MSL	mean sea level
μg	microgram(s)
μmhos	micromhos
N	nitrogen
NANPCA	Nonindigenous Aquatic Nuisance Prevention and Control Act
NEPA	National Environmental Policy Act
NFIP	National Flood Insurance Program

NGO	nongovernmental organization
NGP	Northern Glaciated Plains
NHD	National Hydrography Dataset
NISA	National Invasive Species Act
NLA	National Lakes Assessment
NLF	Northern Lakes and Forests
NOAA	National Oceanic and Atmospheric Administration
NOBOB	"no ballast on board"
NO_2	nitrogen dioxide
NPDES	National Pollutant Discharge Elimination System
NRCS	Natural Resources Conservation Service
NSMPP	Nonpoint Source Management Program Plan
NWI	National Wetlands Inventory
NWS	National Weather Service
Obs.–P	observed phosphorus
OHWL	ordinary high-water level
OP	orthophosphorus
P	phosphorus
PFC	perfluorinated chemical; perfluorochemical
PFOS	perfluorooctane sulfonate
PM	particulate matter
ppb	part(s) per billion
RAPP	Refuse Act Permit Program
RCWD	Rice Creek Watershed District
RIM	Reinvest in Minnesota
RRBC	Red River Basin Commission
RRWMB	Red River Watershed Management Board
SARA	Superfund Assessment and Reauthorization Act
SLICE	Sustaining Lakes in a Changing Environment
SRF	state revolving loan fund
SRWD	Sauk River Watershed District
SU	standard unit(s)

SWCD	soil and water conservation district
SWINS	Storm Water Infrastructure Needs Survey
SWPPP	stormwater pollution prevention plan
TALU	Tiered Aquatic Life Use
TCE	trichloroethylene
TCMA	Twin Cities metropolitan area
TMDL	total maximum daily load
TP	total phosphorus
TSS	total suspended solids
UMRBA	Upper Mississippi River Basin Association
UMRS	Upper Mississippi River System
USACE	U.S. Army Corps of Engineers
USAID	U.S. Agency for International Development
USCG	U.S. Coast Guard
USDA	U.S. Department of Agriculture
USGS	U.S. Geological Survey
USMA	U.S. Maritime Administration
VGP	vessel general permit
WCA	Wetlands Conservation Act
WCP	Western Corn Belt Plains
WINS	Wastewater Infrastructure Needs Survey
WMO	watershed management organization
WQT	water quality trading
WRDA	Water Resources Development Act

PART I

INTRODUCTION

CHAPTER 1

Time for Action

K. William Easter and Jim Perry

T his book is one in a series of volumes Resources for the Future (RFF) is
publishing on water policies for various countries and U.S. states. Minnesota
is representative of states that historically have had adequate water supplies except
during drought periods, which may occur every 5 to 10 years. In such states, water
problems tend to center on quality, flooding, and having enough storage so that in
drought years, adequate water exists to meet basic domestic and industrial
demands. For Minnesota and other Great Lakes states, drought periods can also
cause serious navigation problems on rivers, such as the Mississippi and Illinois, as
well as on one or more of the Great Lakes. Minnesota also represents a situation
found to some degree in many states: after a drought ends, interest in water
conservation rapidly wanes. Yet Minnesota historically has had an innovative
institutional structure and an engaged populace that highly values its natural
resources. Our water resources increasingly are being subjected to new demands
for multiple services, and research frequently detects "new" water quality
problems, such as the presence of pharmaceuticals. Complicating matters, the
public tends to hold explicit expectations that there should be quick, clear, and
fairly painless solutions to these problems and new demands.

ISSUES FACING MINNESOTA POLICYMAKERS

As Minnesota enters the second decade of the 21st century, several critical water
resource issues need to be addressed. First, the state is in a unique position in the
North American continent, situated at the headwaters of three continental-scale
river basins. This means Minnesota has special responsibilities, particularly for
water quality affecting those downstream in both the United States and Canada.
The strongly binational nature of Minnesota water issues requires greater

innovation in policy and institutional arrangements. For example, how does the state develop an effective agreement with Canada to prevent the spread of invasive species through ballast water from lake shipping?

A second, clearly related issue is the growing demand within the state for cleaner water, as highlighted by the state vote for an increase in sales tax to improve water quality through passage of the Clean Water, Land, and Legacy Amendment to the state constitution. As incomes have gone up, so have demands for recreation and improved environmental quality. Minnesota's lakes and rivers play a critical part in meeting this growing demand for an improved environment. No other state can offer its residents more than 12,000 lakes, headwaters of the largest U.S. river, and the shoreline of the lake with the largest surface area in the world. The open question is, will Minnesota be able to sustain or improve the quality of its vast water resources as an increasing number of people demand more services?

A third issue involves the direct effects that land use changes have on the quality of water resources. These impacts include those from agricultural drainage as well as the placement of houses, shopping centers, and towns along almost all major rivers and lakes, such as the cities of Minneapolis and St. Paul on the Mississippi, Duluth on Lake Superior, Mankato on the Minnesota River, and houses along the bluffs of the St. Croix River. The state has managed to drain more than 50% of its original wetlands, and this loss exceeds 90% in the southern and western regions (Gernes and Norris 2006). Because of such losses, the speed at which water and associated contaminants leave farmland and reach the rivers has increased. All of these changes have helped lower water quality throughout the state. These changes also affect the routes and speed of water leaving the land surface. Minnesota has floods more often than was historically the case, due in part to landscape changes such as drainage, wetland loss, and urbanization. Flooding as a water resource issue in the Red River Valley is more complex because the river flows north; the headwaters melt before the downstream reaches, leading to ice jams and floods. This problem will become exacerbated as climate change warms the headwaters of the Red more quickly than it affects downstream reaches in the north.

A fourth issue that has limited Minnesota's effectiveness in managing its water resources is "agency dilution," the division of responsibilities among numerous agencies, including the Department of Health, Pollution Control Agency, Department of Natural Resources, and Board of Water and Soil Resources. No single state government agency has the responsibility for coordinating or overseeing management of Minnesota's more than 12,000 lakes (EQB 2003). Not surprisingly, the role each agency has played has changed in response to new problems. This leads to the Minnesota water imperative: the state must develop a more unified institutional approach that is more holistic and effective.

A fifth issue involves the growing water resource problems that arise from increasing energy demand, such as competition for groundwater between a town and a new ethanol plant or greater quantities of water needed to cool new electric power plants. At the same time, demand is rapidly increasing for energy to manage water resources, such as to transport water for public needs and irrigation, to improve water quality, and to maintain river channels for navigation. Not only is

Minnesota's growing population demanding more water services, but it is also demanding more energy, which in turn requires more water.

A sixth issue is that Minnesota, like most other states, treats its water like a free good with few restrictions on its use, even as the state's population grows and demands more clean water. The state pays the costs of treating and providing the water, but it does not charge consumers for the water it delivers or extracts. According to the Minnesota Environmental Quality Board's 2003 *Biennial Report,* "The appropriation of water, use of water, discharge of wastewater and pollution of water from nonpoint runoff or infiltration are currently free 'uses' of Minnesota's water resources" (EQB 2003, 16). A looming question for Minnesota policymakers is how to change valuation such that the state's growing population treats water as a valued resource and not as a free good.

USING INCENTIVES TO INCREASE THE VALUE OF WATER

With a growing demand in Minnesota for water for a variety of uses, along with an increased concern about water quality, one option is to provide more effective economic incentives to conserve water and protect its quality. Four different types of economic incentives can motivate consumers to cut back on the use of water and pollute less: charges or fees; quotas or rationing; subsidies for water-saving technologies, such as low-flow showers; and markets. These mechanisms can encourage consumers to use less water and create less water pollution, or pay for any water they use or pollute. In Minnesota, the most frequently used mechanisms to reduce water consumption have been water quotas, restrictions on certain uses during drought periods, and subsidies for water-saving technologies. However, more recently, a number of Minnesota towns and cities have changed their policies such that water charges are now more closely related to the quantity of water used. Those who consume 20,000 liters (L) per month pay less per liter than those who use 40,000 L per month.[1] Yet, as EQB points out, there still is no value-based fee for water use (EQB 2003, 16).

There are different ways to price or charge for water. Some provide an incentive to use less water, but others do not. Six of the more common methods are a fixed charge per month or quarter; a constant rate per unit of water used; a decreasing block rate per unit of water used; an increasing block rate per unit of water use; a two-price method, which includes a fixed charge plus a charge per unit of water used; and peak-load or seasonal pricing (Table 1.1). The fixed charge per month is found mostly in small towns and rural water systems that have not invested in water meters and just want to cover basic operating costs. This method provides no incentive to use less water, however. Consumers that use 150,000 L per month pay the same as those that use 5,000 L.

The five methods other than the fixed charge provide varying degrees of incentives to conserve water. The constant rate (e.g., $2 per 1,000 L used) provides the same incentive to conserve no matter how large the user. If consumers use 150,000 L they pay $300, but if they use only 5,000 L they pay $10. The decreasing block rate provides a smaller incentive the more the user consumes; for example,

Table 1.1 *Methods of water pricing for urban uses*

Method	Price per month	Water use (L/month)	Water bill ($/month)
Fixed charge	$50	5,000	$50
		150,000	$50
Constant rate	$2/1,000 L	5,000	$10
		150,000	$300
Decreasing rate ($/1,000 L)	$2 up to 100,000 L +	5,000	$10
	$1 over 100,000 L	150,000	$250
Increasing rate ($/1,000 L)	$2 up to 100,000 L +	5,000	$10
	$5 over 100,000 L	150,000	$450
Two-price method	$30 + $2/1,000 L	150,000	$330
Peak-load pricing			
Summer	$5/1,000 L	150,000	$750
Winter	$1/1,000 L	150,000	$150
Fall/spring	$3/1,000 L	150,000	$450

the rate may drop from $2 to $1 per 1,000 L once a consumer uses 100,000 L during the month. Consumers that use 150,000 L per month would pay $250. This rate structure is frequently used to give industrial plants an incentive to locate in a particular city. The increasing block rate is just the opposite: the incentive to conserve goes up as the consumer uses more. For example, the rate could go from $2 per 1,000 L up to $5 per 1,000 L once consumers use 100,000 L during the month. In this case, consumers of 150,000 L per month would pay $450. The increasing block charge generally provides the best incentive to reduce water use and tends to be employed in cities with limited water supplies and growing water demands.

In the two-price method, the fixed charge ($30) is to pay for those recurring costs that are not related to the quantity of water consumed. The second charge is a variable one based on the amount of water used and can be a constant, increasing, or decreasing rate. It can also include a specific charge for water itself, in addition to delivery and treatment costs, based on the scarcity value of the water. In other words, the greater the scarcity of water, the higher the charge. Finally, seasonal or peak-load pricing can be used to increase the charge during periods of high water use (summer), when conservation is particularly important and the marginal value of water is high because of high water demand. However, as we discuss below, high prices may not be enough to curtail profligate water use because of leaks, large lawns, or undisciplined consumption.

Because the demand for a number of water uses, such as drinking and cooking, tends to be quite steep (inelastic), large price increases will likely bring about only small changes in the quantities used for these purposes, particularly for those with higher incomes. When this is the case, higher water prices may not achieve all the water conservation needed, even though some have reduced their lawn watering and car washing. A better alternative in such cases may be to combine higher water prices with other demand management strategies. One strategy would be to combine high water prices with subsidies for the adoption of water-saving technology, such as low-flow toilets or drip irrigation for landscapes. During periods of drought, water quotas or bans on certain uses, such as watering lawns and washing cars, can also be used to supplement higher water prices. If a household quota is used, it will probably have to vary based on the number of people in the household. This can be hard to determine and will not remain constant over time. In Minnesota cities, the most common strategy has been to restrict selected water uses such as lawn watering.

An alternative to pricing that may become important as future water supplies drop relative to demand is use of a market to allocate water. This, if organized effectively, would price water and provide incentives to use less. There have been few attempts to establish water markets in municipal water systems, although Haddad (2000) proposed one for Monterey, California, after a particularly dry period in the 1990s. The one case where the water market is effective is for bottled water. In a number of Minnesota households, bottled water is the major source of drinking water, and some people even use it for cooking. Water markets have been used in other states and countries to trade irrigation water among farmers and to transfer water between sectors, particularly from agricultural to domestic uses (Easter et al. 1998).

In the future, we may want to consider the use of markets to allocate Minnesota's groundwater supply as demands exceed recharge rates. As is pointed out in Chapter 12, the recharge of groundwater may drop significantly with climate change while demand for water for irrigation and other uses increase significantly. The end result may be growing water scarcity. This may require us to cap the use of water, particularly groundwater, and require all increases in water use to be purchased from existing users who have reduced their water use. The price paid for the water would reflect the cost of conserving water and the scarcity value of water. It will take time to develop the information and enforcement mechanisms needed to operate such a market. Water rights would have to be developed and allocated to users based on water availability and, most likely, historic water use. Water rights and trades would have to be recorded. Pumping records would have to be kept and checked to make sure users did not exceed their allocations, unless they purchased additional rights.

Another key requirement for such an allocation system to work effectively is accounting for the interconnectedness of groundwater and surface water. Drawing down the groundwater will, in many cases, reduce streamflows and lake levels, as well as wetland areas. Water rights cannot be established without fully accounting for these interconnections and possible downstream effects.

Minnesota has already considered the possibility of using markets to help manage water quality improvement, primarily through pollution permit trading. Trading will allow owners of point sources of pollution (primarily factories and municipalities) who have discharge permits to trade a share of their permitted load. In this case, those with low treatment cost are able to treat more and sell the unused portion of their permits. Buyers would be owners of point sources with higher costs of treatment. Minnesota and other states are now considering the use of such markets to reduce the level of discharge of nitrates and phosphorus into water bodies by extending trading to nonpoint sources of pollution such as agriculture. However, as Chapter 11 points out, this will be difficult. The large number of farmers and the wide variation of impacts among farms and even among individual fields make it very difficult to predict impacts on any given water body.

This difficulty in determining the likely impacts highlights one of the most significant challenges to market trading: our limited understanding of the receiving water resource. The logic of trading is that we are able to describe ecosystem services desired by society and quantify trade-offs in services as water quality changes. This is expressed quantitatively as a load (often a total maximum daily load, or TMDL) that represents a cap, the total quantity of impact that can be allocated among users. Yet our ability to precisely relate ecosystem services to instream loads is quite immature. In addition, nonpoint sources do not currently have a discharge limit as do point sources. As a result, trading between point and nonpoint sources has been tried only rarely so far in Minnesota. And as Fang et al. (2005) point out, the Minnesota Pollution Control Agency (MPCA) had to play a major role in facilitating the trades that have been accomplished. Thus it appears that if we are to achieve an active market for phosphorus or nitrate pollution permits, restrictions will have to be placed on nonpoint-source pollution.

A further issue that influences our ability to manage waters precisely is the transaction cost of such management. The economic incentives described above require a support structure that conducts a rigorous assessment, monitoring, and enforcement program. Minnesota has assessed approximately 20% of its waters to date, making it difficult to believe that all waters are protected and managed in the public's best interests. Furthermore, we require a policy mechanism in which ecosystem services are fairly and reasonably assigned to each water body and a modeling effort sufficiently generic that transaction costs for assigning contaminant loads can be reasonably borne by society.

Prices can also be used to reduce water pollution by charging firms and municipalities for the pollutants they discharge into water bodies. For example, the Twin Cities Metropolitan Council uses a "strength charge" on wastewater discharges; this charge is based on the level of oxygen demand, total suspended solids, and the total volume discharged. These per-unit strength charges for waste discharge should increase per unit when the discharge per month increases and should vary among contaminants based on the relationship between a contaminant and ecosystem services. In doing so, the charges would provide additional incentive to reduce pollution. These charges do not vary in that manner today. If the charges were set high enough, some high water-using firms might begin recycling their water.

The degree to which Minnesota will use these four different types of economic incentives to manage its water resources, particularly water quality, will depend a great deal on demand for the range of uses and services that water provides. Clearly, as Minnesota's water resources face increased demands, economic incentives will have to play a larger role in allocating use of our water resources, as evidenced by the new (January 2010) conservation-pricing water law. People will have to pay more for the services water provides. As with anything else in a market society, as the demand goes up for a resource relative to its fixed supply, we have to pay more to obtain each unit of the resource. In the process of responding to higher prices for water, we will discover that we can conserve more and use less, resulting in less wastewater being discharged into our rivers, streams, and lakes.

OVERVIEW

This book addresses the six critical issues framed above, along with others, through a series of chapters broken down into five major sections. Part II discusses Minnesota's water resources, their distribution, and the laws and policies that govern our relationship with them. Part III takes a biophysical approach, examining water issues from the perspective of lakes, wetlands, and groundwaters. Part IV discusses the ways that societal uses of the waters compete. Some uses (e.g., large-scale agriculture) compete with others (e.g., wetland ecology and hydrology). Part V examines emerging issues such as climate change, flooding, and invasive species, seemingly discrete issues that are capturing significant amounts of public attention and resources. It becomes apparent that each of these is closely related to all the other issues we face in water resource policy. Finally, Part VI discusses central themes in Minnesota water policy: themes, ideas, and potential solutions on the path to a sustainable water future.

In Chapter 2, Kent Lokkesmoe and Jay Leitch discuss the policies that affect Minnesota's transboundary waters, those that flow into and out of Canada and adjacent states. The Red River Basin Commission was developed to foster discussion among Minnesota, North Dakota, South Dakota, and Manitoba. The International Joint Commission has wide-ranging authority and programs to address water management throughout the Great Lakes, including eight U.S. states and two provinces in Canada. The St. Croix Basin Team, linking Minnesota and Wisconsin, has been one of the most innovative water resource institutions. In spite of apparent opportunities, no specific institutions assist in comanaging water between Minnesota and Iowa. In Chapter 3, Steve Heiskary discusses the nature of Minnesota's landscapes and how landscape character influences the flow of water resources. Nearly all of Minnesota was glaciated in the late Pleistocene, a process that shaped its topography and structured most of its hydrology. Glaciations in three-quarters of the state and the lack of recent glaciation in southeastern Minnesota resulted in three major river basins and a driftless subregion in the southeast where stream valleys are relatively steep and geologic substrates are karstic. All future water decisions must take into consideration the subregional nature of the landscape and what that means for water resources.

That regional nature will represent a challenge to effective policy, especially in the ways the state addresses impaired waters, as Rob Johansson and Faye Sleeper discuss in Chapter 4. Minnesotans want to think of themselves as innovative and egalitarian. New water laws, such as the Clean Water, Land, and Legacy Amendment, put Minnesota in a position of national leadership in providing resources to address impaired waters. Minnesota policymakers, legislators, and members of the public shared a strong and innovative vision for the future of their waters and have followed that vision to pass major funding legislation. However, implementation of this legislation, including the disparity between expectations and possible solutions, has posed problems. For example, the public and many legislators expect that this major legislative commitment will "solve" the impaired waters problem. In fact, only a small percentage of the state's waters has been assessed, and "solutions" are far distant. In Chapter 5, Brad Karkkainen discusses how federal water resource laws, such as the Federal Water Pollution Control Act, are interpreted and applied in Minnesota and where state laws complement federal legislation. He documents that Minnesota has been innovative in several aspects of water policy, ranging from water quality concerns to efforts to limit new well construction. This legal innovation has not been limited to within-state institutions.

As Louis Smith and John Kolb discuss in Chapter 6, many approaches have been taken to local-scale water management for the state's lakes, but few of them have been effective in maintaining lake quality over time. They review these approaches and attempt to answer an important question for water resource management: which suite of local water management institutions is most effective and why. They find that the watershed approach, sometimes combined with a lake association augmented with strong community involvement, seems to be the most effective model for local water management. In Chapter 7, Jay Leitch and Gyles Randall discuss wetlands as they used to occur in the landscapes of Minnesota and as they occur today. The authors point out that more than 50% of the historic wetlands in southwestern Minnesota have been drained, often explicitly following government policy and using government financial support. The lack of storage causes waters to leave the landscape more quickly, exacerbating downstream flooding as well as causing the loss of many ecosystem services on the land. These policies have been changed, and Minnesota is now trying to maintain or even increase its acreage of wetlands.

Similarly, past government policy has done little to protect groundwater. Bob Tipping, Scott Alexander, and Calvin Alexander point out in Chapter 8 that groundwater has become depleted and degraded in many areas of the state, often because it has been treated as an open-access or free resource. When aquifers are poorly bounded, such as in fractured geology or karstic situations, substances may move readily from the surface into an aquifer and from one aquifer to another. Such movements often cause contamination, posing a threat to public health and ecosystem services. Historically, the state's policy environment has not been effective in addressing groundwater problems involving discrete aquifers. Minnesota also has not done well in considering surface water and groundwater together, or in framing integrated policies that recognize the intricacies of multiple overlain aquifers and their connection to surface water.

In Chapter 9, Greg Lindsey discusses the state's urbanization, the policies it has followed that have resulted in this demographic, and the policy innovations that will be needed to protect its waters as Minnesota becomes more urbanized over the next 50 years. He finds this particularly challenging given the projected growth in the urban population and the financial requirements needed to meet the increasing infrastructure needs, both to replace old systems and to build new ones. Jerry Fruin and Richard Stewart suggest in Chapter 10 that federal navigation policy, implemented by the U.S. Army Corps of Engineers (USACE), has played a key role in the state's economy and water resources. For example, the USACE maintains a 9-foot channel in the Mississippi River from the Twin Cities to the Iowa border and then south to New Orleans, allowing low-cost shipping to the Gulf of Mexico. Dredging, disposal of dredge spoil, and impacts of navigation practices all have significant effects on water resources in the state. The authors also discuss the importance of shipping on the Great Lakes and how this has changed since 1854. Today ocean ships with cargoes of more than 25,000 tons sail as far inland as the Duluth–Superior Harbor, resulting in impacts from ships and channel management. U.S. water policy follows the federal-floor, state-ceiling approach: states can enact policies and standards more, but not less, strict than federal levels. As a result, federal rules and policies imposed on the state create a mosaic of laws and institutions that often are confusing. For example, Jay Coggins and Steve Taff discuss in Chapter 11 how federal policies (e.g., subsidies, restrictions) in both agriculture and energy have controlled the ways Minnesotans choose landscape practices. That in turn has controlled many instream conditions. Yet the authors find that farm policy over the past 20 years has not had much of an impact on water quality. To improve water quality, they think changes in federal water policy are needed, along with changes in farm policy.

Among the most significant influences on Minnesota landscapes and their waters over the next 50 years will be climate change, invasive species, and flood control. As Chuck Dayton and Don Pereira discuss in Chapter 12, Minnesota's position at the junction of three biomes is largely a function of climate differences. The midcontinental region is an area subject to almost certain climatic changes. Although timing of future climates cannot be predicted with much certainty, we have a high degree of certainty that significant increases in droughts, floods, and temperature extremes will occur over the next 50 years. Those changes will likely result in alterations in hydrology and the ecology of Minnesota's waters. The condition of the landscape will strongly influence how those climatic influences are expressed. Minnesota demographics are strongly bimodal now, with 71% of the state's population living in urban areas and 29% in rural areas (Discovery Communications 2010). These urban landscapes are comparatively green, with a relatively high percentage of parks and protected areas. However, urban settings, by their very nature, increase the temperature of adjacent waters and release waters much more quickly than do places with less impervious area.

One of the most disconcerting and troublesome aspects of water policy, in Minnesota as elsewhere, is the way policy does or does not deal with waters that are difficult for the public to understand. Wetlands were regarded as wastelands for nearly a century. Groundwaters are hidden assets, flowing through substrates distant from human eyes. As a result, policies for managing these two resources have long been at variance with expressed goals for water resources. Among the contaminants impacting Minnesota's waters and managed by its policies, invasive species are one of the most damaging and most difficult to address. In Chapter 13, Frances Homans and Ray Newman discuss invasive species policies and the degree of success the state has had in dealing with invasives. They report that a range of actions designed to fulfill societal needs, such as the release of ballast water from shipping, have often led to the introduction of invasive species and a resulting loss of ecosystem services. Scientists have a relatively solid understanding of the characteristics that allow a species to be invasive in a new landscape or water body. There would seem to be an opportunity for proactive policies that attempt to prevent new invasives rather than simply focus on control of current ones. However, this may be difficult when authority is limited by state boundaries. Here greater federal action is needed to stop damaging invasions. Even with federal action, Minnesota may need to take a more punitive approach to reducing the spread of invasive species and, as the authors suggest, consider invasive species as biological pollution.

Invasives are only one aspect of the state's overall water resource management problems. Recreation, especially water-based, is very important to the people of Minnesota. The number of recreational properties along lakes and rivers has increased geometrically in the last 25 years. That increased utilization has resulted in many changes in ecosystem services. For example, fishing pressure has increased, fish management practices have changed, boat wake zones have been established, and nutrient management zones have been proposed. All of these landscape practices, such as drainage, dredging, and agricultural and forest management, influence the speed and timing of water leaving the landscape. Ken Brooks, James Fallon, David Lorenz, James Stark, and Jason Menard discuss these relationships in Chapter 14 and explain how the historic definition of recurrence interval, using terms such as 100- or 500-year flood, has caused confusion and misunderstanding in communicating with the public. Global climate change is increasing the severity and frequency of storms and droughts; landscape practices such as wetland loss and agricultural drainage are facilitating water movement off the landscape. Floods are becoming more frequent and more difficult to predict, yet they are having greater economic impact as humans increase the value of infrastructure in proximity to water bodies. Our policy mechanisms for management of the relationships between infrastructure development and flooding remain ineffective. John Helland draws these thoughts together in Chapter 15 by discussing the politics of water resources in Minnesota. He highlights the very individualized and subjective nature of politics and the ways that all policies are the result of the political process, which is inherently individual. That strong role of personal values adds unpredictability to an already complex policy landscape, while ensuring a process through which stakeholders can express their views and influence outcomes. The latter, of course, is the core of the American representative democratic process.

CONCLUSIONS

The final two chapters synthesize and pose general questions to be addressed as we move forward. In Chapter 16, John Wells examines sustainability and identifies five key principles—interconnectedness, informed decisionmaking, precaution, transparency, and accountability—that advance sustainability. He applies these themes to Minnesota water policy through example and analysis, leading to core principles that must be addressed in Minnesota's water future. In Chapter 17, we examine the strengths and weaknesses of Minnesota water policy, as developed in the preceding chapters, applying a gap analysis to those policies. We conclude that some major gaps exist in both Minnesota and U.S. water policy.

First, Minnesota fails to take a holistic approach to its use and management of water, with too many state agencies in charge of different aspects of the state's water resources. Second, states and local governments fail to adequately use economic incentives to encourage water conservation. The main emphasis of most towns and cities is cost recovery. Third, the use and conservation of groundwater will become more critical in the future, with recharge rates likely to be reduced as a result of climate change and urbanization, while demands for clean groundwater will increase. This means Minnesota's policy that allows most landowners to pump whatever amount of groundwater they want will have to change. Fourth, the state's water policy needs to clearly recognize that surface water and groundwater are, in most cases, interconnected. High levels of pumping are likely to reduce both streamflows and lake levels in many areas. Fifth, the state will need to change how both state and local policies address drought. With climate change likely to mean longer and more frequent droughts, just banning lawn watering and car washing will not be enough. New laws such as the 2010 law establishing conservation rate pricing will be required to create adequate water use incentives and achieve effective water management.

Finally, federal policy needs to change if Minnesota water policy is going to be more effective. For example, as highlighted in Chapter 11, we are not likely to have much success in reducing nonpoint-source pollution unless both federal water and farm policies are changed so that they specifically address nonpoint-source pollution, particularly from runoff from farms and urban areas. Another area where federal policy needs to be strengthened is with invasive species. States such as Minnesota that try to prevent the introduction and spread of invasive species are limited in what action they can take, because their authority stops at the state border. The Asian carp is a good example of a situation where the states tried to do something to keep a species out of the Great Lakes, but the federal government blocked their action rather than taking a lead role in stopping the fish from spreading.

It is clear that if Minnesota is to continue to have adequate and cleaner water, changes in policy will be needed at the federal, state, and local levels of government. Fortunately, Minnesotans, if not their elected officials, seem to understand that water resources cannot be improved on the cheap with no new taxes. They quite willingly voted in 2008 for an increase in the sales tax to help clean up Minnesota's water by passing a Constitutional Amendment to ensure that water resource management improves.

NOTE

1. 1liter = 0.26 gallons.

REFERENCES

Discovery Communications. 2010. HowStuffWorks: Geography of Minnesota. http://geography.howstuffworks.com/united-states/geography-of-minnesota.htm (accessed December 1, 2010).

Easter, K. William, Mark W. Rosegrant, and Ariel Dinar. 1998. *Markets for Water: Potential and Performance*. Boston: Kluwer Academic Publishers.

EQB (Environmental Quality Board). 2003. Biennial Report: Minnesota Water Priorities 2003–2005. St. Paul: Minnesota Planning.

Fang, Feng, K. William Easter, and Patrick Brezonik. 2005. Point-Nonpoint Source Water Quality Trading: A Case Study in the Minnesota River Basin. *Journal of the American Water Resources Association* 41 (June):645–658.

Gernes, Mark, and Doug Norris. 2006. A Comprehensive Wetland Assessment, Monitoring and Mapping Strategy for Minnesota. St Paul: Minnesota Pollution Control Agency.

Haddad, Brent M. 2000. *Rivers of Gold, Designing Markets to Allocate Water in California.* Washington, DC: Island Press.

PART II

OVERVIEW OF MINNESOTA'S WATERS AND LAWS

CHAPTER 2

Interstate and International Water Management

Kent Lokkesmoe and Jay Leitch

M innesota, the land of more than 10,000 lakes and over 92,000 miles of streams, lies at the headwaters of three major continental watersheds: the Mississippi River flowing to the Gulf of Mexico, the Red River flowing to the Hudson Bay, and the Great Lakes. Although the state shares some border waters, only a few minor tributaries flow from other states into Minnesota. Thus Minnesota is upstream and relies on precipitation in the form of rain or snow for surface water and groundwater replenishment. Being upstream brings an added dimension to the state's water management policies.

Formal water management began in Minnesota following the drought in the 1930s and has expanded in scope and purpose as a result of concerns about potential impacts to waters within the state and impacts that waters leaving the state might have on others. The state's water issues and resulting policies fall into three areas: intrastate, interstate, and international. Whereas the balance of this book addresses intrastate water issues, this chapter explores the transboundary, interstate and international issues Minnesota faces and the innovative management approaches the state has developed or adopted, including cooperation and compromise, compacts, legal actions, and treaties.

States generally have sole management authority over waters within their boundaries, with some exceptions, such as the federal Clean Water and Endangered Species Acts, that limit or compel state action. Likewise, political subdivisions within the states, also referred to as local units of government (LUGs), including counties, townships, municipalities, and water management boards, have local responsibility but are bound by state laws and state agency policies. A clear and distinct chain of statutory and government authority exists through which to address intrastate water management.

The chain of statutory authority over transboundary waters that either form state boundaries or flow out of the state is far less clear. Such waters are subject to extraterritorial jurisdiction from the federal government, formal agreements among states, and often international treaties. Although transboundary water issues are similar to intrastate issues (e.g., flooding, water supply, invasive species, water quality), their management is complicated by less formalized and less clear authorities and responsibilities. In addition, the number of stakeholders increases substantially when water issues cross state or national borders.

Minnesota's 1,919-mile border is 67% surface water. The state shares this largely wet border with five states and two Canadian provinces (Figure 2.1). The surface water border includes north- and south-flowing rivers, the largest freshwater lake in the world, remote wilderness areas, urban rivers, prairie lakes, and large reservoirs. As expected, this variety of situations has led to a plethora of "transjurisdictional cooperation" arrangements (Foster 1990, 11), some with authority and others with only the powers of persuasion.

INTERSTATE AND TRANSBOUNDARY ISSUES

States deal with other states via formal and informal agreements, interstate compacts, and judicial solutions. Within the United States, the states are more or less autonomous political units constrained only by the U.S. Constitution, which identifies the distinct roles of the federal and state governments. Congress has passed bills and federal agencies have engaged in rulemaking that further direct how states interact with respect to interstate water management. Because of Minnesota's upstream setting and its largely wet border, the state has a rich transboundary water policy history with its five neighboring states of Michigan, Wisconsin, Iowa, South Dakota, and North Dakota. Because Minnesota and Michigan share only a short border through Lake Superior, separating Minnesota from Lake Michigan's Isle Royale, the issues with Michigan are included below in the discussion about the Great Lakes. Along the 263-mile Iowa border, which includes no surface water except a few small lakes, there are no known transboundary issues with these waters. Thus Iowa is included only in the discussions about the Upper Mississippi River Basin.

Wisconsin

Most of Minnesota's border with Wisconsin is surface water. From northeast to southeast, the statutory boundary runs through Lake Superior (110 miles); follows a local river, the St. Louis, east from Duluth for 21 miles; follows north-south along 98°18′ longitude for about 40 miles; then follows a regional river, the St. Croix, for 129 miles until its confluence with the Mississippi, a continental river, which forms the 140-mile southeastern border between Minnesota and Wisconsin. Along the nonwater portion of the Minnesota–Wisconsin border, several small watersheds rise in Wisconsin and flow into Minnesota, creating exceptions to Minnesota's status as an upstream state.

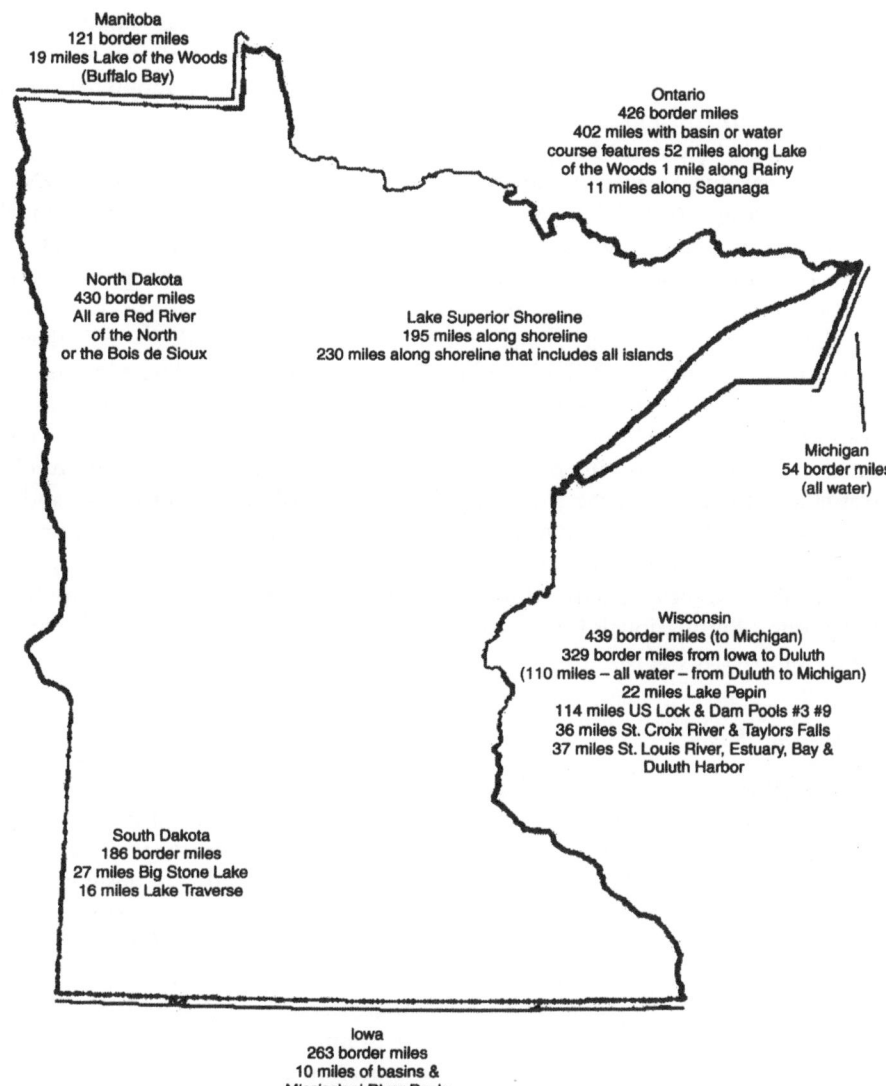

Figure 2.1 *Minnesota's basin and watercourse border features*

Note: Basin and watercourse features along the border are indicated by the heavier lines. They are based on 1:24,000 scale digital data.

St. Croix River

The Wilderness Act (1964) and the National Wild and Scenic Rivers Act (1968) identified the Upper St. Croix River (above Taylor's Falls) as nationally significant, setting it aside as one of the first segments in the federal Wild and Scenic Rivers System. In 1972, the Lower St. Croix River became the 10th river segment in that system. The 52-mile segment from Taylor's Falls to the confluence with the

Mississippi River has been designated a recreational riverway. The National Park Service manages the river above Taylor's Falls, while Minnesota and Wisconsin cooperatively manage the Lower St. Croix. Both states have adopted land use regulations that are administered by local government.

The Lower St. Croix Management Commission consists of representatives from the National Park Service, Minnesota Department of Natural Resources (DNR), and Wisconsin DNR. The commission meets as needed, but at least annually, to provide oversight, discuss mutual issues, and recommend action to state- or federal-level decisionmakers. It has a technical team that reviews proposals and provides input to the commission.

The Lower St. Croix Management Commission owed its effectiveness primarily to a jointly funded Minnesota-Wisconsin Boundary Area Commission. The Boundary Area Commission had three staff members dedicated to coordination of St. Croix River issues, who arranged meetings, kept minutes, developed background information, and conducted studies. When state funding was eliminated in 1999, the Boundary Area Commission ceased to exist, and the states were unable to continue to provide the same level of support. At present, the Lower St. Croix Management Commission provides only very general coordination. Its principal functions are handled by the St Croix Basin Team, an ad hoc group of professionals from state and federal agencies who meet regularly and guide a proactive approach to management of the river.

Mississippi River

The Mississippi River forms Minnesota's southeastern border with Wisconsin, downstream of the confluence with the St. Croix River at Hastings. The Upper Mississippi River Basin Association (UMRBA), a regional interstate organization, was formed in 1981. Illinois, Iowa, Minnesota, Missouri, and Wisconsin joined to coordinate the states' river-related programs and policies and to work with the appropriate federal agencies. UMRBA is the successor to the former federally authorized and funded Upper Mississippi River Basin Commission. When that commission was terminated by a presidential executive order in 1981, the five states' governors signed a joint resolution calling for "the continuation of an interstate organization to maintain communication and cooperation among the states on matters related to water planning and management" (UMRBA n.d.).

UMRBA is involved with programs related to ecosystem restoration, hazardous spills, water quality, floodplain management and flood control, commercial navigation, and water supply. Eight federal agencies participate as advisory members in the association:

- U.S. Army Corps of Engineers (USACE)
- Department of Agriculture: Natural Resources Conservation Service (NRCS)
- Department of Homeland Security: Coast Guard (USCG) and Federal Emergency Management Agency (FEMA)
- Department of the Interior: U.S. Fish and Wildlife Service (FWS) and U.S. Geological Survey (USGS)

- Department of Transportation: U.S. Maritime Administration (USMA)
- U.S. Environmental Protection Agency (EPA).

Although the federal members are nonvoting, they are active participants in UMRBA activities. With their participation, UMRBA is able to serve as a regional forum in which the states and their federal counterparts discuss policy and management issues related to the river and its riparian zone. UMRBA's purpose is to facilitate dialogue and cooperative action regarding water and related land resource issues. More specifically, the association aims to do the following:

- serve as a regional interstate forum for discussion, study, and evaluation of river-related issues of common concern to the states of the Upper Mississippi River Basin;
- facilitate and foster cooperative planning and coordinated management of the region's water and related land resources (e.g., creating a coordinated spill response system);
- create opportunities and means for the states and federal agencies responsible for management of water resources in the Upper Mississippi River Basin to exchange information; and
- develop regional positions on river resource issues and serve as an advocate of the basin states' collective interests before Congress and the federal agencies. UMRBA has been successful in obtaining federal authorization and funding (approximately $20 million a year) for the USACE Environmental Management Program. This funding supports the long-term Resource Monitoring and Habitat Restoration Project.

UMRBA engages in discussion, evaluation, and study of topics ranging from policy and budget matters to specific resource management concerns. The association has addressed a wide range of issues, including nonpoint-source pollution, water quality planning and management, interbasin diversion, cost-sharing strategies, water project financing, sediment and erosion, hazardous spills, toxic pollution, habitat restoration, navigation capacity, channel maintenance, flood response and recovery, floodplain management, wetland protection, hydropower development and licensing, and drought planning. The member states pay dues to support an executive director. State member representatives are upper-level managers from resource agencies and are able to provide staff to support necessary activities.

South Dakota

Only about 25% of the 186-mile border with South Dakota is surface water, including Big Stone Lake, Lake Traverse, Mud Lake, and the Bois de Sioux River. Big Stone Lake is a 26-mile-long, 1-mile-wide, 12,601-acre reservoir and the headwater of the Minnesota River. Lake Traverse is a 16.5-mile-long, narrow, 11,200-acre reservoir that empties into Mud Lake. Mud Lake is a 7.5-mile-long, narrow, 3,850-acre reservoir and the headwater of the Bois de Sioux River, which becomes the Red River of the North after its confluence with the Otter Tail River in Breckenridge, Minnesota. The Continental Divide runs through Traverse Gap between Lake Traverse and Big Stone Lake. Several local tributary watersheds along the Minnesota–South Dakota border have headwaters in South Dakota, further exceptions to Minnesota being upstream.

The Lake Traverse and Mud Lake reservoirs are managed by the USACE primarily for flood control and secondarily for water supply. Water management issues are usually related to maintenance of the pool levels. The two states' fisheries agencies have devoted time and energy to synchronizing fishing and boating regulations on these border lakes.

Big Stone Lake is managed cooperatively by state agencies. This lake is a good example of transboundary multijurisdictional cooperation. In the early 1980s, citizens of South Dakota and Minnesota requested assistance from both states and EPA to begin an effort to restore Big Stone Lake. The primary concerns were poor water quality, excessive algal blooms, sedimentation, rooted aquatic vegetation, and reduced recreation potential. A series of EPA Section 314 and 319 grants, beginning in 1983, have provided funding for lake and watershed restoration projects; the most recent Section 319 grants were awarded in 1996 and 1999. Currently, U.S. Department of Agriculture (USDA) and Environmental Quality Incentives Program funding is being used to implement additional conservation practices in Roberts and Marshall Counties. The key partners in the Big Stone Lake Restoration Project are watershed landowners, lake residents, Upper Minnesota River Watershed District, Citizens for Big Stone Lake, South Dakota DNR, Minnesota Pollution Control Agency (MPCA), EPA, NRCS, FWS, and local counties, conservation districts, and municipalities (EPA n.d.).

North Dakota

The entire length of Minnesota's 420-mile border with North Dakota is a north-flowing, interstate river, starting with about 25 miles of the Bois de Sioux River, followed by 395 miles of the Red River of the North. The Red River is both an interstate and an international river, as it continues north of the U.S. border another 150 miles through southern Manitoba to Lake Winnipeg. This western boundary of Minnesota also represents a change from the eastern to the western water law philosophy. The eastern water law philosophy is based on riparian rights and a reasonable use doctrine, whereas the western philosophy is "first in time, first in right," a prior appropriation doctrine. This change in water law, upstream–downstream conflicts, and the north-flowing, low-relief, international river have spawned scores of issues. Recurring transboundary problems along the Red include dueling agricultural dikes, when one party builds dikes to force more water toward a neighbor; invasive species; interbasin transfers; allocation of water supply; water quality issues, such as algal blooms resulting from phosphorus loadings in Lake Winnipeg; and flooding, such as occurs in downstream regions as ice in southern, upstream areas melts first.

Red River

The Red River could be the poster child for transjurisdictional cooperation to resolve water management issues, with its unique geographic and political setting. Water is the number-one resource management issue in this part of the country, and every LUG, from general government—including township, county, and city—to special districts (watershed management boards), gets in the game. LUGs have some authority in water management, watershed-based management boards have considerable authority, and groups of LUGs organized around water issues have little authority but considerable influence. Here, as with other wet border areas in Minnesota, nongovernmental organizations (NGOs) have emerged to help solve problems.

Red River management also includes a state agency group: the Red River Water Resources Council. This group has existed for more than 25 years and consists of agency representatives from the Minnesota DNR, MPCA, Minnesota Board of Water and Soil Resources, North Dakota Water Commission, North Dakota Department of Health, and Manitoba Water Stewardship. Federal agencies are ex officio. The primary focus of this group is to share information concerning water issues of mutual concern. The council has no statutory authority but exists to provide state-agency-level coordination that is more specific than could be provided by the many other Red River groups.

Several Red River NGOs have been created, the most recent being the Red River Basin Commission (RRBC 2010).[1] The RRBC was formed in 2002 to continue a grassroots effort to address land and water issues in a basinwide context. The commission was formed from a merger of the Red River Basin Board and the International Coalition, with some modifications to the Red River Water Resources Council to accommodate the new commission. The RRBC is a chartered, not-for-profit corporation under the provisions of Manitoba, North Dakota, Minnesota, and South Dakota law. It maintains offices in Moorhead, Minnesota, and Winnipeg, Manitoba, and is dedicated to innovation in the management of the Red River Basin's water resources. The members pay dues to support the commission's activities.

The RRBC is made up of a 41-member board of directors, composed mainly of representatives of LUGs and NGOs: cities, counties, rural municipalities, watershed boards, water resource districts, joint powers boards, First Nations and Native American representation, a water supply cooperative, a lake improvement association, environmental groups, and four at-large members. The governors of North Dakota and Minnesota and the premier of Manitoba have appointed members to the board. While the size of the commission provides most groups with representation, it also presents challenges.

The RRBC has adopted a vision, a mission statement, and a set of guiding principles, based on input from basin residents, to guide its future activities. Although general in nature, these documents serve as the foundation on which to develop goals and objectives for water management in the basin. These goals and objectives, along with the mission statement and guiding principles, provide a framework for the board to conduct business. However, the RRBC has no statutory authority to resolve water issues and receives no direct tax dollars. It is a classic example of an organization created to address issues that scores of other LUGs and agencies could not or would not address in a comprehensive scope, but that has no authority or funding.

Depending on the scope of the water management issue, Red River issues are ameliorated by cooperation among LUGs in the two states, among special districts or regional organizations, with NGOs, or with international organizations. The potential downside to too much representation and too many groups is that it can often lead to protracted deliberation, diffused messages, and dilution of funding available for water management in the region.

INTERNATIONAL TRANSBOUNDARY WATER MANAGEMENT

Minnesota's 547-mile northern border with Canada is nearly all surface water, and much is remote wilderness. The border with the province of Ontario includes the 2,000-square-mile Lake of the Woods (about 52 miles of border), the Rainey River (98 miles), a long string of border lakes and rivers that separate the Boundary Waters Canoe Area (in Minnesota) from Quetico Provincial Park (in Ontario), and the 50-mile-long Pigeon River, which empties into Lake Superior. Only a small portion of the 121-mile border with Manitoba is surface water, about 19 miles across Buffalo Bay of Lake of the Woods. U.S. states cannot enter into formal agreements with Canada, so water management issues along this northern border become issues between the United States and Canadian federal governments. However, the same group of NGOs, LUGs, and state and provincial agency players identified above is usually involved in the deliberations.

Recognizing the potential for ongoing issues related to U.S.–Canadian boundary waters, the two countries signed the Boundary Waters Treaty of 1909, which authorized the International Joint Commission (IJC 2010) as the umbrella organization for management of international border waters, with a broad coordinating role in administering the treaty. Issues along this largely wet northern Minnesota border include Red River water quality, interbasin transfer of aquatic species, commercial and recreational fisheries management, reservoir levels, and a host of issues with the Great Lakes, including out-of-basin diversions. The IJC includes Canadian and U.S. commissioners appointed to make recommendations to their respective federal governments. The commission has limited direct authority:

> The foundation upon which the basin's existing water management mechanism rests is a body of law created by state, provincial, and national legislative bodies to address water and related land resources. Added to that are the rules and regulations developed and used by dozens of government agencies to implement these laws. Zoning regulations, city ordinances, and a host of other actions by local governments likewise affect how water resources are managed. (Krenz and Leitch 1993, 99–100).

When border issues cannot be resolved through cooperation and compromise, the IJC is invited to help. The commission has established numerous issue- or water-body-specific boards and commissions to advise it, including the following (IJC 2000b):

- International Red River Board
- International Rainy Lake Board of Control
- Great Lakes Water Quality Board
- Great Lakes Science Advisory Board
- Great Lakes Fishery Commission
- International Lake Superior Board of Control

Manitoba

The Red River of the North, which flows from Minnesota into Manitoba, has been the subject of numerous water management issues, including water quality, invasive aquatic species (Leitch and Tenamoc 2001), interbasin water transfers, and flood fighting (IJC 2000a). A seemingly ongoing issue is North Dakota's desire to transfer Missouri River waters into the Red River watershed either directly or via Devil's Lake. Minnesota has often sided with Manitoba with respect to these interbasin transfers and has opposed the proposed diversions.

Lake of the Woods, a 344,000-acre transboundary lake, lies within Minnesota, Manitoba, and Ontario. The northernmost community in the Lower 48 states, Angle Inlet, is on the Northwest Angle, an area of Minnesota cut off by the lake and accessible by road only through Manitoba. Concerns over fluctuating water levels in Lake of the Woods resulted in formation of a Canadian Lake of the Woods Control Board in 1919 (IJC 2000b). The 1925 Lake of the Woods Convention and Protocol, a treaty between the United States and Canada, set elevation and discharge requirements and established the IJC International Lake of the Woods Control Board. In short, the IJC is the "go-to" agency for water issues along Minnesota's northern border.

Ontario

The U.S. Congress was involved with flood control issues along the Minnesota–Ontario border in the early 20th century (U.S. House Committee 1971), when it got involved with regulating Rainy River flows and levels. Most broad-based, substantive transboundary water issues have been or continue to be addressed in some form or another by the IJC. Because of their size and significance, the Great Lakes present a host of water management concerns. Just a few-mile stretch of Minnesota's northeast border through Lake Superior is common with Ontario, a bit more with Michigan, and still more with Wisconsin.

Great Lakes

In the northeastern part of Minnesota, the Great Lakes watershed drains to the east. The most significant recent water management development was the passage of the Great Lakes–St. Lawrence Seaway Compact. Great Lakes concerns began in earnest in 1981, when a coal slurry pipeline was proposed to take water from Lake Superior to Wyoming. Because of this issue, the eight Great Lakes states (including Minnesota and its bordering state of Michigan) and two provinces entered into a nonbinding agreement, the Great Lakes Charter of 1985, signed by the 10 Great Lakes governors and premiers. The charter is a good-faith agreement in which these leaders committed to do the following:

- give prior notice to, and consult with, each other before approving any new or increased diversions or consumptive uses over 5 million gallons per day, averaged over any 30-day period;
- manage and regulate all new withdrawals that result in a new or increased diversion or consumptive use of Great Lakes water over 2 million gallons per day averaged over any 30-day period; and
- collect and share comparable information on all Great Lakes water withdrawals of 100,000 gallons per day averaged over any 30-day period.

A year later, in 1986, the U.S. Congress passed the Water Resources Development Act (WRDA), which contained a provision requiring unanimous approval of all Great Lakes governors for any diversion of water out of the Great Lakes basin. The WRDA provisions did not include a minimum threshold, so the law applies to a diversion of any size. Some states did not follow through with earlier commitments under the charter, because the WRDA provides veto power over any diversions out of the basin. Because unanimous approval is needed from all eight governors, any Great Lakes state could veto a diversion request.

In 1998, the Ontario-based Nova Group requested and received a permit from that province's Ministry of the Environment to ship, in bulk containers, approximately 160 million gallons per year of raw water from Lake Superior for the purpose of selling the water "in Asia." Because the amount of water withdrawn would be less than 5 million gallons per day averaged over any 30-day period, and because the proposal was in Canada, neither the Great Lakes Charter's prior notice and consultation requirements nor the WRDA provisions were applicable. This permit raised significant concern among the Great Lakes states and other interested parties.

Additionally, a growing concern has been that the federal WRDA could be successfully challenged because there were no standards or processes to guide decisions. These concerns started a seven-year process of developing both a binding compact between the eight states and a nonbinding agreement between these states and the two provinces. The negotiations were complicated, because each state was starting with very different regulatory procedures and because all of the governors and premiers changed during this process. Two agreements were needed because a binding agreement between the United States and Canada would be a treaty, and only the U.S. State Department can enter into treaties. A compact among states must be passed by each state, using identical language, and then be approved by Congress. The resulting compact met these requirements and went into effect as of December 8, 2008.[2] Most states had to pass new laws to meet the compact requirements. Minnesota's existing water appropriation permit program addressed the water withdrawal requirements of the compact, however, and new authorities were not required.[3]

The compact prohibited diversions to any location outside the Great Lakes Basin, with two very specific exceptions: a community that straddles the watershed boundary or a community within a straddling county may apply for an exception to the prohibition. The compact established a council that includes the governor or designee from each state, and the agreement established a regional body that includes council members and representatives from Ontario and Quebec. These entities provide specific administrative review functions and have review or approval authority for new or increased diversions and consumptive use proposals. They review and approve exceptions to the prohibition on diversions and review and comment on consumptive uses over 5 million gallons per day. The Council of Great Lakes Governors' staff serves as secretariat for both groups. The members pay dues to support the activities.

CONCLUSIONS

A persistent problem, a common threat, and the appearance of unilateral decisionmaking over a bilateral, transboundary water resource have all been common threads that have led to cooperation, compacts, and treaties among states and provinces. An additional method of resolution—using the court system—is not addressed in this chapter, because most court findings lead to forced cooperation and compromise, interstate compacts, or international treaties, which fall outside the purview of this discussion.

One key ingredient in sustaining interstate and international, transjurisdictional cooperation through a committee of affected groups is having adequate financial support to provide staff to organize and facilitate discussion. Another key is including the right levels and expertise of NGO, LUG, and agency representatives. Leadership and a clearly defined mission are also important to reaching decisions regarding interstate and international water management issues. Finally, statutory authority or substantial influence is needed to affect real policy changes.

The proliferation of multijurisdictional groups using old-fashioned politics of cooperation and compromise has enabled Minnesota to collectively manage resources it shares with its neighbors. Having one state or nation make unilateral decisions is not good for the overall system, and a system that can ensure broad participation is desirable. Minnesota has the unique position of being upstream, but shared decisionmaking with downstream neighbors results in better, more enforceable overall policies. For any group to function well, it needs to recognize a common threat and be willing to compromise and to pay to support the effort needed to address that threat.

However, old-fashioned water management politics also has its downsides. The time it takes to reach decisions can be greatly lengthened by the grassroots democracy of inclusion, whereby no stakeholder, NGO, or LUG is left out of the process. The already large number of represented groups interested in an issue can double when state or international borders are crossed. While each stakeholder

(NGO, LUG, regional or state agency) may be quick to take credit for successes, it can be equally quick to deny responsibility or withhold funding. Further, with such a wide host of players, partners, stakeholders, and decisionmakers, it is easy to see how the sum of well-intentioned small choices can lead to an unintended overall outcome, an example of what has been termed the tyranny of small decisions (Kahn 1966; Leitch 2003). Also, money to help solve water management problems is not unlimited, and each additional group involved dilutes the total financial pool available. Finally, another downside of the old-fashioned politics of grassroots democracy is that many individual board members, especially of ground-level groups, have little or no appropriate expertise to bring to the table. For example, a local township board member may vote on an issue that ultimately affects residents of another country hundreds of miles downstream, when his or her political perspective is the immediate neighborhood. In some instances, where water management issues persist, overall social well-being might best be served with a watershed-wide authority led by an independent, objective decisionmaker.

NOTES

1. As Brown (2001, 265) put it, "Few areas of human activity have been so dominated by NGOs as the environmental movement [or water management in the Upper Midwest]. Broadly speaking, NGOs evolve to fill gaps by government and the business sector. Literally thousands of such groups have been formed in both industrial and developing societies."
2. Minn. Stat. § 103G.801.
3. The Council of Great Lakes Governors' website (www.cglg.org) has detailed information on the status of ongoing issues within the basin.

REFERENCES

Brown, L.R. 2001. *Eco-Economy: Building an Economy for the Earth*. New York: W.W. Norton & Company.
EPA (U.S. Environmental Protection Agency). No date. Big Stone Lake Restoration Project. www.epa.gov/region8/water/nps/sdbigstone.pdf (accessed January 14, 2010).
Foster, C.H.W. 1990. What Makes Regional Organizations Succeed or Fail? In *Proceedings of the Symposium on International Transboundary Water Resources Issues*. pp. 11–18. Edited by John E. FitzGibbon. Bethesda, MD: American Water Resources and Canadian Water Resources Association.
IJC (International Joint Commission). 2000a. *Living with the Red*. Washington, DC: IJC.
———. 2000b. *Transboundary Watersheds*. Washington, DC: IJC.
———. 2010. Home page. www.ijc.org (accessed December 3, 2010).
Kahn, A.E. 1966. The Tyranny of Small Decisions: Market Failures, Imperfections, and the Limits of Economics. *Kyklos* 19:23–47.
Krenz, G., and J.A. Leitch. 1993. *A River Runs North: Managing an International River*. Fargo, ND: Red River Water Resources Council.
Leitch, J.A. 2003. Floodplains and the Tyranny of Small Decisions. *ASFPM [Association of State Floodplain Managers] News & Views* 15 (5):1, 10–12.
Leitch, J.A. and M.J. Tenamoc, eds. 2001. *Science and Policy: Interbasin Water Transfer of Aquatic Biota*. Fargo, ND: Institute for Regional Studies, North Dakota State University.

RRBC (Red River Basin Commission). 2010. Home page. www.redriverbasincommission.org (accessed December 3, 2010).

UMRBA (Upper Mississippi River Basin Association). No date. About UMRBA. www.umrba. org/aboutumrba.htm (accessed January 17, 2010).

U.S. Congress. House. Committee on Flood Control. 1971. Newlands-Broussard-Rainey River Regulation Bill. 64th Cong., 1st sess., HR V13975.

CHAPTER 3

Minnesota's Landscape Characteristics: The Implication for Water

Steven Heiskary

G lacial activity made Minnesota one of the most water rich states of the Lower 48, with more than 13.1 million acres of water, including lakes, wetlands, and rivers. Actual numbers or miles of lakes, wetlands, and streams may vary depending on map scale or data source. Based on recent National Hydrography Dataset (NHD) estimates, Minnesota has about 105,000 stream miles, which are distributed among 81 major watersheds, and about 9.3 million wetland acres. The newest NHD places the number of lakes at 12,200; of these, 800 are greater than 500 acres, 4,000 are between 100 and 500 acres, and the remainder are between 10 and 100 acres. Minnesota's large number of lakes, relative to other states in the nation, was acknowledged in the 2007 National Lakes Assessment (NLA); Minnesota had more lakes drawn for this statistically based survey of the nation's lakes than any other state in the Lower 48 (MPCA 2009a).

While lakes, rivers, wetlands, and groundwater are intrinsically linked, we address policy and management issues in three separate chapters, with Chapter 7 focusing on wetlands and Chapter 8 on groundwater. This chapter looks at Minnesota's lakes and rivers, examining their spatial distribution, how they are organized for management purposes, regional patterns that influence the quantity and quality of water, and how land management practices have impacted these resources. Policy and management issues are raised in the context of this discussion.

AN OVERVIEW OF MINNESOTA'S LAKE AND RIVER RESOURCES

Minnesota's major basins (Figure 3.1) are a primary basis for broad-scale water resource management and planning. The Red and Rainy River Basins drain north

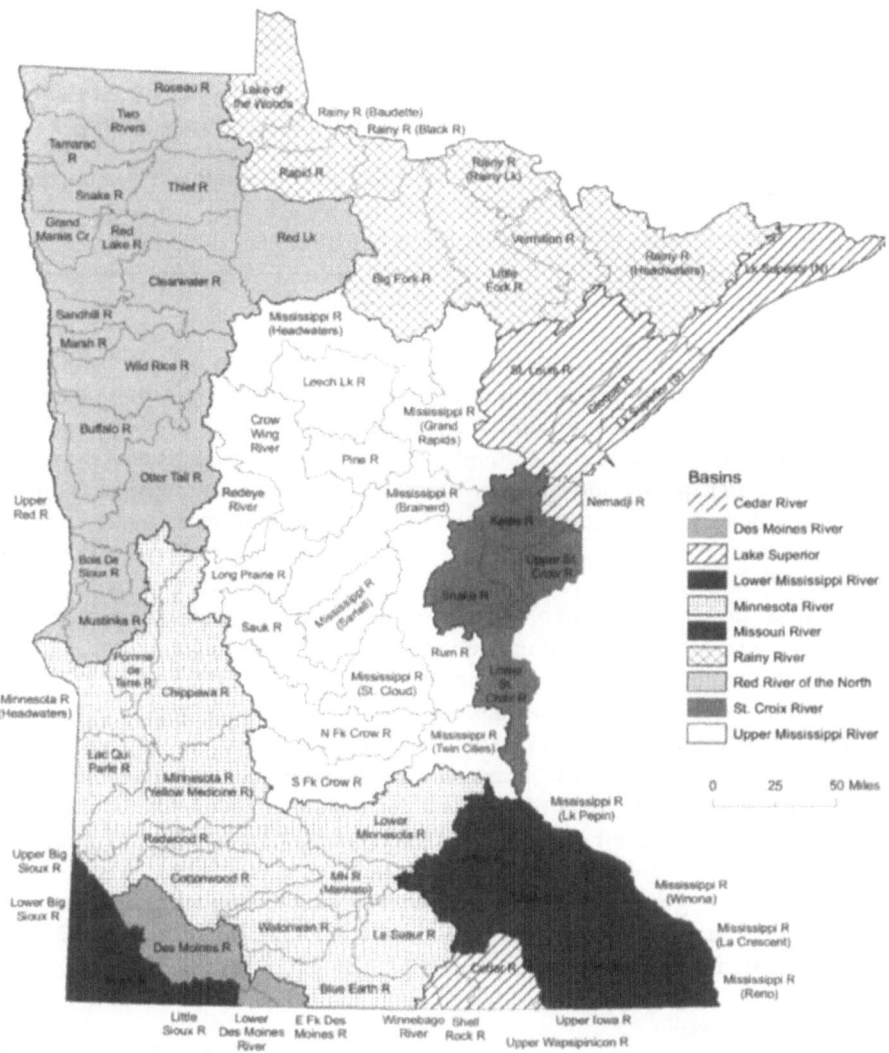

Figure 3.1 *Basins and major watersheds*

to Hudson Bay, while the Mississippi, Minnesota, and St. Croix and the very small basins—the Cedar, Des Moines, and Missouri—drain south to the Gulf of Mexico. The Lake Superior Basin, including the numerous streams along the North Shore of Lake Superior and the St. Louis and Nemadji Rivers, drains to Lake Superior and represents the headwaters of the St. Lawrence, which flows to the Atlantic.

These basins comprise 81 major watersheds (Figure 3.1), which are a primary focus for monitoring and the development of strategies for addressing water quality and quantity issues within each of the basins. Though there are numerous rivers in Minnesota, three—the Mississippi, Red, and Rainy—drain the majority of the state, and each deserves a closer look.

Case Study 1. Minnesota's Major Rivers and Downstream Responsibility

Being at the headwaters of three major rivers, the Mississippi, Red, and Rainy, Minnesota has an important responsibility to downstream water users. While numerous policy issues exist, focusing on one for each of these major basins can help demonstrate this responsibility and show how downstream considerations need be factored into Minnesota's water management.

Mississippi River. The Mississippi River, from its headwaters at Lake Itasca, flows south 2,350 miles to the Gulf of Mexico (MPCA 2009c). It drains over 65% of Minnesota via four basins: the Upper Mississippi, St. Croix, Minnesota, and Lower Mississippi. The states through which the Mississippi River flows face numerous water quality and quantity issues. Minnesota, at the headwaters of the river, has a downstream responsibility to ensure that the water leaving the state is of good quality.

One of the more high-profile issues, raised in recent years, is hypoxia (low oxygen) in the Gulf of Mexico. Various studies have documented how upstream sources of nitrogen and phosphorus contribute to the excessive algal growth that leads to the hypoxic conditions in the Gulf. Hypoxia contributes to reductions, and in some cases elimination, of valuable forms of aquatic life over a large portion of the Gulf. Studies have estimated the relative contributions to this problem from the various states that drain to the Mississippi, including Minnesota (e.g., Robertson et al. 2009). These studies have attributed 1% to 5% of the nitrogen load to Minnesota, and of this, the majority arises as diffuse or nonpoint-source runoff in the Minnesota River (Alexander et al. 2008). Although the state has done a good job at reducing point-source phosphorus at wastewater treatment facilities, much less progress has been made on nitrogen. Efforts to reduce nonpoint-source nitrogen loading must consider the role of wetland removal and tile drainage, both of which allow for efficient downstream transport of dissolved nitrogen.

Currently, various strategies are under discussion to determine how best to reduce the loading of excess nitrogen and phosphorus going to the Gulf. Minnesota has been active in these discussions via the Upper Mississippi River Basin Association (UMRBA) and involvement in task forces. At this point, it is unclear what mandates might be put forth for states to implement, such as load allocations (similar to a large-scale total maximum daily load, or TMDL), water quality standards, or some combination of these. The U.S. Environmental Protection Agency (EPA) has long been urging states to develop nutrient standards in response to a broader call for action on excess nutrients (e.g., EPA 1998). Minnesota has been a leader in developing nutrient standards by promulgating lake nutrient standards in 2008 and drafting river nutrient criteria. Whatever approach is promoted at the federal level, it seems evident that there will be a continued emphasis on addressing the problem of excess nutrients, and to be successful, nonpoint-source nutrient loading must be reduced. It is important that Minnesota remains actively engaged in discussions on Gulf hypoxia, nutrient criteria development, and related issues in both science- and policy-focused forums. For these purposes, continued involvement in UMRBA, the Association of Water Pollution Control Administrators, EPA Regional Technical Assistance Groups, and various EPA-sponsored forums and workgroups will provide access to the current state of the science on the issue, as well as an opportunity to share Minnesota's viewpoint on the best way to address problems.

Red River. The Red River of the North Basin stretches from northeastern South Dakota and west-central Minnesota northward through eastern North Dakota and northwestern Minnesota into southern Manitoba. It ends where the river empties into the southern end of Lake Winnipeg. The Minnesota portion of the Red River Basin covers about 37,100 square miles in the northwestern corner of the state, in all or part of 21 counties (MPCA 2009b).

The basin contains some of the richest and flattest farmland in the world, but this does not come without its problems. Although the ditching and draining of wetlands have enhanced the ability to farm large portions of the basin, this practice has created water quality and quantity problems in the Red mainstem and many of its major tributaries. Excess nutrients and sediments run off the landscape, leading to impaired water quality and aquatic biota downstream. Rapid runoff in the spring and after major rain events often leads to flooding along the river, particularly in the downstream reaches. This can be worsened further in the spring as the ice in the downstream reaches may not have melted or ice floes may cause water to be backed up. In late summer, the opposite problem may occur as rivers drop to very low levels, which can contribute to water quality problems as well (MPCA 2009b).

Because the Red River drains portions of northeastern South Dakota, western Minnesota, eastern North Dakota, and southern Canada, it is clearly a river of regional and international significance. One of the water quality issues raised in recent years is the effect of elevated phosphorus (P) concentrations, or loads, on Lake Winnipeg. One of the largest lakes in the world, Winnipeg periodically experiences severe nuisance blooms of blue-green algae as a result of elevated phosphorus concentrations. A Manitoba Conservation estimate places the U.S. portion of the Red River total phosphorus (TP) contribution to Lake Winnipeg at 43% (Bourne et al. 2002). Minnesota accounts for about 40% to 50% of the U.S. portion of the watershed, but it contributes about 75% of annual flow to the Red (Christensen 2007), along with a significant amount of the U.S. portion of the downstream phosphorus load that impacts Lake Winnipeg. It is important that Minnesota be fully engaged with North Dakota, Manitoba, and the Canadian government in developing strategies to reduce downstream phosphorus loading.

Rainy River. The Rainy River Basin is home to some of Minnesota's finest forest and water resources. Voyageurs National Park and the Boundary Waters Canoe Area Wilderness (BWCAW) are located within the Rainy River Basin, as are several of the state's most famous walleye fisheries and many top-notch trout streams (MPCA 2001). The majority of the land within the basin is forested. Prominent uses of natural resources in the basin are forestry, mining, and various forms of recreation. Although land use composition is not as complex as in the other basins, management of water resources can be equally challenging, given the large amount of land in a federal wilderness area (BWCAW), a national park, eight Minnesota counties, and a shared border and significant shared water resources with the province of Ontario and the Canadian government.

Forest and wetland land uses predominate across the basin and contribute to the generally good water quality found in the basin. However, this basin is not without its water quality challenges. A recent one was the 2008 303(d) listing of Lake of the Woods (LOW) for excess nutrients. The designation "303(d)" refers to the section of the Clean Water Act that requires assessment of the condition of the nation's waters relative to water quality standards. A 303(d) listed water would be considered impaired because it does not meet one or more of these standards. Different portions of LOW have distinct differences in water quality. Much of the northern portion in Canada is oligotrophic to mesotrophic in nature, whereas the Minnesota portion is more eutrophic and has had a history of late-summer blue-green algal blooms. Collaborative monitoring by the Minnesota Pollution Control Agency (MPCA), LOW property owners, the Minnesota Department of Natural Resources (DNR), and the province of Ontario has provided a rich database for the lake. Data from the Minnesota portion were compared with the Northern Lakes and Forests (NLF) ecoregion standards, and levels of phosphorus and chlorophyll-a were found to exceed the standards. As a result of the 303(d) listing, a collaborative study is under way that involves all interested parties on this important border water issue. This study will result in a TMDL that will have policy implications for regulated dischargers in the watershed of the lake. In this case, Minnesota has a responsibility to ensure that its portion of the LOW watershed is not exporting excessive amounts of phosphorus that may be contributing to water quality impairment.

REGIONALIZATION AS A BASIS FOR ASSESSING CONDITIONS AND DEVELOPING WATER QUALITY CRITERIA

Although basin planning and developing monitoring and management strategies on a watershed basis are very useful approaches, they do not, on their own, provide an adequate framework for assessing the condition of Minnesota's streams and lakes or for developing standards for this purpose. This is because of distinct differences in Minnesota's landscape characteristics that exhibit a gradient from the northeastern to the southwestern portions of the state. As these features—land forms, land uses, and vegetation—change, so does the character of lakes and streams in the various regions.

The recognition of regional patterns and their implications for resource management is not new. In mapping the presettlement vegetation of Minnesota, Marschner (1930) demonstrated the distinctly different landscapes that

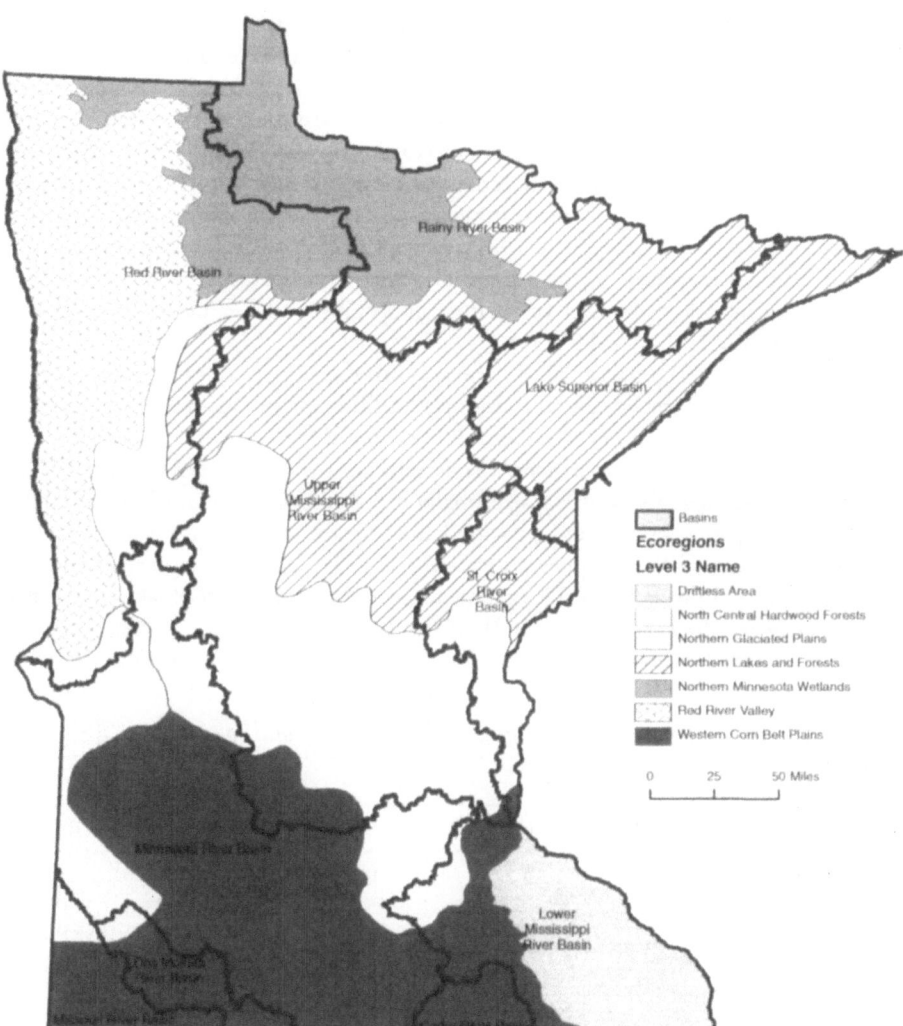

Figure 3.2 *Ecoregions and basins*

characterize the state, with a northeast-to-southwest pattern that still exists today (Tester 1995). Moyle (1956) recognized distinct regional patterns that were generally considered a function of geology, vegetation, hydrology, and land use. These observations helped shape fishery and wildlife management in the state. Zumberge (1952) provided perhaps the most comprehensive overview of the forces that formed many of Minnesota's lakes and in the process also described distinct regional patterns.

Understanding regional patterns is important to the assessment, management, and development of policies for Minnesota's lake and stream resources. There have been many approaches to regionalization, including the Ecological Classification

System and agroecoregions, but the one that has had the most impact on water resource management in Minnesota has been Omernik's ecoregions as mapped by EPA for the contiguous Lower 48 states (Omernik 1987). Ecoregions are intended to provide a spatial framework for ecosystem assessment, research, inventory, monitoring, and management (Omernik and Bailey 1997). Minnesota is characterized by seven Level III ecoregions (Figure 3.2), which have provided a framework for describing patterns in water quality, morphometry, and ecology of Minnesota's lakes (Heiskary et al. 1987). Level III refers to the ecoregion scale used in the original mapping of the Lower 48 states as reported in Omernik (1987). Later mapping efforts sought to identify subregions within Level III, and these were referred to as Level IV.

Analysis of Minnesota's lakes focused on these ecoregions, with the vast majority (98%) of the state's lake resources falling in four ecoregions: Northern Lakes and Forests (NLF), North Central Hardwood Forests (CHF), Western Corn Belt Plains (WCP), and Northern Glaciated Plains (NGP) (Heiskary and Wilson 1989). A detailed analysis of 104 minimally impacted (reference) lakes, distributed across these four ecoregions, provided a basis for describing watershed, lake morphometry, and water quality characteristics for each ecoregion (Table 3.1). This analysis has proved useful for characterizing the condition of lakes and setting goals in watershed projects (Heiskary and Wilson 1989). These same data contributed to the development of regional lake eutrophication criteria that were

Table 3.1 *Typical summer water quality by ecoregion and land use*

Parameter	Ecoregion			
	NLF	CHF	WCP	NGP
# of lakes	32	43	16	13
Area (ha)	61–208	74–297	75–168	198–258
Depth: mean (m)	2.5–10.5	4.8–7.9	1.9–3.4	1.5–1.8
Depth: maximum (m)	9.1–24.1	13.0–22.0	3.0–8.2	2.0–3.0
Watershed land use				
% forested	54–81%	6–25%	0–15%	0–1%
% wetland	14–31%	14–30%	3–26%	8–26%
% cultivated	0–1%	22–50%	42–75%	60–82%
% pasture	0–6%	11–25%	0–7%	5–15%
% developed	0–7%	2–9%	0–19%	0–2%
Water chemistry				
TP (µg/L)	14–27	23–50	65–150	122–160
Chlorophyll-a (µg/L)	4–10	5–22	30–80	36–61
Secchi disk (m)	2.4–4.6	1.5–3.2	0.5–1.0	0.4–0.8
T. Kjeldahl N (mg/L)	0.4–0.75	< 0.60–1.2	1.3–2.7	1.8–2.3
Alkalinity (mg/L)	40–140	75–150	125–165	160–260
pH (SU)	7.2–8.3	8.6–8.8	8.2–9.0	8.3–8.6
Chloride (mg/L)	0.6–1.2	4–10	13–22	11–18
Conductivity (µmhos/cm)	50–250	300–400	300–650	640–900

Notes: ha = hectares; m = meters; µg = microgram(s), L = liters; mg = milligrams; SU = standard units; µmhos = micromhos; cm = centimeters.
Source: Adapted from Heiskary and Wilson (2008).

Table 3.2 *Lake eutrophication criteria defined by ecoregion for specific lake types and uses*

Ecoregion and lake type (official use classification)	TP[a] (μg/L)	Chlorophyll-a[a] (μg/L)	Secchi[b] (m)
NLF: Designated lake trout (Class 2A)	12	3	4.8
NLF: Designated stream trout (Class 2B)	20	6	2.5
NLF: Aquatic recreation use (Class 2B)	30	9	2.0
CHF: Designated stream trout (Class 2B)	20	6	2.5
CHF: Aquatic recreation use—deep (Class 2B)	40	14	1.4
CHF: Aquatic recreation use—shallow (Class 2B)	60	20	1.0
WCP & NGP: Aquatic recreation use—deep (Class 2B)	65	22	0.9
WCP & NGP: Aquatic recreation use—shallow (Class 2B)	90	30	0.7

Notes: Aquatic life and recreation use classes as defined in Minnesota Rules 7050.0140, subpart 3, and 7050.0222 (Minnesota Revisor of Statutes 2007). *Class 2A* is used for waters supporting a cold-water fishery and refers specifically to lakes that support natural populations of lake trout. *Stream trout* refers to all other designated (managed) trout lakes. *Class 2B* is a broad designation that applies to the majority of Minnesota's rivers and lakes and allows for the propagation and maintenance of cool- or warm-water fish and associated aquatic life.
[a]TP and chlorophyll-a should remain below these concentrations.
[b]Secchi should be not less than this value to ensure that the specific use is maintained.
Source: Adapted from Heiskary and Wilson (2008).

promulgated into Minnesota's water quality standards in 2008 (Table 3.2). Details on the approach used to develop these standards may be found in Heiskary and Wilson (2008).

Case Study 2. Regional and Temporal Trends in Lake TP Based on Sediment Diatom Reconstruction

Lake-sediment cores provide a valuable archive of information over a historical continuum of water quality and related environmental factors. For Minnesota, they serve to reaffirm the existence of regional patterns, provide a basis for describing natural background conditions, and give insight into how land use changes influence lake water quality. In particular, fossil diatoms, a form of algae preserved in the cores, have allowed researchers to reconstruct historic phosphorus, chloride, pH, and other measurements.

Based on sediment cores from 55 lakes distributed across Minnesota (Heiskary and Swain 2002), distinct differences among regions were evident in the comparisons of pre-European and modern–day TP concentrations (Figure 3.3). For the NLF lakes as a group, there was no significant difference in pre-European versus modern–day TP, and in general, land uses (historically and today) are predominantly forests and wetlands. Significant increases in TP were noted for the CHF lakes, and these increases can be attributed to agricultural and urbanized land uses that are now common across the region (Ramstack et al. 2004).

Shallow lakes were underrepresented in the original study, and this led to additional collection of cores from shallow lakes. The resulting 79-lake model (Heiskary et al. 2004) exhibited a significant modern–day TP increase relative to

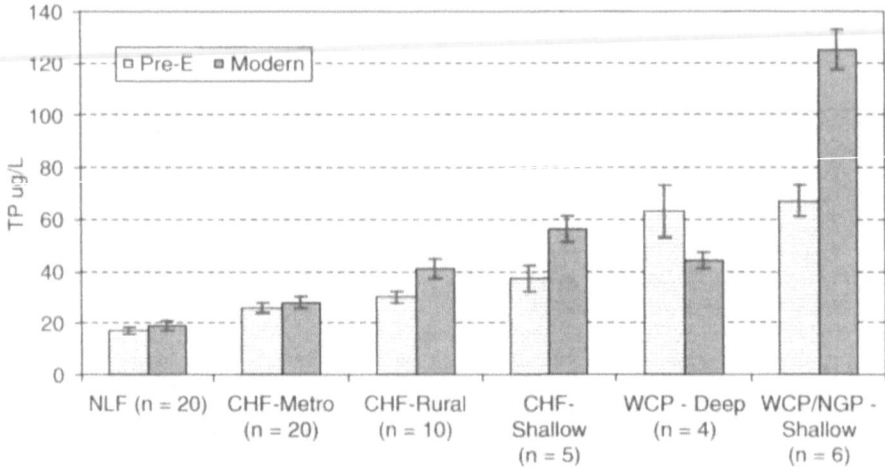

Figure 3.3 *Diatom-inferred TP concentrations: Pre-European versus modern-day*

pre-European values for shallow lakes (Figure 3.3). It was also evident that pre-European TP in shallow CHF and WCP lakes was higher than in their deeper counterparts. Collectively, the sediment core data reaffirmed the need to regionalize lake eutrophication standards, demonstrated differences between deep and shallow lakes, and provided a basis for defining "background" concentrations—all of which contribute to the development of sound management strategies (such as specifically developing shallow lake standards) for Minnesota's lakes.

The combination of land use and other factors that account for the differing ecoregion characteristics strongly influence stream water quality and quantity as well. In a statistical assessment of water quality data from minimally impacted stream sites, McCollor and Heiskary (1993) characterized the "typical" water quality of various ecoregions (Table 3.3).

Table 3.3 *Stream water quality by ecoregion*

Percentile/region	Total phosphorus (mg/L)			Total suspended solids (mg/L)			Nitrate-N (mg/L)		
	25%	50%	75%	25%	50%	75%	25%	50%	75%
NLF	0.02	0.04	0.05	1.8	3.3	6.0	0.01	0.03	0.09
NMW	0.04	0.06	0.09	4.8	8.6	16.0	0.01	0.02	0.08
CHF	0.06	0.09	0.15	4.8	8.8	16.0	0.04	0.10	0.26
NGP	0.09	0.16	0.25	11.0	34.0	63.0	0.01	0.14	0.51
RRV	0.11	0.19	0.30	11.0	28.0	59.0	0.01	0.07	0.21
WCP	0.16	0.24	0.33	10.0	27.0	61.0	1.40	3.90	7.40

Notes: mg/L = milligrams per liter; interquartile range based on data from representative minimally impacted streams (1970–1992).
Source: McCollor and Heiskary (1993).

Understanding these patterns in stream water quality and the factors that contribute to elevated concentrations is important to policy and management efforts designed to improve stream water quality and protect downstream receiving waters. TP and total suspended solids (TSS) often covary, with phosphorus adhering to soil particles that are conveyed to lakes and rivers via surface runoff, gully erosion, and streambank sloughing. The increased intensity of land uses (Table 3.1) combined with soil characteristics results in higher TP and TSS concentrations in CHF streams as compared with NLF ecoregion streams (Table 3.3). This is even more evident in the WCP and NGP ecoregions, where elevated TP and TSS are common across most streams in these ecoregions. Nitrate-N, in contrast to TP, is delivered principally in a dissolved form and is readily transported to streams via surface water, subsurface runoff, and groundwater (Tesoriero et al. 2009). Subsurface delivery is most common in highly drained (tiled) landscapes, which are common throughout agricultural lands in the WCP ecoregion. This results in some of the highest nitrate-N concentrations in the state (Table 3.3), and as noted previously, this source category must be addressed in any policy discussions that seek to find solutions for minimizing the downstream transport of N.

Ecoregions are used in conjunction with watershed-based approaches for assessment and management in Minnesota. For example, watersheds (basins) that straddle two adjacent ecoregions (e.g., the Upper Mississippi River Basin; see Figure 3.2) reflect characteristics of both, and the ecoregion framework can be used to explain differences and develop meaningful water quality standards. In Minnesota, Level III ecoregions (Figure 3.2) provide a basis for regionalizing river nutrient criteria (Heiskary et al. 2010) and will also be used in the development of regional TSS standards, which will replace the existing turbidity standard for streams. This particular example of regionalization is quite significant from a policy standpoint, because prior to the development of the new TSS standards, all streams across the state were expected to meet the same turbidity standard.

Up to this point, we have emphasized broad spatial patterns and temporal changes at the statewide and ecoregion level. To bring the policy and management discussion to a more local scale, we now offer a couple of lake-specific examples to further demonstrate regional differences and the impact of land use changes on lake water quality. Although specific to these two lakes, the management issues addressed affect numerous lakes throughout the state.

Case Study 3. Impact of Urbanization on Water Quality of an Urban Lake

Fish Lake, like many lakes in the Twin Cities metropolitan area (TCMA), had minimal development in its watershed up until the late 1960s. From 1971 to 1975, the residential area around the lake was developed (Heiskary and Swain 2002), resulting in increased runoff entering the lake. A series of events from 1980 to 1987 resulted in an expansion of the watershed entering the lake. These efforts increased the watershed of Fish Lake from about 261 acres to more than 3,300 acres (116:1 watershed-to-lake ratio). A significant increase in TP (Figure 3.4) and extraordinary increase in chloride from the 1970s to the 1990s can be attributed to the expansion and highly urbanized nature of its watershed (76% of the watershed area is built up or roads). Recently collected data indicate that TP remains elevated above that observed prior to the expansion of the lake's watershed and above that which is needed to meet the water quality standard (40 µg/L; Table 3.2).

The management implications from this extensive urbanization and linking of water bodies in this urban area are significant. Fish Lake was placed on Minnesota's 2006 list of impaired waters because of elevated nutrients. A TMDL is under development, and the city of Eagan is faced with making substantial reductions in nutrient loading to the lake, which is complicated by the large watershed and limited area available for treatment of stormwater. While this is one specific example of the impact of urbanization on lake water quality and quantity, the general theme is not unique to Fish Lake, but rather is shared by many in the rapidly urbanizing TCMA as well as other such areas in Minnesota. It is quite likely that had current stormwater management policies been in place in the 1980s, they would have minimized the increase in TP in Fish Lake and perhaps numerous other TCMA lakes that experienced rapid development in the 1980s and 1990s.

Case Study 4. Impact of Agricultural Practices on a Rural Lake

Lake Shaokotan in Lincoln County, with a maximum depth of about 12 feet and a predominantly agricultural watershed, is fairly typical of lakes in this ecoregion. The lake has a history of water quality problems, including severe nuisance blooms of blue-green algae, summer and winter anoxia, and periodic fish kills. A detailed diagnostic study was initiated in 1989, and restoration efforts were under way by 1991. Implementation included rehabilitation of three animal feedlots, four wetland areas, and shoreline septic systems.

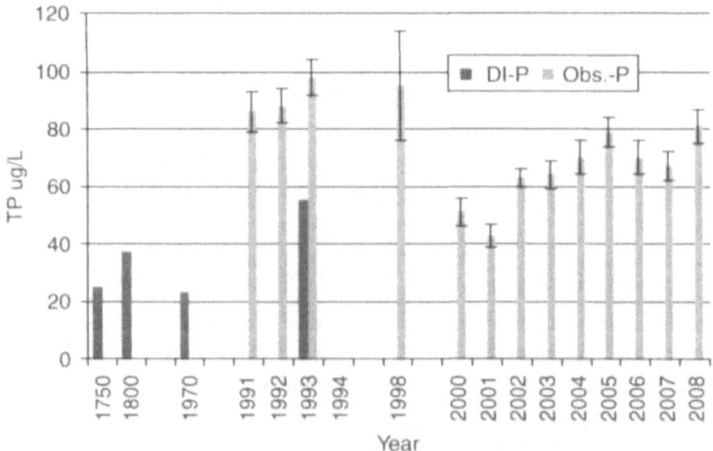

Figure 3.4 *Land cover composition and drainage network for Fish Lake in Eagan*

Notes: DI-P = diatom-inferred phosphorus; Obs.-P = observed phosphorus.

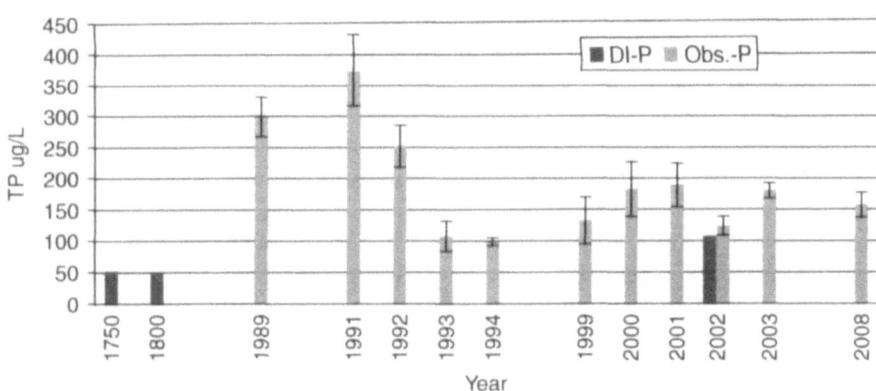

Figure 3.5 *Summer-mean total phosphorus and sediment diatom-inferred phosphorus for Lake Shaokotan*

Notes: DI-P = diatom-inferred phosphorus; Obs.-P = observed phosphorus.

By 1994, significant reductions in in-lake phosphorus were realized, with concentrations approaching the ecoregion-based standard of 90 µg/L (Figure 3.5). Nuisance algal blooms were less frequent, and anecdotal evidence in 1999 suggested that rooted plant populations were increasing. Subsequent plant surveys in 2000 and 2002 found essentially no rooted plants, however, and increases in TP and chlorophyll-a were noted (Figure 3.5). This increase was largely attributed to an abandoned feedlot operation in the near-shore area of the lake and storm events that flushed runoff from this area to the lake. TP and chlorophyll-a remained above the water quality standards for the NGP ecoregion (Table 3.2) Hence the lake was included on the 2002 303(d) list, and a TMDL study was completed in 2010.

At this point, the ecoregion-based eutrophication criteria appear to be reasonable and achievable for the lake, given that pre-European diatom-reconstructed TP (Figure 3.3) was well below the criteria (Table 3.2) and in light of the progress that had been attained as a result of watershed efforts in the early 1990s. From a management and policy standpoint, this case study indicates the difficulty of preserving water quality gains when nonpoint sources are the primary source of excessive nutrients and when reliance is strong on implementation of best management practices. This suggests there may be a need to modify policy on how nonpoint sources are addressed in the context of pollution abatement efforts, in particular when reductions in these sources are needed to meet a TMDL.

CONCLUSIONS

Minnesota is blessed with a wealth of water resources. These resources are distributed among 10 major basins and 81 major watersheds. While the basins and major watersheds serve as a good framework for monitoring and management, the underlying regional patterns must be considered as well in the overall assessment and water quality standards development process. For Minnesota, a combination of the watershed approach and the ecoregion framework has provided a good basis for designing monitoring strategies, assessing the condition of the waters, developing protection and restoration strategies, and drafting basin and eventually watershed-based plans. Although much progress has been made in recent years, the need continues to assess the condition of Minnesota's waters, evaluate the impacts of land use and climatic changes, address impairments, and develop plans to protect the quality and quantity of water. In all cases, understanding and accounting for the underlying regional patterns in water resource quantity and quality will be important. Several new or evolving efforts should be valuable in this regard and serve to further shape water resource management and policy in Minnesota. Some pertinent examples follow.

River eutrophication standards are under development as part of EPA's nationwide effort to encourage states to develop standards for lakes, rivers, wetlands, and estuaries. Minnesota has developed lake nutrient standards on an ecoregion basis (Table 3.2). River eutrophication standards will have a similar region-based approach that will consider regional patterns in stream water quality (Table 3.3) but will also require some adaptations that take into account that rivers flow from one region to the next. Regionalized TSS standards will replace the existing turbidity standard, which will be an advancement as well and should allow for improved assessments and development of more realistic TMDLs. Background on Minnesota's approach to river eutrophication standards may be found in Heiskary et al. (2010).

The DNR, with assistance from the MPCA, is leading a new sentinel lakes program called Sustaining Lakes in a Changing Environment (SLICE) (DNR 2010). The focus of this interdisciplinary effort is to improve understanding of how major drivers of change, such as development, agriculture, climate change, and invasive species, can affect lake habitats and fish populations, and to develop a long-term strategy to collect the necessary information to detect undesirable changes in Minnesota lakes. To increase the state's ability to predict the consequences of land cover and climate change on lake habitats, SLICE uses intensive lake-monitoring strategies on 24 regionally representative Minnesota lakes (Figure 3.6). This includes analyzing relevant land cover and land use, identifying climate stressors, and monitoring the effects of these factors on the lake's habitat and biological communities. Further, cooperation with multiple entities charged with aquatic resource management ensures rigor (e.g., the right people doing the job), efficiency, relevance, and shared ownership in commonly held goals.

The Tiered Aquatic Life Use (TALU) framework will be a significant revision to the Water Quality Standards of Minnesota's aquatic life use classification. The TALU framework refines and expands existing water quality standards, with a goal of improving how water resources are monitored and managed. One of the strengths of the TALU approach is its ability to take into account natural variations in water resources. Because of Minnesota's diverse water resources, rigid standards and beneficial uses lead to errors in assessment and management. For example, within rivers and streams, natural factors like water body size, geographic location, hydrology, water temperature, and stream gradient influence chemical, physical,

Figure 3.6 *Map of the SLICE program's sentinel lakes as of 2008, with major land types noted*

Source: Map developed by the Minnesota DNR.

and biological composition. As a result, different expectations are needed for these water resources. Without setting appropriate expectations, chemical and biological goals may be underprotective of the highest-quality resources and overprotective of others. TALU allows for better goal-setting processes through the application of a framework that recognizes tiers of aquatic life use based on a stream's type and potential and should serve to refine existing statewide or regionally based criteria. By accounting for natural variation in these systems and setting appropriate goals for water resources in Minnesota, TALU will set biological and chemical goals that are protective yet attainable.

The MPCA's watershed-based approach for biological and chemical monitoring, water resource assessments (e.g. 303(d)), and development of TMDLs should lead to a coordinated effort to address the many water quality issues Minnesota confronts. As this strategy is implemented over the next several years, it will make use of ecoregion-based criteria, indices of biotic integrity, and other tools and approaches, such as TALU, to ensure the proper assessment of Minnesota's water resources and the development of appropriate implementation strategies to protect or restore the quality of the waters.

REFERENCES

Alexander, R.B., R.A. Smith, G.E. Schwarz, E.W. Boyer, J.V. Nolan, and J.W. Brakebill. 2008. Differences in Phosphorus and Nitrogen Delivery to the Gulf of Mexico from the Mississippi River Basin. *Environmental Science & Technology* 42 (3):822–830.

Bourne, A., N. Armstrong, and G. Jones. 2002. *A Preliminary Estimate of Total Nitrogen and Total Phosphorus Loading to Streams in Manitoba, Canada.* Manitoba Conservation Report No. 2002-04. www.lakewinnipeg.org/web/downloads/LakeSci_NutriantLoading.pdf (accessed January 10, 2011).

Christensen, V. 2007. *Nutrients, Suspended Sediment, and Pesticides in Water of the Red River of the North Basin, Minnesota and North Dakota, 1990–2004.* Scientific Investigations Report 2007-5065. Reston, VA: U.S. Geological Survey.

DNR (Minnesota Department of Natural Resources). 2010. Sustaining Lakes in a Changing Environment (SLICE). www.dnr.state.mn.us/fisheries/slice/index.html (accessed December 4, 2010).

EPA (U.S. Environmental Protection Agency). 1998. *Clean Water Action Plan.* Washington, DC: Office of Water.

Heiskary, S., W. Bouchard, and H. Markus. 2010. Minnesota Nutrient Criteria Development for Rivers. St. Paul: Minnesota Pollution Control Agency. www.pca.state.mn.us/index.php/water/water-permits-and-rules/water-rulemaking/proposed-water-quality-standards-rule-revision.html (accessed January 5, 2011).

Heiskary, S., and E. Swain. 2002. Water Quality Reconstruction from Fossil Diatoms: Applications for Trend Assessment, Model Verification and Development of Nutrient Criteria for Minnesota USA Lakes. St. Paul: Minnesota Pollution Control Agency. www.pca.state.mn.us/index.php/water/water-types-and-programs/surface-water/lakes/lake-water-quality/lake-water-quality.html (accessed January 5, 2011).

Heiskary, S.A., E.B. Swain, and M.B. Edlund. 2004. Reconstructing Historical Water Quality in Minnesota Lakes from Fossil Diatoms. Minnesota Pollution Control Agency Environmental Bulletin No. 4 (September). www.pca.state.mn.us/index.php/view-document.html?gid=11453 (accessed January 5, 2011).

Heiskary, S., and C.B. Wilson. 1989. The Regional Nature of Lake Water Quality across Minnesota: An Analysis for Improving Resource Management. *Journal of the Minnesota Academy of Science* 55 (1):71–77.

———. 2008. Minnesota's Approach to Lake Nutrient Criteria Development. *Lake and Reservoir Management* 24:282–297.

Heiskary, S.A., C.B. Wilson, and D.P. Larsen. 1987. Analysis of Regional Patterns in Lake Water Quality: Using Ecoregions for Lake Management in Minnesota. *Lake and Reservoir Management* 3:337–344.

Marschner, F.J. 1930. *The Original Vegetation of Minnesota.* Washington, DC: U.S. General Land Office.

McCollor, S., and S. Heiskary. 1993. *Selected Water Quality Characteristics of Minimally Impacted Streams from Minnesota's Seven Ecoregions* St. Paul: Minnesota Pollution Control Agency. www.pca.state.mn.us/publications/tdr-g1-03.pdf (accessed November 8, 2009).

Minnesota Revisor of Statutes. 2007. Minnesota Administrative Rules: Chapter 7050, Waters of the State. www.revisor.mn.gov/rules/?id=7050 (accessed December 4, 2010).

Moyle, J.B. 1956. Relationships between the Chemistry of Minnesota Surface Waters and Wildlife Management. *Journal of Wildlife Management* 30 (3):303–320.

MPCA (Minnesota Pollution Control Agency). 2001. Rainy River Basin Information Document. www.pca.state.mn.us/index.php/water/water-types-and-programs/surface-water/basins-and-watersheds/rainy-river-basin/rainy-river-basin.html (accessed January 5, 2011).

———. 2009a. National Lakes Assessment Project (NLAP). www.pca.state.mn.us/water/nlap.html (accessed November 30, 2009).

———. 2009b. Red River Basin. www.pca.state.mn.us/water/basins/rainy/index.html (accessed November 28, 2009).

———. 2009c. Upper Mississippi River Basin. www.pca.state.mn.us/water/basins/uppermiss/index.html (accessed November 28, 2009).

Omernik, J.M. 1987. Ecoregions of the Conterminous United States. *Annals of the Association of American Geographers* 77 (1):118–125.

Omernik, J.M., and R.G. Bailey. 1997. Distinguishing between Watersheds and Ecoregions. *Journal of the American Water Resources Association* 33 (5):935–949.

Ramstack, J.M., S.C. Fritz, and D.R. Engstrom. 2004. Twentieth-Century Water-Quality Trends in Minnesota Lakes Compared with Pre-settlement Variability. *Canadian Journal of Fisheries and Aquatic Sciences* 61:561–576.

Robertson, D., G. Schwarz, D. Saad, and R. Alexander. 2009. Incorporating Uncertainty into the Ranking of SPARROW Model Nutrient Yields from Mississippi/Atchafalaya River Basin Watersheds. *Journal of the American Water Resources Association* 45 (2):534–549.

Tesoriero, J. Duff, D. Wolock, N. Spahr, and J. Almendinger. 2009. Identifying Pathways and Process Affecting Nitrate and Orthophophorus Inputs to Streams in Agricultural Watersheds. *Journal of Environmental Quality* 38:1892–1900.

Tester, J. 1995. *Minnesota's Natural Heritage: An Ecological Perspective.* Minneapolis: University of Minnesota Press.

Zumberge, J.H. 1952. *The Lakes of Minnesota: Their Origin and Classification.* Bulletin 35. Minnesota Geological Survey. Minneapolis: University of Minnesota Press.

CHAPTER 4

Implementing the Federal Water Pollution Control Act and Minnesota's Clean Water, Land, and Legacy Amendment

Rob Johansson and Faye Sleeper[1]

*T*his chapter presents a brief history and overview of the Federal Water Pollution Control Act (FWPCA) Amendments of 1972, now known as the Clean Water Act (CWA), including recent legislative and judicial actions. Balancing the efforts of domestic industries to keep pollution control costs at a minimum and the needs of public health and welfare has generally marked federal control of water pollution. As with most federal efforts at controlling pollution, the process has been iterative: public opinion sparks legislative proposals; negotiations in the legislative process narrow the scope of the legislation; rulemaking and guidance by federal agencies seek to interpret and implement congressional intent and scope while considering public comments; lawsuits provide tests of the scope and intent of Congress; public opinion sparks new legislative efforts.

A main characteristic of the CWA is that it encourages states to develop their own regulations and guidance in order to ensure compliance with the clean water provisions. Although states can pursue more stringent water quality objectives than those set out in the CWA, the federal law establishes baseline standards and processes for maintaining and improving water quality in U.S. water resources. Such a federal-floor, state-ceiling policy has resulted in a number of different approaches to achieving clean water across the nation. Minnesota has developed and employed several policy alternatives to federal policies to customize and locally adapt federal guidelines. For example, targeting and emissions trading have been employed as innovative ways of achieving water quality standards.

The most forward-looking aspect of Minnesota water resource management has been the Clean Water, Land, and Legacy Amendment (CWLA), passed in 2006. The CWLA provides a vehicle for allocating significant resources (currently about $80 million per year) for addressing impaired waters, the only such mechanism in the nation. This chapter discusses the vision and politics that led to the CWLA, its implications and limitations, and opportunities it presents.

SETTING THE STAGE FOR THE CLEAN WATER ACT

The federal Clean Water Act sets the baseline for all state water programs. States are allowed to set programs that are more stringent or in addition to the Clean Water Act, but not less stringent. Most states, like Minnesota, and some tribes have been delegated to carry out the requirements of the Clean Water Act. Knowing the basics of the federal Clean Water Act is essential context for understanding state water law and policy.

Human Health Concerns

Initial efforts at controlling water pollution in the United States date back to the early days of public health efforts at controlling waterborne diseases, such as typhoid, cholera, and yellow fever. Lack of a public sewer system and subsequent contamination of residential wells in the city of New York were partially responsible for yellow fever and cholera outbreaks throughout the 1800s. As many as 5,000 city residents perished from the cholera outbreak of 1848–1849 (Wynne 1852). As a result, the first federal public health officials in the United States sought to limit the spread of such diseases partially through the construction of sewer systems (Andreen 2003). Initial wastewater treatments for the city of New York, for example, were initiated at several sites in 1890 and the early 1900s in Queens and Brooklyn (New York City [2007?]).

During this period, public health and sanitation concerns continued to motivate state and local efforts at keeping raw sewage out of waters. In 1899, Congress passed the first federal water pollution control legislation in the form of the Rivers and Harbors Act (Section 13 of which is known as the Refuse Act) to regulate the discharge of "any refuse matter of any kind or description" without a permit into navigable waters.[2] That act set the precedent for federal control of wastes discharged into U.S. water resources and remains in force today. The U.S. Army Corps of Engineers (USACE) is responsible for enforcing the Rivers and Harbors Act and historically has limited enforcement actions for refuse that impedes or obstructs navigation, rather than more broadly applying them to "activities in navigable waters" (U.S. Senate Committee 2000). Next, Congress broadened the scope of water pollution control to ban discharge of oil from ships into coastal waters in the Oil Pollution Act of 1924. In the enactment of both the Refuse Act and the Oil Pollution Act, more ambitious attempts to control other sources of water pollution were bypassed in order to attract sufficient congressional support to enact the legislation. For example, as initially proposed, the Oil Pollution Act of 1924 would have regulated oil discharged in inland navigable waters and penalized accidental spills.[3]

Following a period of urbanization and rapid economic growth in the United States, public concern over the consequent increase in water pollution prompted Congress to enact the landmark Federal Water Pollution Control Act (FWPCA) of 1948. The law balanced pollution control costs, enforcement mechanisms, and public health and welfare concerns. States had responsibility for water pollution controls with funding from the federal government to help build municipal sewer systems. The scope of the law was limited to interstate waters. Direct federal actions were authorized only when polluted waters "endangered the health or welfare of persons in a State other than that in which the discharge originates," and only with the explicit permission of the state where the pollution originated (Powers 2004). In part because of those constraints, amendments were enacted in 1961, which expanded the jurisdiction of the FWPCA to the discharge of pollution in all "navigable waters" that endangered health or welfare.

A Change in Focus

Just prior to the passage of the Federal Water Pollution Control Act Amendments of 1972, two important developments in pollution control efforts laid the groundwork for the future. First, the Water Quality Act of 1965 signaled a change in focus from preventing water pollution that might threaten human health directly to preventing water pollution that might damage aquatic systems and might hinder recreational activities, such as swimming and fishing (Poe 1995). Next, the Supreme Court found in *United States v. Republic Steel* in 1960 and *United States v. Standard Oil* in 1966 that the Refuse Act applied to industrial pollution discharge, and that such discharge required permits. The resulting upsurge in citizen lawsuits concerning industrial discharge prompted President Richard Nixon to issue an executive order in 1970 creating the Refuse Act Permit Program (RAPP). That program required facilities discharging wastes into public waters to apply for a permit from the USACE. However, the executive order made it clear that no enforcement action would be taken against those applying for a permit (Van Wye 2004). Ambient water quality standards based on ecological health and permitting provided the framework for the 1972 FWPCA Amendments.

PASSAGE OF FEDERAL WATER POLLUTION CONTROL ACT AMENDMENTS OF 1972

The signature U.S. water pollution control legislation was passed under the title of the Federal Water Pollution Control Act Amendments of 1972 by the 92nd U.S. Congress. On October 4, 1972, Congress passed the bill, but President Nixon vetoed the legislation on October 17, calling it "a bill whose laudable intent is outweighed by its unconscionable $24 billion price tag" (Van Wye 2004). Congress overrode the veto the next day, so the FWPCA were enacted and became effective on October 18, 1972. The development of that legislation represented many compromises among industries' desire to minimize liability, the Nixon administration's desire to limit expenditures, and public interest. It nevertheless represented a much stronger law governing benchmarks for rehabilitating U.S. water resources.

Technology and Ambient Water Quality Standards

In general, the goals of the FWPCA Amendments of 1972 were to eliminate the discharge of pollutants into navigable waters by 1985 and to attain, when possible, waters deemed "fishable and swimmable." Those goals continue today.

One key component of this law was to replace the measure of endangerment to public health or welfare with specific technology-based standards, which effectively made enforcement actions much easier to pursue. Another notable component of the 1972 amendments was that it was the first time agriculture was explicitly recognized in federal law as a source of pollution (Ribaudo and Caswell 1999).

Some key provisions of the 1972 FWPCA Amendments were as follows:

- The amendments defined liability for discharges of oil and hazardous substances and the federal role in cleanup operations.
- They created a Clean Lakes Program.
- They established a system of construction grants for municipal water treatment facilities.
- Section 208 required states to assess water quality impairments and develop programs to control them.
- Section 301 required water treatment facilities to implement best practicable control technology and industrial facilities to employ best available technologies by 1983.
- Section 303 established listing requirements for impaired waters and development of total maximum daily load (TMDL) standards for controlling pollutant discharge into those waters.
- Section 402 established the National Pollutant Discharge Elimination System (NPDES) to authorize the issuance of discharge permits by the U.S. Environmental Protection Agency (EPA).
- Section 403 stipulated guidelines for EPA to issue permits for discharges into the territorial sea, the contiguous zone, and ocean waters farther offshore.
- Section 404 authorized the USACE to issue permits for the discharge of dredged or fill material into navigable waters at specified disposal sites. (EPA was authorized to prohibit the use of a site as a disposal site based on a determination that discharges would have an unacceptable adverse effect on municipal water supplies, shellfish beds and fishery areas, wildlife, or recreational uses.)

1977 Amendments: The Clean Water Act

Further amendments to the Federal Water Pollution Control Act enacted in 1977 rebranded the FWPCA as the Clean Water Act (CWA). These amendments specified that standards for nonconventional pollutants, such as phosphorus, nitrogen, and ammonia, would be promulgated by EPA by 1983, and that industrial sources were to achieve compliance with technology standards by 1987. Some of the earlier deadlines for water treatment facilities and industrial dischargers were extended under certain conditions. Most important, however, was that the 1977 amendments provided exemptions to the dredge and fill permitting requirements under Section 404. Most of these exemptions pertain to normal farming, ranching, and silviculture activities.

1987 Amendments: The Water Quality Act

Additional amendments in 1987 also required a congressional override of a president's veto—President Reagan vetoed a bill in 1986 because of concerns about increasing taxes and federal spending that could result from the proposed provisions for improving the water quality in the nation's lakes, a program to clean up the nation's estuaries, and regulation of stormwater discharge (Shabecoff 1987). As enacted, the Water Quality Act of 1987 contained many of these provisions:

- The grant system for wastewater treatment plants was converted to a revolving loan system.
- NPDES permits were now required for some stormwater discharge.
- Antibacksliding provisions were attached to the NPDES permitting process.
- States were required to develop strategies for toxics cleanup in waters where the application of best available technology would not be sufficient to meet state water quality standards and support public health.
- Section 319 was established, a $400 million program to support state development and implementation of nonpoint-source management programs where Section 208 plans were not successful.

REMAINING CHALLENGES TO REALIZING THE GOALS OF THE CWA

Several issues remain in meeting the stated goals of achieving "fishable and swimmable" waters. First, the intent of Congress regarding the scope of the CWA remains unclear following two Supreme Court rulings, known as *SWANCC* and *Rapanos*. Second, federal funding for secondary treatment of wastewater and for nonpoint-source controls is unlikely to meet necessary wastewater and stormwater management capital costs. And finally, a large number of current water impairments are linked to nonpoint-source discharge from agricultural activities, which are exempt from most CWA requirements.[4]

Legal Challenges

Congress can exercise authority over states so long as that authority does not violate the 10th Amendment: "The powers not delegated to the United States by the Constitution, nor prohibited by it to the States, are reserved to the States respectively, or to the people." The most relevant authority for regulating pollution granted to Congress by the U.S. Constitution is that listed under the Commerce Clause (Article 1, Section 8, Clause 3), which says that Congress shall have power to "regulate commerce with foreign nations, and among the several states, and with the Indian tribes." This allows Congress to write laws regulating pollution in waters, because of the nexus with interstate commerce.

Two recent Supreme Court decisions have affected the implementation of permitting provisions in the CWA. In its 2001 decision in *Solid Waste Agency of Northern Cook County (SWANCC) v. U.S. Army Corps of Engineers*, the Supreme Court argued that permitting authority granted by the USACE does not extend to "isolated waters" and wetlands that are not "adjacent" to navigable waters, interstate waters, or their tributaries. However, in 2006, the Supreme Court split in its ruling in *Rapanos v. United States*, finding both that the "waters of the United States" include wetlands only when they are "adjacent," connected by a "continuous surface connection" as opposed to an intermittent hydrologic connection (as written by Justice Scalia for the plurality), and that a wetland is adjacent if it affects the chemical, physical, and biological integrity of a navigable water (as written by Justice Kennedy in his concurring opinion) (Meltz and Copeland 2009).

The application of the plurality test or the ecological test of "significant nexus" has left permitting of dredge and fill disposal somewhat unclear (Copeland 2009b). EPA indicates that the *Rapanos* decision has "created a lot of uncertainty with regards to EPA's compliance and enforcement activities," and "processing enforcement cases where there is a jurisdictional issue has become very difficult" (EPA 2009a). An initial analysis by the USACE indicated that it anticipated that its permitting workload would increase "dramatically," in part to process a large backlog of jurisdictional determinations and to conduct "significant nexus" determinations (USACE 2007). Several legislative proposals have been introduced into Congress since the *Rapanos* decision to more clearly define what it means by "waters of the United States."

Two other large CWA issues are currently being litigated through the courts, both relevant to agricultural production. First, the rules describing how NPDES permitting provisions are to be applied to concentrated animal feeding operations (CAFOs) has been challenged by industry and environmental groups for a variety of reasons.[5] Second, EPA determined that pesticide applications over waters of the United States do not need NPDES permits, because relevant provisions controlling that application were already required as part of the Federal Insecticide, Fungicide, and Rodenticide Act. However, the Sixth Circuit Court found that such a finding was erroneous, and consequently EPA will begin to move forward with developing permit requirements for pesticide applications.

Funding Challenges

The state revolving loan program currently distributes funding for municipal water treatment based primarily on a state's population and the estimated capital costs necessary for the state to comply with secondary treatment as well as more stringent treatment requirements in the CWA and some sewer capacity investments. Congress appropriated $689 million in 2009 for state revolving fund grants, but the current formula meets only between 1% and 11% of eligible surveyed needs (Copeland 2009a). EPA's 2004 survey of total needs, which include the estimated capital cost of meeting new and maintaining existing water treatment infrastructure, showed that the total needs exceeded $200 billion over 20 years (EPA 2008a).

The Association of State and Interstate Water Pollution Control Administrators argues that a majority of water body impairments are the result of nonpoint-source pollution, such as stormwater runoff and agricultural activities, and that current management approaches have largely been a failure (ASIWPCA 2010). Funding to implement necessary nonpoint-source controls in these impaired waters is estimated to be approximately $1.9 billion annually (EPA 2008b); recent funding for Section 319 grants has averaged $205 million annually (Copeland 2009b).

Challenges Posed by Agriculture and Nonpoint-Source Pollution

In most cases, nonpoint-source discharge into water bodies is dispersed across the landscape, making attribution, permitting, and enforcement actions unwieldy. Controlling nonpoint sources of water pollution has generally been approached at the federal level by use of voluntary programs, such as agricultural conservation programs funded by the Farm Bill or through Section 319 grants under the CWA.

In some cases, agricultural activities are treated as point sources. Since 1972, CAFOs have been defined as point sources and are required to comply with the NPDES permitting provisions (EPA 2003). Further, beginning in 1987, some nonpoint sources were required to apply for NPDES permits—the Water Quality Act required permits for discharges of industrial stormwater and from municipal separated storm sewer systems from cities of 250,000 persons or more. Nevertheless, regulatory approaches to controlling the bulk of agricultural and nonpoint-source discharge using the Clean Water Act or other authorizing federal law has generally proved to be difficult and controversial and could infringe on state and local jurisdictions (Hecox 2010).

As an example, consider a recent case heard by EPA's Office of Administrative Law on June 8, 2009. EPA asserted that the farmer was liable for a fine of $157,500 for failing to apply for a NPDES permit 180 days prior to feedlot discharge occurring. The evidence of discharge was given as follows: "manure and other feedlot pollutants would leave Respondent's feedlot when a sufficient rain occurred, ... those pollutants would travel down drainage paths created during such rains and make their way across a cornfield, eventually arriving at an unnamed tributary. From there such pollutants would then flow to Elliot Creek. Both the unnamed tributary and Elliot Creek are waters of the United States" (EPA 2009d, 3).

The administrative law judge found, however, that EPA's assertion was not supported by actual water quality sampling at the site, upstream and downstream. As such, the assertion of discharge could not be established. Although the case in question concerned agricultural activities at a point source (a large animal feedlot) and not a nonpoint source (e.g., a field of corn), it highlights the importance of having actual monitoring data to enforce permitting requirements, demonstrating the difficulty of permitting agricultural nonpoint sources. As a result, a good deal of the effort to limit nonpoint-source discharge into waters has been left to the states. States are required to set water quality standards and develop TMDLs and remediation plans for waters that are listed as impaired, which includes a determination of the nonpoint-source component of the impairment. When states are delinquent in these responsibilities, the federal government can step in and develop the TMDL.

MINNESOTA AND THE CLEAN WATER, LAND, AND LEGACY AMENDMENT OF 2006

Minnesota's approach to implementation of the Clean Water Act for impaired waters and development of TMDLs evolved from a long history within the state of establishing environmental programs that are innovative and often more effective than those in other parts of the nation. Even before the enactment of the Clean Water Act, Minnesota was already addressing water quality issues, initially driven by human health concerns.

Early Control Efforts in Minnesota

Like those in the rest of the nation, Minnesota health officials had long been concerned about diseases, such as typhoid, that spread through contaminated drinking water. In the 1800s, the focus was on filtering the water for drinking, rather than protecting water supplies from raw sewage and other effluent. Rochester, St. Paul, and Minneapolis all developed water filtration plants in the early to mid-19th century. In Duluth in 1891, only 10% of residences had a sewer connection to dispose of waste a safe distance from drinking-water sources; by 1914, such disposal was mandatory. Like many cities, Minneapolis and St. Paul used the Mississippi River for industrial and slaughterhouse wastes, as well as the sewage of 680,000 people. Construction of the Ford Dam in 1914 for navigation purposes stopped the flow of wastes south of Minneapolis but caused a "sludgepool" to form behind the dam, killing all aquatic life in that part of the Mississippi except tubificid worms. Construction of sewer and treatment plants to transport the sewage south of the Twin Cities culminated with the completion of the Pig's Eye Sewage Treatment Plant in 1938 (Scherkenbach 2007).

In 1945, the Minnesota Water Pollution Control Commission, consisting of four staff members, was created within the Department of Health. The focus during these years was to encourage upstream cities to treat sewage well enough that downstream users could disinfect the stream water for potable use. For example, in 1948, the Water Pollution Control Commission raised a concern in a state Board of Health Survey of Pollution of the St. Louis River, which drains into Duluth Harbor, Lake Superior: "Industries at Cloquet use the river as a source of process water and then finally, along with the municipalities, for the dilution and disposal of wastes and sewages. To remove all pollution from the river would not be economically feasible, but some steps toward reducing the pollution, particularly in regard to sludge-forming wastewater from industries, are possible and should effect considerable river improvement" (MSBH 1948).

Minnesota Pollution Control Agency

The Minnesota Pollution Control Agency (MPCA) was established in 1967, prior to passage of the federal Clean Water Act, in response to an outcry against the use of waters as conveyances for waste and in response to several large oil spills in the Minnesota River, which feeds into the Mississippi. These oil spills caused 10,000 ducks to die and made national news and the cover of *Audubon* magazine. In many ways, this was Minnesota's equivalent of a river catching on fire (Lee 2002).

Unlike EPA, which shared institutional roots with health and interior programs, the early MPCA grew out of the previous Water Pollution Control Commission, based in the Minnesota Health Department, and it continued to focus on treating and diverting wastes from the rivers and lakes, with a new emphasis on aquatic life and protecting water resources for recreational uses. Also unlike EPA, the MPCA was never combined with the Department of Natural Resources. This separation has allowed the environmental concerns to stand alone within a state agency rather than competing within a natural resources department. That has been important in the state's environmental record, as the drivers for these two departments are different, albeit perhaps equally important (Scherkenbach 2009).

Implementing the CWA in Minnesota

Since the passage of the CWA, Minnesota has been innovative in its implementation. Some have suggested that Minnesota's bounty of water resources, a citizenry that cares deeply about water resources, the diversity of landscape and land uses, and the presence of strong environmental champions explain why the state occasionally approaches environmental protection programs in a different manner than the rest of the nation. Minnesota's natural diversity spawns a mix of interests that actively engage in environmental issues ranging from agriculture to mining, forestry to urban living, industrial interests to environmental advocates, and hunters and fishers to cabin and resort owners. The MPCA is unique among state environmental agencies in that its top decisionmaking authority rests in a citizens' board, appointed by the governor on a rotational basis, which enhances its accountability to citizen demands (Scherkenbach 2009).

An example of Minnesota's unique approach to water pollution control is how the state addressed separation of the combined sanitary and storm sewers from 1985 to 1994. Although other large cities also began to address this issue, most at a much later date, the Twin Cities project was the largest in the nation at that time (EPA 1999). The combined sewers regularly exceeded capacity during storm events, because they carried both stormwater and sewage to the sewage treatment facilities. The facilities did not have the capacity to treat storm flows, resulting in bypasses of untreated sewage into lakes and rivers. In 1985, the Minnesota legislature, with the support of the MPCA and the metropolitan council, tackled this issue by funding one-third of the cost of separating these sewers. Minnesota was able to obtain additional funds from federal sources and the communities of Red Wing, St. Paul, Minneapolis, and their suburbs (EPA 1999).

This early attention to ensuring that sufficient funds were available to improve the state's waters in an effective manner is a common thread in Minnesota's environmental programs. For example, Minnesota passed and funded a companion program to the Federal Conservation Reserve Program, called Reinvest in Minnesota. This state–federal partnership has resulted in a high level of federal Conservation Reserve Program funds being allocated to Minnesota. Minnesota's legislature and citizens, once convinced of the value of an activity, often take action to ensure that the staff and financial resources are provided to address the concerns.

Nonpoint-Source Pollution

An important component of addressing impaired waters both nationally and in Minnesota is reduction of pollution from nonpoint sources. This is another example of the state's innovative implementation of CWA provisions. Section 319 of the CWA provided funding to assist in implementation of nonpoint-source controls. States were required to provide a funding match in order to obtain this funding. Rather than just providing such matching funds, as did many states, the MPCA developed a more comprehensive stand-alone program that served as a companion to CWA Section 319. This program, the Clean Water Partnership (CWP), relies on local entities to lead nonpoint-source projects with technical assistance and oversight from the MPCA and other state agencies. The CWP has three phases: Phase I consists of assessment of the problem and development of a detailed implementation plan, Phase II is implementation, and Phase III is for continuation of those projects that need additional funds to complete implementation. This third phase remains an important feature of the companion nonpoint-source efforts, as mitigating nonpoint-source pollution is sometimes inexact and complicated.

For more than 20 years, this companion program has helped coordinate organizational and local infrastructure in order to more effectively address water quality issues. In addition, the program laid a foundation very similar to that of the Section 303 impaired waters program: identifying the causes of water body impairment, developing a corrective action plan, and implementing the cleanup plan through various land and water practices. The impaired waters approach, including TMDLs, follows the same approach as a CWP project but incorporates both point and nonpoint sources. TMDLs are mandated through the Clean Water Act, whereas the CWP has been a voluntary program. Nevertheless, CWP provided an effective foundation for Minnesota to meet its impaired waters and TMDL responsibilities under the CWA, whereas in other states, such a foundation was not readily available.

Development of TMDLs in Minnesota

Before 1990, both in Minnesota and the nation at large, there was limited focus on Section 303(d) of the Clean Water Act, which required states to identify waters that did not meet water quality standards, provide a list of these waters to EPA, include a public notice of the list, and complete TMDLs for these waters. TMDLs were used primarily to determine effluent limits for direct point-source dischargers. The MPCA and agencies of other states submitted lists of waters not meeting quality standards, but for this early effort, compliance was not rigorous (Hora 2009). In 1993, Minnesota faced a legal challenge by the Minnesota Center for Environmental Advocacy for noncompletion of TMDLs (EPA 2009b). The state prevailed in that lawsuit. However, beginning in 1997, environmental advocacy groups around the nation began to successfully sue EPA over the agency's failure to enforce Section 303(d).

As lawsuits mounted nationally, EPA increased pressure on states not only to complete their listings, but also to implement the TMDL provisions. In many states with litigation, the courts set stringent schedules for completing TMDLs. As a result, many of these TMDLs were not effective mechanisms for addressing water quality issues and have not led to restoring water to applicable quality standards.

Minnesota, like other states, was required to complete TMDLs with existing resources and very little direction. In the early years, the MPCA also faced the belief by many opponents that TMDLs, as often was the case with other initiatives, were a temporary concern and would eventually fade away. The program began with a few staff members, tasked with transferring the philosophies and tenets of the Clean Water Partnership program to the TMDL process. During those early years, some "early adopters" among local officials stepped forward and agreed to help pilot the new Section 303(d) process.

EPA Region 5 continued to apply pressure on the MPCA to meet the obligations of Section 303(d) by completing TMDLs more quickly. The MPCA had been developing TMDLs at a slower pace than other states, because it wanted to ensure that these studies would be adequate and result in water quality improvement. The Minnesota Center for Environmental Advocacy (MCEA) and other Minnesota environmental advocacy groups agreed with this approach, having seen numerous TMDLs from other states that likely would never be implemented. However, EPA indicated that it could withhold the nonpoint-source funds that Minnesota receives annually through CWA Section 319, a significant source of funding, and that it could actually do the TMDLs for Minnesota if the state took too much time to do so.

In response, Minnesota took a unique approach. Between 1997 and 2002, the state's leading environmental group focused on this issue, the MCEA, partnered with the Minnesota Chamber of Commerce (MCC) to place pressure on the MPCA to approach TMDLs in a transparent manner. Environmental organizations, industries, and municipalities were concerned about the impact on business and economic growth. The MCEA wanted to ensure that TMDLs were done well and would lead to improvements in impaired waters. To that end, the MCEA and MCC engaged a group of business and environmental advocacy interests to meet monthly with representatives of the MPCA to help shape how Minnesota would implement the program. An important backdrop to those discussions was a set of proposed federal regulations for implementation of CWA Section 303(d) that raised concerns across the nation.

The result of these meetings over several years was that the MPCA became more transparent in its process for identifying and listing impaired waters. This was especially true for the narrative water quality standards (Robertson 2009), statements that prohibit unacceptable conditions in or on the water, such as floating solids, scum, visible oil film, or nuisance algae blooms. Narrative standards are sometimes called "free froms," because they help keep surface waters *free from* visible and basic types of water pollution. The association between a narrative standard and beneficial use is less well defined than it is for numeric standards. Because narrative standards are not quantitative, determining that one has been exceeded typically requires a "weight of evidence" approach to data analysis, showing a pattern of violations (MPCA 2010). These early discussions were an important first step in identifying how Minnesota would implement Section 303(d) of the CWA and establish new unique policies for the state.

Critical Estimates of TMDL Implementation Costs

In 2002, the legislative auditor completed a report called Minnesota Pollution Control Agency Funding, based on interviews with staff and management responsible for completing TMDLs. The report reviewed the MPCA's water quality monitoring program, listing process, and completion of TMDLs, and concluded with the recommendation that the "MPCA should provide the 2003 Legislature with a multi-year TMDL implementation and financing plan, outlining 1) what mix of existing and new resources would be needed to meet federal requirements, 2) specific strategies the agency will use to assess water quality statewide, and 3) the types of strategies the agency will likely pursue to clean up impaired waters" (MPCA 2003). The MCC successfully lobbied to add a rider to the 2002 budget bill requiring that the MPCA respond to the legislative auditor in a report to the legislature by early 2003. This rider not only forced a necessary analysis, but also highlighted the issue of addressing impaired waters for the Minnesota legislature.

Minnesota changed administrations in January 2003, so the response was initiated under Commissioner Karen Studders and completed under Commissioner Sheryl Corrigan. Commissioner Corrigan assumed her position with a desire to tackle the impaired waters issue and move that issue higher on the agendas of both the MPCA and the state. She also believed that stakeholders should be involved in shaping the program. In its report, the MPCA made a commitment to the legislature to convene a stakeholder process that included key sector representatives. The stakeholder process was to begin in spring 2003 and finish with a set of recommendations in fall 2004 (MPCA 2003).

The response also developed funding estimates for statewide monitoring and assessment, completion of TMDL studies, and restoring impaired waters. Based on knowledge gained through its monitoring and assessment program, the MPCA estimated a cost of $8.2 million per year to complete a statewide assessment every 10 years as required by the CWA. Funding available at that time was $1.1 million annually, resulting in a funding gap of about $7.1 million per year. The MPCA also estimated the cost of completing TMDLs to be $8.9 million per year. The MPCA had available at that time approximately $3.1 million annually of state general fund and Section 319, leaving a gap of $5.8 million per year. The most difficult costs to estimate were those for restoring impaired waters. The Clean Water Partnership provided the basis for estimating restoration costs for nonpoint-source impairments: $45 million to $230 million per year. The range was large due to the uncertainty and lack of actual data from the TMDL program at that point (MPCA 2003).

The response accomplished two important things: it served as an educational tool for the legislature on the program and its needs, and it elevated the agency's engagement of key stakeholders. Minnesota was poised to develop new policies that went beyond the CWA and to identify the true costs of identifying and restoring impaired waters.

The Clean Water, Land, and Legacy Amendment (CWLA) of 2006

The MPCA contracted with the Minnesota Environmental Initiative (MEI) to facilitate development of the new law. After discussing and evaluating several approaches to a stakeholder process, the MEI, MPCA and MCC agreed on a nested approach (Robertson 2009). The core working group was called the Group of 16, still referred to as the G16. This diverse group consisted of representatives from farming organizations, local government organizations, industry and business interests, environmental advocates, and state agencies. The group met monthly, first learning about the various aspects of the TMDL program and other work that could inform their deliberations, and eventually hammered out agreements on a policy framework and funding. The second tier, referred to as the G40, met three times during the process and provided feedback to the G16. This group included staff from the G16 organizations that worked on the water program, as well as other entities such as academic institutions, tribal governments, and a broader group of stakeholders and citizens. Finally, a third, unlimited group was invited in several times to learn about the work and offer input: the MEI had two open forums of more than 150 people, one at the beginning of the process and another near the end, as another way to gain input from a broader group of stakeholders (MPCA 2004).

During this time, the MPCA staff continued to develop TMDLs as required, while also providing input and background information for the G16 to consider. The G16 continued to keep the legislature informed of its progress throughout the process, which served the group well during the 2005 legislative session. The G16 worked through issues on a consensual basis. The very different groups involved recognized the need for full consensus in order for their work to be successful. The result of the work was a package that included two companion parts: a policy framework to guide the future of the program from monitoring through restoration; and a financial needs assessment and proposed method of financing.

The policy framework laid out how the program and its phases—monitoring and assessment, TMDLs, and implementation—would be conducted; who was responsible for each phase; and how stakeholders were to be engaged throughout the process. The policy framework had several key provisions that were unique for state impaired waters programs:

- a strong engagement of citizens in the process, from water quality monitoring through completion of TMDLs and land and water restoration activities;

- involvement and coordination of four state agencies: the Board of Water and Soil Resources, Department of Agriculture, Department of Natural Resources, and Pollution Control Agency;
- recommendations to move to a geographic approach rather than address individual pollutants for individual water bodies;
- prioritization of all phases of the program and targeted implementation activities for highest effectiveness;
- attention to protecting waters not yet impaired;
- strong public participation throughout the process, rather than only at the time of the official public notice;
- an emphasis on restoration of water bodies from both nonpoint and point sources using existing laws and incentives; and
- establishment of a Clean Water Council for continued stakeholder involvement and oversight of the program.

The comprehensive law was unique in the nation in that it developed a model for completing TMDLs and restoring waters. This law underscored Minnesota's intent not only to comply with the Clean Water Act, but also to establish new policies regarding public participation, prioritize impaired waters activities, distribute the nondelegated activities among several state agencies, and place protection activities on an equal plane with restoration activities.

Funding the CWLA

The G16 also developed a funding proposal, which drew heavily from the MPCA's response to the legislature. However, the group refined the estimates for restoration funds, as shown in Table 4.1.

The G16 agreed on a range of $75 million to $100 million annually. Because of the diversity of the group, it reviewed a list of more than 50 funding options, using the following criteria:

- connection among pollution source, funding source, and use is logical;
- bigger is better;[6]
- does not encourage negative environmental behavior;
- recognizes larger societal benefits of clean water;
- is easy to administer;
- creates equity between point and nonpoint sources;
- has legislative viability;
- has stability and longevity;
- funding source is secure; and
- is acceptable to current administration.

Governor Timothy Pawlenty had taken a pledge against new taxes, which eliminated any tax options or fees that might be interpreted as new taxes. The G16 members concurred that asking for $100 million per year from the existing state general fund was not likely to be successful (Robertson 2009). The funding

Table 4.1 *Estimated annual need for impaired waters*

Assessment	$8.2 million
TMDL report	$8.9 million
Nonpoint-source restoration	$46 million–$230 million
Point-source restoration	$200 million
Estimated annual need for impaired waters	$263.1 million–$447.1 million

Source: MPCA (2004).

ultimately proposed by the G16 was a fee on residential and commercial water hookups. Although this did not meet all the above criteria, it did generate enough funds, was relatively low cost to administer, and was acceptable to all 16 members. Residences were to be charged $3 per month, and commercial establishments would pay $12.50 per month (State of Minnesota 2007b).

The G16 proposed this legislation in 2004, 2005, and 2006. Each year, an increasing number of legislators supported the plan by signing onto the legislation as coauthors. As with any new major legislation, unexpected concerns were raised over the three years it was debated by the legislature. It was clear from 2004 on, that the funding level was unacceptable to both the governor and the legislature. During the 2006 legislative session, the state had a budget surplus, so the governor and the legislature agreed on a onetime funding package of $26 million per year for the total program for fiscal years 2007 and 2008. This allowed the MPCA to begin ramping up its monitoring and assessment and its TMDL programs, and also provided a small amount of funding for implementation. While this onetime funding was a start, the G16 and legislators believed in the need for permanent or long-term funding (Robertson 2009). Representative Dennis Ozment and Senator Dennis Fredrickson were able to garner support of their respective chambers after two sessions to pass the policy framework into Minnesota Statute 114D as the Clean Water, Land, and Legacy Amendment (CWLA) in 2006 (State of Minnesota 2007b).

During this time, several events took place that were important to decisions made in the 2006 legislative session. First, the MPCA undertook a large TMDL based on turbidity and eutrophication impairments of Lake Pepin. Lake Pepin is a widening of the Mississippi River at Lake City, Minnesota, a water body with a drainage basin of approximately 55% of the state. All parties were watching this TMDL very closely because of its reach.

Then, in 2005, the MPCA issued a permit to a new facility that would combine the discharges of Maple Lake and Annandale, located in the Lake Pepin drainage basin. The MCEA sued the MPCA over this permit because of a regulation that limits discharges to waters if the discharger will contribute to a violation of water quality standards.[7] The MCEA argued that the MPCA wrongly allowed a new discharge of phosphorus to waters that did not meet water quality standards for phosphorus.

The lower courts found in favor of the MCEA, but the MPCA appealed to the Minnesota Supreme Court, which accepted the case. The lower court case was in process during the 2006 session, and the supreme court case was pending during the 2007 session. Legislators and members of the G16 felt a new urgency because of this court case, and this helped elevate the need for additional funding. Ultimately, on May 17, 2007, the supreme court found in favor of the MPCA (State of Minnesota 2007a). However, legislators were convinced of the need for this program and its ability to improve and protect Minnesota's waters.

For many years, hunting and fishing conservation groups had been proposing a constitutional amendment to dedicate three-eighths of 1% of Minnesota's sales tax for habitat restoration. Momentum began to build in 2006 to meld this effort with that of the G16 and other clean-water advocates. Ultimately, the legislature passed a bill that allowed voters to decide on a constitutionally dedicated increase in sales tax, to be divided among the following four efforts: 33% for water quality, 33% for wildlife, 19.75% for arts funding, and 14.25% for parks. Citizens voted on the following language:

> Shall the Minnesota Constitution be amended to dedicate funding to protect our drinking water sources; to protect, enhance and restore our wetlands, prairies, forest, and fish, game and wildlife habitat; to preserve our arts and cultural heritage; to support our parks and trails; and to protect, enhance, and restore our lakes, rivers, streams, and groundwater by increasing the sales and use tax rate beginning July 1, 2009, by three-eighths of one percent on taxable sales until the year 2034? (State of Minnesota 2008)

In spite of a state and national economic downturn, Minnesotans passed this new sales tax by a significant margin on November 4, 2008.

Implementing the CWLA: Opportunities and Challenges

Implementation of the CWLA was challenging in the first legislative session after passage of the constitutional funds. The clean water fund was directly overseen by the legislature, which set up a special legislative committee to develop a recommendation for allocation of the funds. Although the Clean Water Council (CWC) was established by law, the legislature did not link the council to the funds. Thus the CWC made recommendations to the legislature but had no authority over the funds. The CWC still has authority to oversee implementation of the Clean Water, Land, and Legacy Amendment and has worked to guide the agencies involved in the implementation of the program.

Several significant directions have emerged from the CWC. First, the CWC has supported the initial split of responsibilities among agencies. Monitoring and assessment are primarily the responsibility of the MPCA, with assistance from the Departments of Natural Resources and Agriculture. TMDLs are also the responsibility of the MPCA. The responsibility shifts to the Board of Water and Soil Resources for distributing implementation funds. This division of responsibilities required additional coordination among agencies. It appears that this coordination has been successful to date, but maintaining shared responsibility for the program will be a challenge and will require vigilance by both the CWC and the legislature. It also remains unclear that this is the most effective model.

Another direction embraced by the CWC is a shift away from monitoring in a random fashion across the state and toward a more intensive watershed-based monitoring scheme, based on a 10-year rotation. This will allow TMDLs to be completed within a watershed at one time, rather than continually doing studies in the same watershed and in some cases the same water body. This approach was recommended in the Clean Water, Land, and Legacy Amendment. It also allows for completing the restoration at the same time that protection activities take place. Minnesota joins only a few other states in establishing a policy of approaching its waters more holistically.

The CWLA requires prioritization, but often a pure prioritization to meet the goals is proposed, and areas of the state that are not receiving funding successfully challenge the priorities. As a result, effort and resources are realigned to ensure geographic distribution. One aspect of this prioritization is targeting of practices, in both urban and agricultural areas. Existing incentive programs are open to any landowner who meets basic criteria, but they do not target optimal locations for practices. The Minnesota legislature dedicated a portion of the first funds from the Clean Water Fund to get statewide Light Detection and Ranging (LIDAR) coverage, which makes this targeting relatively easy to accomplish (DNR 2010). The challenge will no longer be the technology for targeting, but the political will to implement a valid targeting plan that may not distribute the funds from incentive programs equally across the state.

The G16 and legislature both supported the notion of using existing programs to channel the funds to local units of government, to avoid setting up new governmental infrastructure. Although this prevented the creation of new infrastructure, it has dispersed funds through 12 distinct funding programs at the Minnesota Board of Water and Soil Resources, which seems unnecessarily complex and in fact has contributed to a more cumbersome bureaucracy. The combination of these factors could result in ineffective implementation.

Finally, there is a huge rift in Minnesota and other parts of the nation over the role of agriculture in protecting water resources. Historically, farmers have been viewed as stewards of the land, but as society as a whole has become more fractured and litigious, so has the debate over the role of agriculture in environmental protection. Federal law has fueled this split by regulating a portion of both urban and rural nonpoint sources. This has led to some interests being defensive, feeling they are under continual attack, while others believe agriculture is doing nothing about water quality and is to blame for much of the problem. This rift likely will limit the success of the Clean Water, Land, and Legacy Amendment. Alternatively, Minnesota may find a way to bridge this schism, allowing the best minds across the continuum of citizens and landowners to work more cooperatively in addressing the water issues. Natural and social scientists cannot rehabilitate Minnesota's impaired waters without the cooperation of landowners and policymakers.

NEXT STEPS IN ACHIEVING "FISHABLE AND SWIMMABLE" WATERS

The Minnesota experience with water pollution shares many of the same characteristics as national experience. Table 4.2 summarizes both federal and Minnesota key policy and legislative actions in water pollution control. Initial efforts at controlling water pollution were motivated by human health concerns. Lack of adequate sewage treatment resulted in contaminated drinking water and widespread disease. Following significant economic growth and urbanization during and after World War II, the focus of water pollution control shifted from preventing adverse human health impacts to the broader consideration of maintaining the health of aquatic ecosystems. That shift in focus was maintained by several court rulings that supported the new water pollution control legislation. Those rulings were accompanied by institutional development to help manage funding for water treatment, monitor aquatic ecosystem health, and enforce technical standards.

Despite the similarities, for a variety of reasons, Minnesota's clean water efforts have sometimes deviated from the approaches that the federal government and other states have taken. Because of the importance of maintaining and improving water quality to its citizens, due to the abundance of lakes, streams, and wetlands, Minnesota has clearly set policies that go beyond the Clean Water Act. The CWLA provides funding for developing TMDLs and implementing real solutions to meet those targets. In addition, the CWLA requires state agencies to prioritize not only the studies, but also the monitoring, which is the first step in the process. The CWLA also establishes a more rigorous public participation process than is required by the federal level—in Minnesota, agencies must incorporate citizens and stakeholders throughout the entire process, rather than just provide a public comment period after the TMDL has been written. Stakeholders are part of monitoring, developing the TMDL, and identifying and implementing solutions. Furthermore, the MPCA as a separate agency has some freedom to determine its course, and with legislative support and funds, it has been able to successfully carve out a path of environmental protection that is more effective in solving water quality issues. Minnesotans are not satisfied to just let government dictate programs and activities, but actively engage in development of policy and implementation of programs. Thus state departments must work in partnership with the citizens. Finally, Minnesota has benefited from having strong environmental government officials, congressional representation, and advocacy groups.

Yet many challenges remain. Industry and environmental advocates continue to pursue legal actions at the federal and state levels to limit or expand the authorities under the Clean Water Act to control discharge into U.S. waters. The needs of municipal water treatment systems far exceed federal resources available. Similarly, federal authority to control nonpoint-source pollution by regulatory means and funding to implement controls on these sources remain limited relative to the challenges posed by such discharges.

Table 4.2 *Timeline of federal water pollution legislation and Minnesota actions*

Law	Year	Key federal provisions	Minnesota actions
River and Harbors Act	1899	Regulated the discharge of any nonliquid refuse matter of any kind or description into navigable waters. Covered construction of all bridges, docks, piers, etc.	Early water quality concerns focused on health and water contamination.
Oil Pollution Act	1924	Prohibited intentional release of oil into navigable coastal waters.	Mississippi River contamination behind the Ford Dam in 1914.
Rivers and Harbors Act	1938	Provided "due regard" to wildlife conservation in permitting construction.	Completion of the Pig's Eye Sewage Treatment Plant in 1938 in St. Paul.
Federal Water Pollution Control Act	1948	Provided funding for state and local water treatment. Provisional authority given to surgeon general for polluted interstate waters that endangered public health.	Minnesota Water Pollution Control Commission created in 1945.
Federal Water Pollution Control Act	1956	Extended federal financial support of wastewater treatment plants.	
United States v. Republic Steel	1960	Supreme Court ruled that Refuse Act applied to industrial discharge of wastewater.	
Federal Water Pollution Control Act	1961	Increased federal support for water treatment. Allowed federal actions against polluters with the consent of the state's governor.	Several large oil spills in the Minnesota River (1962 and 1963), with high duck mortality.
Water Quality Act	1965	Required ambient water quality standards for the first time to protect health and welfare. Increased federal supports for water treatment.	
Clean Water Restoration Act	1966	Increased funding for water treatment construction.	
United States v. Standard Oil	1966	Supreme Court ruled that Refuse Act applied to industrial wastewater.	Minnesota Pollution Control Agency created in 1967. State water quality standards adopted in 1968.
Water Quality Improvement Act	1970	Added liability provisions for cleanup of oil spills.	Minnesota adds state water quality standards.
Kalur v. Resor	1971	Federal District Court ruled the Refuse Act Permit Program violated the newly enacted National Environmental Policy Act provisions requiring environmental impact statements.	

Table 4.2 *Timeline of federal water pollution legislation and Minnesota actions*

Law	Year	Key federal provisions	Minnesota actions
Federal Water Pollution Control Act Amendments	1972	Section 303 required states to develop a list of impaired waters and set TMDLs detailing allowable discharges into those water bodies. Technology-based standards set for point-source discharges, known as the NPDES. If water quality did not meet ambient standards, additional point-source controls were authorized. Permitting requirements set for discharge of dredged or fill material into U.S. waters.	
United States v. NRDC	1976	Consent decree requiring EPA to address 129 "priority pollutants" through setting water quality standards, upgrading best available technology standards, issuing pretreatment standards, and creating new source performance standards.	
Indictment of Allied Chemical Corporation	1976	First criminal FWPCA case, which resulted in fines of $13.2 million for contamination of the James River by discharge of toxic materials.	
Clean Water Act	1977	Exempted most farming activities from Section 404, which authorized the USACE to permit "dredge and fill" activities.	
Water Quality Act	1987	Moved grant system for municipal sewage treatment to a revolving loan program under Title VI. Created stormwater regulations for nonpoint-source discharge from urban areas. Section 319 established funding for nonpoint-source projects.	Clean Water Partnership program created to augment Section 319 grant program. In 1985–1994, combined sewer system in Twin Cities was separated to reduce combined sewage–stormwater overflow.
TMDL Guidance	1992	Allows cap-and-trade systems for meeting performance standards.	Minnesota Center for Environmental Advocacy sued Minnesota in 1993 over failure to complete TMDLs.
Minnesota Clean Water, Land, and Legacy Amendment	2006		Constitutional amendment to fund Clean Water, Land, and Legacy Amendment. Establishes policies and process for Minnesota to more comprehensively monitor and assess waters of the state, complete TMDLs, restore impaired waters, and protect waters not yet impaired.
Coeur Alaska Inc.	2009	Supreme Court ruled that discharge of mining waste can be permitted under Section 404 rather than Section 303.	

Federal Directions

Recent legislation proposed by Congress primarily attempts to resolve the uncertainty regarding the scope of the CWA following the *SWANCC* and *Rapanos* rulings; increase funding provisions for municipal water treatment and nonpoint-source controls; or increase application of CWA requirements to underregulated sources of pollutant discharge.

The Water Quality Investment Act (HR 1262) was sponsored by Minnesota representative James Oberstar and was introduced in the U.S. House of Representatives on March 3, 2009. The bill passed the House on March 12, with a vote of 317 to 101, and was referred to the Senate Committee on the Environment and Public Works on March 16. This bill authorizes appropriations for financial years 2010–2014 to provide grants and programs relating to the causes, prevention, reduction, and elimination of water pollution. Construction of new systems is eligible for such assistance, provided that the replacement or rehabilitation of the existing or new collection system is to address an existing adverse environmental condition, and the project otherwise meets the requirements of the CWA. The bill also amends the definition of "treatment works" to include acquisition of lands and interests in land that are necessary for construction.

The Clean Water Restoration Act (S 787) was sponsored by Senator Russell Feingold on April 2, 2009. This bill would amend the CWA to replace the term "navigable waters" with "waters of the United States," defined to mean all waters subject to the ebb and flow of the tide, the territorial seas, and all interstate and intrastate waters and their tributaries, including lakes, rivers, streams (including intermittent streams), mudflats, sandflats, wetlands, sloughs, prairie potholes, wet meadows, playa lakes, natural ponds, and all impoundments of the foregoing, to the fullest extent that these waters, or activities affecting them, are subject to the legislative power of Congress under the Constitution.

A recent executive branch action has drawn new attention to the Clean Water Act. On May 12, 2009, President Barack Obama issued the Chesapeake Bay Protection and Restoration Executive Order. This order established a committee to be chaired by EPA, tasked with defining the "next generation of tools and actions to restore water quality in the Chesapeake Bay and to describe the changes to be made to regulations, programs, and policies to implement these actions" (White House 2009). More recent legislation echoes the goals in that executive order. For example, Senator Cardin has published initial legislative language that would amend the CWA to include specific provisions for reducing pollutant discharge into the Chesapeake. Those provisions include establishment of a Chesapeake Bay-wide TMDL; no net increases of nitrogen, phosphorus, or sediment loads above those required to meet water quality standards; and binding load allocations for all nonpoint sources. To support these goals, bay-wide nutrient cap-and-trade programs for nitrogen and phosphorus would be required by 2011, as well as separate permitting programs for Chesapeake Bay tributaries (Walitsky 2009).

Minnesota's Directions

Minnesota continues to improve its impaired waters approach by implementing components of the Clean Water, Land, and Legacy Amendment. A dependable source of income is allowing the MPCA and its partners to move to a watershed approach; over 80% of TMDLs are led by local watershed specialists, state agencies have continued to improve their coordination, and Minnesota is leading the nation in restoration.

The state is still challenged by basic tenets of the Clean Water Act, however. The opening phrase of the CWA talks about chemical, biological, and physical approaches. Minnesota, like other states, has focused primarily on the chemical aspects, but it has begun to emphasize biological aspects of water quality as well, assessing and restoring biological communities. The comprehensive system to remove both urban and agricultural runoff poses additional challenges, because the physical alteration of these runoff patterns has affected water quality as well as quantity. As the world population increases, so does demand for food, fiber, and biofuels, requiring more land and more intensive yields from the land. These efforts could be at odds with CWA and CWLA requirements, which seek to limit nutrient, sediment, and chemical runoff into Minnesota's water bodies.

The state also will be affected by national actions and by the impact of Minnesota's activities on waters not within or adjacent to its borders. The hypoxic zone in the Gulf of Mexico or impaired waters in other states could force Minnesota to look beyond phosphorus, the nutrient of concern in its fresh water, to nitrogen, which is the nutrient of concern in salt water and in some other freshwater bodies. The state's contribution of nitrogen and other pollutants to water resources outside its borders via the Mississippi River could increase the need for state efforts to limit nutrient loads. Minnesota also is at the heart of the debate over how to balance the needs for fuel, food, and fiber with the need to protect its water resources. So even with its forward-looking approach, Minnesota shares many of the same future challenges as the rest of the nation. And if history provides any indication of the future, it is likely that the road to meeting these challenges will be an iterative collaboration among the public, and federal and state legislators and regulators.

NOTES

1. The views and findings presented in this chapter are those of the authors and should not be interpreted as those of the USDA.
2. March 3, 1899, Ch. 425, Sec. 9, 30 Stat. 1151. 33 U.S.C. § 401.
3. The Clean Water Act of 1972 later repealed the Oil Pollution Act of 1924. However, in 1990, Congress signed into law the Oil Pollution Act (OPA) of 1990—largely in response to the Exxon *Valdez* incident in 1989. The new OPA set up a national Oil Spill Liability Trust Fund to provide financial resources to respond to oil spills and provided new requirements for industry planning. Lastly, the OPA increased penalties for noncompliance (see EPA 2009c).

4. Of the 75,449 reported causes of impaired waters listed on the 303(d) list, approximately 45% are from pollutants linked to agricultural production: pathogens, nutrients, organic enrichment, sediment, pesticides, and excessive algal growth (EPA 2008c and EPA 2010).
5. See Fifth Circuit Court of Appeals, *National Pork Producers Council v. EPA*, CA5, No. 08-61093.
6. The Policy Work Group decided to target fewer but larger sources of revenue in an effort to minimize the number of potential political conflicts. This was only one of several considerations in the group's discussion of funding options (MPCA 2004).
7. The Clean Water Act (40 C.F.R. 122.4) requires the owner or operator of a new source or new discharger proposing to discharge into a water segment which does not meet applicable water quality standards or is not expected to meet those standards and for which the state or interstate agency has performed a pollutants load allocation for the pollutant to be discharged, to demonstrate, before the close of the public comment period, that (1) there are sufficient remaining pollutant load allocations to allow for the discharge; and (2) existing dischargers into that segment are subject to compliance schedules designed to bring the segment into compliance with applicable water quality standards.

REFERENCES

Andreen, W.L. 2003. Evolution of Water Pollution Control. *Stanford Environmental Law Journal* 22:215–294.

ASIWPCA (Association of State and Interstate Water Pollution Control Administrators). 2010. Comments on Coming Together for Clean Water, November 9 Letter to Assistant Administrator Peter Silva. www.asiwpca.org/home/docs/Funding.pdf (accessed January 12, 2011).

Copeland, C. 2009a. *Allocation of Wastewater Treatment Assistance: Formula and Other Changes.* Report No. RL31073. Washington, DC: Congressional Research Service.

———. 2009b. *Water Quality Issues in the 111th Congress: Oversight and Implementation.* Report No. R40098. Washington, DC: Congressional Research Service.

DNR (Minnesota Department of Natural Resources). 2010. Minnesota Elevation Mapping Project. Minnesota Geospatial Information Office, Digital Elevation Committee. www.mngeo.state.mn.us/committee/elevation/mn_elev_mapping.html (accessed December 2, 2010).

EPA (U.S. Environmental Protection Agency). 1999. *Combined Sewer Overflow.* Report No. 832-F-99-041. Washington, DC: EPA Office of Water.

———. 2003. National Pollutant Discharge Elimination System Permit Regulation and Effluent Limitation Guidelines and Standards for Concentrated Animal Feeding Operations (CAFOs); Final Rule. *Federal Register* 68 (29) (February 12).

———. 2008a. Appendix A: Summary of CWNS Cost Estimates. Report to Congress. In *Clean Waters Needs Survey 2004.* Washington, DC: EPA.

———. 2008b. Appendix E: Nonpoint Source Pollution Control: Documented Needs and Modeled Estimates. Report to Congress. In *Clean Waters Needs Survey 2004.* Washington, DC: EPA.

———. 2008c. *Clean Watersheds Needs Survey: 2008 Report to Congress.* Report No. EPA-832-R-10-002. Washington, DC: EPA Office of Water and Office of Wastewater Management. http://epa.gov/cwns/cwns2008rtc/cwns2008rtc.pdf (accessed January 12, 2011).

———. 2009a. *Congressionally Requested Report on Comments Related to Effects of Jurisdictional Uncertainty on Clean Water Act Implementation.* Report No. 09-N-0149. Washington, DC: Office of Inspector General.

———. 2009b. Litigation Status: Summary of Litigation on Pace of TMDL Establishment. www.epa.gov/owow/tmdl/lawsuit.html (accessed April 14, 2010).

————. 2009c. Oil Pollution Act Overview. www.epa.gov/OEM/content/lawsregs/opaover. htm (accessed June 25, 2010).

————. 2009d. *United States Environmental Protection Agency before the Administrator in the Matter of Lowell Vos, d/b/a/ Lowell Vos Feedlot, Woodbury County, Iowa* Docket No. CWA-07-2007-0078. www.epa.gov/oalj/orders/lowell-vos-id-060809.pdf (accessed January 12, 2011).

————. 2010. National Summary of Impaired Waters: Causes of Impairment for 303(d) Listed Waters. http://iaspub.epa.gov/waters10/attains_nation_cy.control?p_report_type=T# causes_303d (accessed June 30, 2010).

Hecox, E. 2010. Water Quality Law Summary: Nonpoint Source Water Pollution and BMPs. Bureau of Land Management. www.blm.gov/nstc/WaterLaws/abstract2.html (accessed June 25, 2010).

Hora, M. 2009. Personal communication between Marvin Hora, manager of assessment and standards, Minnesota Pollution Control Agency, and the authors. November 6.

Lee, S.J. 2002. Operation Save a Duck and the Legacy of Minnesota's 1962–63 Oil Spills. *Minnesota History Magazine* 58 (2):105–123.

Meltz, R., and C. Copeland. 2009. *The Wetlands Coverage of the Clean Water Act Is Revisited by the Supreme Court: Rapanos v. United States.* Report No. RL33263. Washington, DC: Congressional Research Service.

MPCA (Minnesota Pollution Control Agency). 2003. Minnesota's Impaired Waters: Report to the Legislature. www.pca.state.mn.us/publications/reports/lrwq-s-lsy03.pdf (accessed March 15, 2010).

————. 2004. Impaired Waters Stakeholder Process: Policy Framework, July 2003–February 2004. Minnesota Environmental Initiative Report. www.pca.state.mn.us/publications/ reports/lrwq-iw-2sy04.pdf (accessed June 25, 2010).

————. 2010. Water Quality Standards. www.pca.state.mn.us/water/standards/index. html#nnstandards (accessed June 25, 2010).

MSBH (Minnesota State Board of Health for the State Water Pollution Control Commission in collaboration with Wisconsin Committee on Water Pollution and Wisconsin Board of Health). 1948. *Report of the Follow-up Survey of the Pollution of the St. Louis River, 1947–1948.* St. Paul: MSBH.

New York City. [2007?]. New York City's Wastewater Treatment System. Department of Environmental Protection Report. www.nyc.gov/html/dep/pdf/wwsystem.pdf (accessed January 12, 2012).

Poe, G. 1995. *Water Control Policies.* Department of Agricultural, Resource, and Managerial Economics Report No. E.B. 95-06. Ithaca, NY: Cornell University.

Powers, A. 2004. Major Acts of Congress: Federal Water Pollution Control Act (1948). www. encyclopedia.com/doc/1G2-3407400129.html (accessed October 13, 2009).

Ribaudo, M.O., and M. Caswell. 1999. U.S. Environmental Regulation in Agriculture and Adoption of Environmental Technologies. In *Flexible Incentives for the Adoption of Environmental Technologies in Agriculture.* pp. 7–26. Edited by F. Casey, S. Swinton, A. Schmitz, and D. Zilberman. Norwell, MA: Kluwer.

Robertson, M. 2009. Personal communication between Michael Robertson, environmental policy consultant to the Minnesota Chamber of Commerce, and the authors. October 14.

Scherkenbach, T. 2007. The Evolving Landscape of Wastewater Regulation over the Last 35 Years. Presentation to Minnesota Environmental Partnership, Clean Water Action, and WLSSD Clean Water Act 35th Anniversary Celebration. November 2007, St. Paul.

————. 2009. Personal communication between Timothy Scherkenbach, deputy commissioner, Minnesota Pollution Control Agency, and the authors. October 30.

Shabecoff, P. 1987. House Ignores Reagan to Pass Clean Water Bill. *New York Times* Jan. 9.

State of Minnesota. 2007a. *In the Matter of the Cities of Annandale and Maple Lake NPDES/SDS Permit Issuance for the Discharge of Treated Wastewater, and Request for Contested Case Hearing,* Minnesota Supreme Court, Court of Appeals. www.lawlibrary.state.mn.us/archive/supct/ 0705/opa042033-0517.htm (accessed June 25, 2010).

————. 2007b. Minnesota Statutes 2007, 114D: Clean Water, Land, and Legacy Amendment. Minnesota Office of the Revisor of Statutes. www.revisor.mn.gov/bin/getpub.php?type=s&num=114D (accessed June 25, 2010).

————. 2008. Chapter 151—H.F. No. 2285: Proposed Amendment to the Minnesota Constitution. www.revisor.mn.gov/data/revisor/law/2008/0/2008-151.pdf (accessed June 25, 2008).

USACE (U.S. Army Corps of Engineers). 2007. Guidance Highlights for Rapanos and Carabell Decision. www.usace.army.mil/cecw/pages/cwa_guide.aspx (accessed January 12, 2011).

U.S. Congress. Senate. Committee of Environment and Public Works. 2000. Sections 9–20 of the Rivers and Harbors Appropriation Act of 1899, as Amended through P.L. 106-580, Dec. 29, 2000. http://epw.senate.gov/rivers.pdf (accessed January 12, 2011).

Van Wye B., ed. 2004. *The Clean Water Act Thirty-Year Retrospective*. Washington, DC: Association of State and Interstate Water Pollution Control Administrators.

Walitsky, Sue. 2009. Cardin Announces Details of Draft Chesapeake Bay Reauthorization, Including New Funding for States and New Enforcement Provisions. http://cardin.senate.gov/news/record.cfm?id=317548 (accessed December 6, 2010).

White House. 2009. Executive Order. Chesapeake Bay Protection and Restoration. www.whitehouse.gov/the_press_office/Executive-Order-Chesapeake-Bay-Protection-and-Restoration/ (accessed December 6, 2010).

Wynne, J. 1852. Appendix C to the Report of the General Board of Health on the Epidemic of Cholera of 1848 and 1849, presented to Both Houses of Parliament. London.

CHAPTER 5

Minnesota Water Law: A Unique Hybrid

Bradley C. Karkkainen

*W*ater law in the United States is first and foremost a matter of state law, descended from our English common-law heritage and modified over the years by judicial precedents and state legislative enactments, and it varies considerably from state to state. Minnesota's water law has evolved into a complex and interconnected, if not always fully coherent, web of common-law principles, statutes, and regulations aimed at managing and protecting one of the state's most vital and ubiquitous natural resources—water. An exhaustive examination of all the relevant law would form an entire treatise; this chapter undertakes the more modest task of outlining some of the most important legal principles and statutory provisions.

OWNERSHIP OF THE WATER

The background assumption in Minnesota, as in all states, is that both surface waters and groundwater resources are owned by the state in its sovereign capacity as trustee for the benefit of the public (Klein et al. 2009).[1] Individuals, municipalities, corporations, and other legally cognizable entities do not own water outright. Instead, they hold what are known in law as usufruct rights—rights to use water for certain purposes, subject to state supervision and regulation. In some cases these use rights are exclusive, but in other cases they are shared. Some use rights are of limited duration, whereas others are presumed to be permanent, at least absent subsequent changes to the law. Some use rights arise by operation of law without any formal regulatory approval; others are granted through an administrative permitting system. Whatever their scope and origin, all rights to use water are held subject to the state's ongoing power to regulate.[2]

In general, Minnesota is what legal scholars have termed a "regulated riparian" jurisdiction, meaning that it has modified the traditional common law of riparian rights without abandoning that body of law completely (Dellapenna 2010). In a pure riparian rights system, the owners of land adjacent to surface waters— "riparian land," from the Latin *ripa*, or riverbank—hold broad-ranging self-help rights to divert and use those waters on their riparian lands for their own benefit. Riparian rights arise as an incident of ownership of lands along the river or shoreline and generally cannot be severed or transferred apart from the riparian land to which they are attached. Riparian use rights are further limited by the principle that the use must be "reasonable" in light of the correlative rights of other riparian landowners. "Reasonableness," however, is a highly indefinite and context-dependent concept, leaving the scope and character of any individual landowner's use rights also indefinite and subject to change with changing circumstances (Klein et al. 2009). Similarly, under common law, landowners held rights to extract the underlying groundwater for use on their surface lands, again subject to a "reasonable use" limitation.

Like many other traditionally riparian jurisdictions, Minnesota has modified this background riparian rights scenario by overlaying a legislatively enacted administrative permitting and regulatory scheme, qualifying traditional riparian rights without completely abolishing them. In certain areas, however, Minnesota's regulatory scheme represents such a substantial departure from traditional riparian law principles that the "regulated riparian" characterization may not fairly capture its character.

By statute, the Minnesota Department of Natural Resources (DNR) is assigned responsibility to develop and administer a comprehensive water resource permitting and regulatory program to protect designated "public waters" (as defined in the next section below), as well as to regulate appropriations of water resources for public or private use, subject to principles and priorities laid out in the statute. In carrying out this duty, the DNR commissioner is directed to "develop and manage water resources to assure an adequate supply to meet long-term seasonal requirements for domestic, municipal, industrial, agricultural, fish and wildlife, recreational, power, navigation, and quality control purposes."[3] The statute thus presumes Minnesota can "have it all," sustainably supporting all major water uses over the long term. This assumption may have been justified in the past, given the relative abundance of water resources in most parts of the state, but it could come under increasing pressure in the future as population growth, increased competition for water, and climate change put additional stresses on Minnesota's water resources.

While the DNR has principal responsibility for managing water allocations and use, including maintenance of adequate quantities of water in the natural system, two other state agencies play major roles in protecting water quality. The Minnesota Pollution Control Agency (MPCA) has primary responsibility for regulating and permitting water pollution and maintaining water quality in the natural system. The Minnesota Department of Health has primary responsibility for safeguarding drinking-water supplies through well permitting and management, source water protection, and public water supply regulatory programs.

PUBLIC WATERS

Minnesota law distinguishes between "public" and "nonpublic" waters. Notwithstanding the terminology, this distinction is *not* based on ownership of the lake- or stream-bed. Like other states, Minnesota claims public ownership of the beds of lakes or streams that are determined to be "navigable-in-fact"[4]—that is, they follow the federal definition of being "used, or capable of being used, in their ordinary and natural condition, as highways of commerce over which trade and travel are or may be conducted."[5] Public ownership of the lands beneath the navigable waters extends up to the ordinary natural low water level (DNR 2010b). In contrast, the beds of non-navigable bodies of water are owned by the adjacent landowners under well-established common-law principles, with boundaries set at the "thread" (or center) of the stream or, in the case of lakes, by lines extending from each riparian tract to the lake's center (DNR 2010b).[6]

As early as 1897, however, the state of Minnesota decided that the public interest extended beyond protection of navigable-in-fact waters to certain other significant water bodies that were "non-navigable" according to the federal legal definition, and therefore their beds were not state-owned. These "public waters" originally included meandered lakes (surveyed lakes with boundaries recorded on federal plat maps), as well as streams capable of beneficial public uses like fishing, waterfowl hunting, boating, or water supply—whether the lake- or streambeds were publicly or privately owned. From an early date, these waters have been subject to heightened standards of legal protection (DNR 2009a).

By statute, the DNR commissioner is responsible for designating and maintaining an inventory of the state's public waters[7] and for regulating any activities that obstruct or "change the course, current, or cross section of public waters," such as the construction, removal, alteration, or change in ownership of dams, reservoirs, control structures, or waterway obstructions.[8] Under the current statutes, public waters include all navigable waters, meandered lakes, waters designated as trout or game lakes, designated trout streams, water basins designated as scientific or natural areas, waters whose beds or shores are owned in whole or in part by the state or federal government, and water basins with designated public access points, as well as any water basin draining an area of more than 2 square miles.[9] In addition, wetlands of 10 or more acres in unincorporated areas or 2.5 or more acres in incorporated areas are designated "public waters wetlands."[10] They are subject to the same protections as other public waters—a source of much controversy, as the regulatory definitions of what counts as a "wetland" tend to be imprecise and the relevant facts are often in dispute, making the delineations of specific sites as wetlands highly contestable. In general, however, all but the smallest streams, ponds, and wetlands fall into one or more of the protected categories and consequently are deemed public waters.

By statute, for all public waters in the state, the DNR commissioner is required to establish permissible lake or stream levels that will "conserve or utilize" those resources,[11] and to manage the state's water resources so as to maintain those target lake and stream levels while also maintaining public waters wetlands as functioning wetlands.

To implement these provisions, the DNR has established a Public Waters Work Permit Program. Before undertaking any work affecting the "course, current, or cross section" of designated public waters (including public waters wetlands), the party undertaking the work must secure a permit from the DNR.[12] This provision applies to such activities as diking, dredging, draining, filling, excavating, or placing structures in public waters or public waters wetlands. The DNR has broad discretion to approve, deny, or place conditions on a permit, based on its determination whether the proposed work is "reasonable, practicable, and will adequately protect public safety and promote the public welfare."[13] DNR regulations provide for categorical treatment of certain types of activities—for example, removable docks are generally permissible if specified conditions are met—and set detailed standards for other types of work.[14]

WATER USE PERMITS

The DNR commissioner also administers a statewide water use, or water appropriation, permitting program. By statute and implementing regulations, any groundwater or surface water user withdrawing more than 10,000 gallons per day or 1 million gallons per year is required to obtain a water use permit from the DNR.[15] Smaller withdrawals, such as wells for rural domestic use, do not require permits, even though their cumulative impacts may be significant. The largest category of permitted water use in the state is for electric power generation, which relies primarily on surface waters to cool thermoelectric power generating plants. Public water supply systems are the second-largest category of permitted users, relying on a mix of surface water and groundwater, depending on local hydrologic and water quality conditions. Industrial processes such as mining, papermaking, and food processing also use large volumes of water, mostly from surface water sources, and represent the third-largest category of permitted water users. Finally, irrigation and other agricultural uses represent the fourth major category of water use, drawing mostly on groundwater resources, although surface water withdrawals are also significant in some parts of the state (DNR 2000).

Water use permits are durational, limited to five years for most purposes but subject to "cancellation by the commissioner at any time if necessary to protect the public interests."[16] Permit applications, including renewal applications, are considered on a case-by-case basis in the order in which they come in, and the commissioner has broad discretionary authority to approve, deny, require modifications to, or impose conditions on any permit, based on standards set out in the statutes. In issuing permits, the commissioner must follow a statutorily established order of water use priorities, with the top priority awarded to domestic water supplies, second to consumptive uses of less than 10,000 gallons per day, third to irrigation and food processing, fourth to power production, fifth to consumptive uses in excess of 10,000 gallons per day, and sixth to nonessential uses.[17] In addition, in deciding whether to award permits, the commissioner must adhere to the conservation standards set out in the statute, including maintenance of public waters at target levels, prevention of harm to wetlands, maintenance of groundwater aquifers at "safe yield" levels, and special protections for trout streams, calcareous fens, and endangered species habitat. To that extent, in situ conservation of water in the natural system may be fairly said to take precedence over all ex situ uses as a matter of Minnesota law, at least in principle, if not always in practice.

Minnesota law allows local units of government, including watershed districts, soil and water conservation districts, and municipalities, an opportunity to review and comment on any permit application within their jurisdiction before the DNR makes its permitting decision.[18]

All active water appropriation permit holders are required to monitor their water use, employing an approved flow-monitoring device, and to report their water use annually to the DNR.[19] In addition, the DNR may require water appropriators to monitor and report the levels of groundwater or surface waters in the sources from which they are drawing.[20] Changes to the permitted water use—including the number or type of diversions, the rate or volume of withdrawals, or the number of acres irrigated—require DNR approval, as do transfers of permits to new owners. The DNR has authority to suspend permits during periods of drought, low water levels, or other shortages.[21] Currently, about 7,000 water appropriation permits are in effect statewide; of these, about 900 are held by public water supply systems (DNR 2009b).

The water appropriation permitting system represents a substantial departure from common-law riparian rights principles. Whereas the common law typically allowed riparian landowners to apply surface waters only to adjacent riparian lands and generally prohibited out-of-basin diversions (Klein et al. 2009), the water permitting system contains no such geographic limitations. Whereas common-law riparian use rights are indefinite, correlative, and subject to change under the malleable "reasonable use" standard, water permits are awarded on an individual case-by-case basis and are quantitatively specific, stating the precise volumetric quantities, flow rates, and seasonal patterns of water withdrawals that are permitted at specified locations, for specified uses, and by specified means of diversion. Whereas riparian rights arise by law without administrative oversight or prior administrative approval, the water appropriation permitting scheme is entirely a statutory creation and is administered top to bottom by a state agency. Whereas riparian rights are held simultaneously by all riparian landowners (Dellapenna 2010), water appropriation permits are issued individually and sequentially by a state agency in such a way that the amount of water deemed available to a subsequent applicant may be limited by the amounts already appropriated by current permit holders.

In certain respects, then, Minnesota's administrative permitting scheme looks less like a riparian rights approach and more akin to a prior appropriation regime—an alternative common-law approach that developed in the arid American West, in which the first appropriator is entitled to continue to withdraw a specified quantity of water indefinitely so long as the appropriated water is put to a "beneficial use," with all subsequent appropriative rights legally "junior" to the superior claims of prior appropriators. The principal difference, however, is that the limited five-year duration of Minnesota water appropriation permits, coupled with administrative management of the permitting system, gives a state administrative agency far greater control over where, when, for what purposes, and how much water is used than does prior appropriation law, predicated as the latter is on the West's deep traditions of self-help and limited government (Dellapenna 2010).

Yet Minnesota's water law is a curious hybrid, with an administrative permitting scheme for large-scale water appropriators sitting uneasily atop an essentially unmodified riparian rights scheme for small-scale riparian users. Although the animating principles of the two schemes are quite different, they can coexist peacefully so long as water is relatively abundant. Should water become scarcer, either in an absolute sense or relative to growing demand, then the two may come increasingly into conflict, and legal adjustments may be required.

RIGHTS OF ACCESS TO SURFACE WATER

In a state with thousands of lakes and streams and millions of avid recreational users—including boaters, swimmers, anglers, waterfowl hunters, and winter snowmobilers—the question of who has access rights to surface waters is critically important. Here Minnesota generally follows traditional riparian law principles.

Riparian landowners have the legal right to "wharf out"—that is, to build docks or other structures from the shoreline out to a navigable depth, whether the lakebed is privately or publicly owned—subject, however, to the DNR's Public Waters Work Permit requirements.[22] Riparian landowners also have rights to own and use shorelands added to their property by accretion, where shoreline sediment deposits added to the land, or exposed by reliction, the receding of a lake or river.[23] In addition, they have rights to collect ice and to boat, fish, hunt, swim, and otherwise use the water body in the usual and customary ways, so long as their use is "reasonable."[24] However, riparian landowners may not drain, dike, or fence off their portion of a water body or otherwise deny surface use to other riparian landowners or to other persons who have legally gained access, and all riparian landowners have equal rights to use a lake over its entire surface, regardless of who owns a particular part of the lakebed.[25] In addition to these common-law limitations, the "exercise of riparian rights is subject to state regulation in the public interest."[26] Where a parcel of riparian land is publicly owned, the public holds riparian rights, and any member of the public may gain legal access over the publicly owned land and use the entire surface of the water body for recreational uses such as boating, fishing, or swimming—regardless of who owns the stream- or lake-bed and whether or not the water body has been designated a public water. Moreover, public ownership of a road abutting the water body is sufficient to confer a right of public access.[27] Although it is generally unlawful to cross private land to gain access to a water body without the permission of the landowner, in instances where the public has openly and notoriously used private land to gain access to a water body for a period of 15 years, Minnesota courts have found an implied dedication of a public easement, creating a permanent public access right.[28] Designation of water or wetlands as "public waters" neither grants additional public access rights nor diminishes a private landowner's right to exclude others from entering or crossing his or her land.[29]

WETLANDS

Minnesota has one of the most extensive and complex programs of wetlands protection in the nation. The Wetland Conservation Act of 1991 established a policy of "no net loss in the quantity, quality, and biological diversity of Minnesota's existing wetlands."[30] Toward that end, the statute prohibits the draining or filling of wetlands "unless replaced by restoring or creating wetlands of at least equal public value under [an approved] replacement plan." Authority to approve replacement plans is delegated to local units of government, which for this purpose are under the supervision and oversight of the state Board of Water and Soil Resources.[31] Local governments must establish appropriate wetlands permitting procedures and must apply substantive standards set out in the state statute and regulations. The statute authorizes exemptions from the replacement requirement in specified circumstances, such as for certain agricultural and forestry activities, utility projects, work by public drainage systems, and *de minimis* activity. Replacement plans may be approved only after the landowner has first taken all steps to avoid and minimize adverse wetland impacts. Replacement requirements may be met by purchasing credits from an approved wetlands "bank" at a replacement ratio specified by regulation, based on the locations and types of wetlands involved.[32]

The Wetlands Conservation Act nominally applies to all wetlands in the state. However, wetlands designated as "public waters wetlands" are entitled to additional protections, including Public Waters Work Permit requirements, which involve an additional layer of project approval by the DNR.[33] In addition, the dredging or filling of wetlands subject to federal jurisdiction under Section 404 of the Clean Water Act requires a permit by the U.S. Army Corps of Engineers, adding a third tier of regulatory approval.[34] Thus wetlands projects may be subject to one, two, or three layers of permit approval, depending on the kind of project, the location, and the type of wetlands involved. Another federal program, the "Swampbuster" provision of the Farm Bill, does not directly impose additional regulatory prohibitions or permitting requirements, but instead creates an additional disincentive to wetlands conversion by making farmers who convert wetlands to agricultural use ineligible for crop subsidies or other federal payments.[35] A few municipalities have added even more complexity by enacting their own local wetlands ordinances, but in many cases these are administered by the same local governments that administer the Wetlands Conservation Act.

DRAINAGE AUTHORITIES

Throughout most of its history, Minnesota actively encouraged the development of public drainage systems with the goal of removing "excess" water from lands that could be made suitable for agriculture (McCorvie and Lant 1993). This resulted in an extensive network of public ditches (or in some cases, underground tile systems), managed by public drainage authorities for the private benefit of landowners served by the drainage system and supported by assessments on the benefited land. State statutes now regularize the procedures of drainage authorities, as well as the substantive standards by which they decide whether to establish, repair, improve, extend, abandon, or redetermine the benefits of drainage systems.[36] Most important, the drainage authority must engage in an explicit cost–benefit balancing, weighing the environmental and other public harms of a proposed drainage project against the benefits to the public, not merely to private landowners, if the project goes forward.[37]

It is unlawful for a landowner to drain water into a public drainage ditch without a permit.[38] A variety of environmental laws, including the federal Clean Water Act and especially its wetlands provisions, Swampbuster, the state's Wetlands Conservation Act, public waters protections, and statutes protecting specified categories of lands or waters, such as endangered species habitat or designated fish and wildlife habitat, may be implicated by changes to the hydrology brought about by drainage projects. These statutes, along with broader environmental factors such as the effects on water quality, land use, and fish and wildlife, must be considered by the local drainage authority prior to making a decision. Any drainage project affecting public waters requires prior approval by the DNR, and the drainage laws require the DNR to review and make advisory reports on all proposed drainage projects before final decisions are made by the drainage authorities.[39]

WATERSHED DISTRICTS AND MANAGEMENT ORGANIZATIONS

Minnesota Statutes, Section 103D, authorizes the creation of watershed districts to control or alleviate flooding, improve stream channels, regulate streamflow, divert or change all or part of watercourses, regulate development on riparian lands, protect and enhance water quality, and provide water supplies for domestic, industrial, recreational, agricultural, hydropower, and other public uses.[40] Watershed districts are empowered to establish regulatory rules concerning water resource management, carry out conservation projects, and acquire, construct, and operate drainage systems and water supply systems. They have the power to levy property taxes and to issue bonds within state-imposed limits. They also have the power of eminent domain. The courts have ruled, however, that they do not have general police powers comparable to those of a municipality.[41]

Watershed districts are created and their boundaries delineated by the state Board of Water and Soil Resources (BWSR). BWSR relies on a petition process, establishing new watershed districts in response to petitions from citizens or local units of government.[42] Currently Minnesota has 47 watershed districts, each with jurisdiction over a hydrologically defined natural watershed; however, about 70% of the state does not fall within an established watershed district (MAWD 2006).

In contrast to the rest of the state, where watershed-based management is optional, watershed management plans are mandatory for all watersheds lying wholly or partly within the seven-county Twin Cities metropolitan area, which is divided into hydrologically defined watershed management organizations

(WMOs).[43] There are three kinds of WMOs: joint powers agreements among the cities and townships within a watershed; watershed districts, as in the rest of the state; and county governments that carry out watershed management functions through their planning departments. The statute and implementing regulations require that each watershed management plan must include an enforceable implementation program, including regulatory controls to protect wetlands, prevent erosion and sedimentation, and manage stormwater and drainage consistently with the watershed management plan goals.[44] The regulations further state that all building, driveway, and grading permits must be issued under standards consistent with the watershed management plan. They also call for the establishment of water quality and quantity monitoring and assessment programs to gauge the effectiveness of plan implementation and to guide the periodic revision of plans in light of changing conditions. Watershed management plans must be revised at least once every 10 years.

PROTECTED LAKES AND STREAMS

Minnesota has designated more than 150 trout lakes and more than 650 trout streams, subject to special fishing rules and regulatory protection as public waters.[45] Access rules are the same as for other public waters. Additionally, the DNR has purchased public easements over about 250 miles of privately owned shoreline to allow public access to high-quality trout-fishing waters (DNR 2003).

The DNR also manages a Shallow Lakes Program, consisting of several categories of specially protected shallow lakes managed to protect habitat for migratory waterfowl. These include 33 migratory waterfowl feeding and resting areas, where the use of motors is prohibited during migration seasons, although the use of electric motors is grandfathered on 13 of these lakes.[46] Another 40 lakes are designated as wildlife lakes (DNR 2010a); here the DNR develops an appropriate management plan, which typically includes artificially lowering lake levels during certain seasons to optimize waterfowl habitat value and may include regulation of the use of motorized watercraft and recreational vehicles.[47] Another 116 lakes are designated as waterfowl refuges or sanctuaries, some actively and others passively managed. Those designated as state game refuges prohibit the hunting of some or all wild game.[48] An even stronger form of protection is afforded by designation as a migratory waterfowl refuge, state duck refuge, or state wildlife sanctuary. Under these designations, public access is prohibited seasonally or, in the case of state wildlife sanctuaries, year-round.[49]

SHORELAND MANAGEMENT AND FLOODPLAIN REGULATION

Minnesota counties and municipalities exercising land use regulatory authority are required to enact and enforce shoreland management ordinances incorporating or exceeding state standards or, alternatively, to incorporate such standards into their general land use planning, zoning, and subdivision ordinances.[50] The Shoreland Management Program is applicable to lakes, ponds, or flowages of 10 or more acres in incorporated areas and 25 or more acres in unincorporated areas. Local ordinances, which have to be approved by the DNR Division of Waters, must set standards for sewage and septic systems and establish setback requirements, minimum lot sizes, and density limits for buildings in the shoreland area, defined for lakes as 1,000 feet from the shore and for rivers as 300 feet from the river or the landward side of the floodplain, whichever is greater. Separate standards are established for existing resorts, aimed at allowing improvements while protecting water quality.[51]

Counties and municipalities in flood-prone areas must also incorporate state floodplain standards into local floodplain management ordinances, and local officials must inventory and map floodplains within their jurisdiction and submit all relevant data, maps, and floodplain ordinances to the DNR for approval.[52] Floodplain regulations apply within the 100-year floodplain and require that the lowest floor of any residential, commercial, or industrial building built within the flood fringe be at least 1 foot above the 100-year flood elevation, after adjusting for any increase in the flood stage due to filling in the floodplain.[53] Only land uses that have a low flood damage potential are permitted within the floodway, and uses that are likely to become polluting or hazardous in the event of a flood are generally prohibited there.[54]

WILD, SCENIC, AND RECREATIONAL RIVERS

Segments of six rivers—the Kettle, Rum, Mississippi (from St. Cloud to Anoka), North Fork of the Crow (in Meeker County only), Minnesota (from the Lac qui Parle Dam to Franklin), and Cannon—are designated as wild, scenic, or recreational rivers under Minnesota's Wild and Scenic Rivers Program, designed to preserve the "outstanding scenic, recreational, natural, historical, scientific, and similar values" of designated rivers and their adjacent lands.[55] "Wild" rivers are defined by free flow, excellent water quality, and adjacent land in an "essentially primitive" condition. "Scenic" rivers are also free-flowing and largely undeveloped, but may have some adjacent development. "Recreational" rivers may have some prior impoundment or diversion and considerable adjacent development, but nonetheless are suitable for protection to advance the purposes of the statute.[56]

Each of these protected river segments has an approved management plan outlining goals and rules for the waterway and adjacent lands. By law, local zoning ordinances must be in conformity with the management plan, and lands owned by the state and its political subdivisions must be managed in accordance with the plan and may not be transferred to any party whose use of the land would be inconsistent with the management plan.[57] The DNR is empowered to acquire fee title or scenic easements to lands adjacent to designated rivers, and to develop recreational facilities (such as "water waysides") compatible with the management plans, although in fact few such acquisitions have been made.[58]

In addition, the 200-mile stretch of the Upper St. Croix River above Taylor's Falls (lying partly in neighboring Wisconsin) is federally designated as a wild and scenic river under the federal Wild and Scenic Rivers Act,[59] one of the eight original rivers to receive such a designation. It is managed by the National Park Service to protect its wild and scenic values. The federal statute prohibits all federal agencies from engaging in any loan, grant, license, or other assistance to any project that would adversely affect the river's wild and scenic values. In addition, the federal government may acquire scenic easements. The Lower St. Croix, consisting of 52 miles from Taylor's Falls south to the river's confluence with the Mississippi at Point Douglas, was added to the national Wild and Scenic Rivers Program in 1972, with the upper 10 miles of this reach designated as scenic and the remaining 42 miles designated as recreational.[60] The Lower St. Croix is jointly managed by the National Park Service, the Minnesota DNR, and the Wisconsin DNR under a cooperative management plan developed by the agencies in consultation with interested citizens and stakeholders, with the federal government assuming primary management responsibility for the stretch between Taylor's Falls and Stillwater, and the states managing the lower and more intensively developed portion of the river (NPS 2001).

MISSISSIPPI RIVER AND RED RIVER MANAGEMENT

As an alternative to formal Wild and Scenic Rivers designation, the first 400 miles of the Mississippi River, from Lake Itasca to a point just south of Little Falls, are managed by the Headwaters Board, a joint powers board consisting of eight northern and central Minnesota counties, working in conjunction with state and tribal authorities and the U.S. Department of Agriculture (USDA) Forest Service in its capacity as manager of the Chippewa National Forest. The Headwaters Board is mandated by statute to enhance and protect the "natural, scientific, historical, recreational, and cultural values" of the headwaters region.[61] It does so by promulgating common land use zoning requirements, which by statute must be incorporated into each county's zoning ordinance,[62] applicable to a corridor generally defined to extend 500 to 1,000 feet on either side of the river. The board has the power to review and certify or overrule certain local land use decisions, including zoning amendments, variances, and subdivision approvals, if they are inconsistent with the corridor plan. It also develops and periodically revises a comprehensive management plan setting forth guidelines for activities on public and private lands along the corridor. These guidelines are incorporated into cooperative agreements with the Leech Lake Tribal Council, the state DNR, and the Forest Service, and they are influential with some private landowners (MHB 2002). Although the plan provisions are not legally binding on federal or tribal authorities or private landowners, state and local governments and their agencies are mandated by statute to "exercise their authorities so as to further the purposes of . . . [the statute] and the plan," and to manage state and local government-owned lands "in accordance with the plan."[63]

The basin of the Red River of the North encompasses the west-central and northwestern portions of Minnesota, northeastern South Dakota, eastern North Dakota, and southern Manitoba up to the river's mouth on Lake Winnipeg. The Minnesota portion covers about 37,000 miles and 10 major watersheds, and it includes some of the richest and flattest farmland in the world (Moehlman 1935). Nonpoint-source agricultural runoff and spring flooding are chronic concerns here, as are shortages of water during the late-summer dry season (Carlyle 1985). The Red River Watershed Management Board (RRWMB) was created by the Minnesota legislature in 1976 to promote coordination of water management in the Minnesota portions of the basin, with a special emphasis on flood control. Currently, eight Minnesota watershed management districts are represented on the RRWMB. The International Joint Commission (IJC) has established an International Red River Board (IRRB), pursuant to the Boundary Waters Treaty between the United States and Canada, to advise and assist the IJC in its dispute avoidance and resolution functions under the treaty, again with respect to both water quality and quantity issues. Finally, a not-for-profit Red River Basin Commission (RRBC), composed mainly of local government officials, was convened in 2002 in an effort to broaden basinwide coordination (Hearne 2007). Efforts are now under way to consolidate and streamline these various (and arguably duplicative) efforts at interjurisdictional coordination through a basinwide integrated water resource management plan (USACE 2008).

GREAT LAKES COMPACT AND AGREEMENT

Minnesota was the first state to formally ratify the Great Lakes–St. Lawrence River Water Resources Compact, a legally binding agreement among the eight states bordering the Great Lakes, which took effect after Congress passed and President George W. Bush signed the necessary federal enabling legislation in 2008.[64] The compact, together with a parallel nonbinding good-faith agreement that includes the Canadian provinces of Ontario and Quebec, generally prohibits out-of-basin diversions of water from the Great Lakes Basin, which holds approximately 20% of the world's fresh surface water. The compact provides limited exceptions for existing diversion points, communities in counties that straddle in-basin and out-of-basin areas, and withdrawals of water in containers of less than 5.7 gallons. The compact and agreement establish a Great Lakes–St. Lawrence River Water Resources Council and Regional Body to create a common database on water resources and uses in the basin, review state and provincial implementation of the compact and agreement, and set basinwide water conservation and efficiency objectives. Each state and province is responsible for developing and administering its own conservation and efficiency program, consistent with regional objectives, and for developing standards, guidelines, and permitting requirements for new and increased water withdrawals, consistent with new basinwide standards calling for environmentally sound and economically feasible water conservation and efficiency measures, and "no significant harm" to the basin's waters and natural resources. New out-of-basin diversions proposed on the basis of one of the enumerated exceptions, as well as large water withdrawals resulting in a consumptive loss to the basin of 5 million gallons or more per day, are also subject to review by the Regional Body prior to final approval by the state or province.

Each state is responsible for enacting the necessary implementing legislation within five years of the effective date of the compact. This may necessitate further action by the Minnesota legislature to put the necessary standards and administrative mechanisms into place to regulate water withdrawals and diversions from that portion of northeastern Minnesota that lies within the Great Lakes Basin.

REGULATION OF WATER QUALITY

Protection of water quality is widely thought to be principally the province of the federal government, operating through the Clean Water Act. In Minnesota, however, as in many other states, protection of water quality is in fact a tripartite affair, with federal, state, and local units of government all playing significant roles.

At the federal level, the Clean Water Act (CWA) generally prohibits the discharge of pollutants into surface waters from point sources—factories, sewer outfalls, pipes, ditches, and other "discrete conveyances"—without a permit.[65] Consequently, each point-source polluter must obtain a National Pollutant Discharge Elimination System (NPDES) permit specifying the permitted levels and kinds of effluent, based on the levels of control achievable by use of the "best available technology" for that category of polluter.[66] In addition, the permitted effluent should not exceed a level that would allow achievement of state-established water quality standards for the water body or segment receiving the pollution. At the state level, the Minnesota Pollution Control Agency (MPCA) administers the NPDES program on behalf of the federal government pursuant to a cooperative agreement between the U.S. Environmental Protection Agency and the state (MPCA 2002). Under the federal statute, the MPCA must exercise its permitting authority according to federal standards, and in addition, it is responsible for setting water quality standards for all the state's surface waters. Local units of government may also play a significant role in protecting water quality on their own initiative or as mandated by state law, as in the Wetlands Conservation Act, the Shoreland Management Program, and the Wild and Scenic Rivers and Mississippi Headwaters Programs, or in the form of drainage authorities, watershed districts, or watershed management organizations.

The CWA does not directly regulate nonpoint-source pollution, primarily polluted runoff from farms, forestry operations, construction sites, highways, urbanized areas, and septic systems. As a consequence, nonpoint-source pollution, especially agricultural runoff, is now the largest contributor to impairments of water quality in Minnesota's waters. Minnesota does have a state Nonpoint Source Management Program Plan (NSMPP), making it eligible to receive federal subsidies under Section 319 of the CWA, which the state uses in turn to subsidize voluntary nonpoint-source pollution reduction projects in some of Minnesota's most polluted waters (MPCA 2008). Like similar programs in other states, however, Minnesota's program has put only a small dent in a very large problem.

The federal statute does contain a backstopping secondary regulatory scheme that is only now beginning to force Minnesota and other states to come to grips with the festering problem of nonpoint-source pollution. The CWA requires states to assess the quality of all waters in the state and publish a list identifying "impaired waters"—those not meeting state water quality standards.[67] For each pollution-impaired water body or segment, the state is further required to develop a total maximum daily load (TMDL)—an allowable pollution budget that would eliminate water quality impairments by bringing the receiving waters up to state water quality standards. EPA regulations further require that the pollution budget established by the TMDL be apportioned among all sources contributing to the impairment, including both point and nonpoint sources, thus indirectly inducing states to address nonpoint-source pollution.[68] As of 2010, the MPCA had identified a total of 3,049 impairments statewide, of which nearly 60% will require TMDLs, although many water bodies have not yet been assessed, and the agency does not anticipate completing these assessments until 2017 (MPCA 2009). The largest numbers of these are in the Mississippi, Minnesota, and St. Croix River Basins, although the Red River, Rainy River, and Lake Superior Basins also contain significant concentrations.

Work on developing TMDLs is well under way in some areas. The largest and potentially most influential of these concerns sediment and excess nutrients (principally phosphorus) in Lake Pepin, a natural lake formed by the widening of the Mississippi River along Minnesota's southeastern border with Wisconsin. Lake Pepin's 48,634-square-mile watershed encompasses a little over half of Minnesota's land area, including the Mississippi, Minnesota, and St. Croix River Basins, as well as a portion of western Wisconsin. As elsewhere, most of the pollutants entering Lake Pepin are from nonpoint sources, especially agriculture in the Minnesota River Valley, but point sources including wastewater treatment plants, urban stormwater discharges, and concentrated animal feeding operations are implicated as well (MPCA 2007). Given its geographic scope, the Lake Pepin TMDL will necessarily become an overarching strategic plan for pollution reduction over this expansive basin, serving as a guiding framework within which smaller-scale and more specific TMDLs for smaller tributaries will be nested. TMDLs in turn rely largely on a strategy of identifying and promoting the adoption of "best management practices" (BMPs) for various categories of pollution sources. Except where the pollution source is already subject to permitting requirements, however, adoption of BMPs is at present voluntary, limiting the effectiveness of both BMPs and the larger TMDL program in meeting surface water quality goals.

The federal Clean Water Act generally does not address groundwater contamination, leaving that matter principally to state jurisdiction. Minnesota's Groundwater Protection Act sets an aspirational goal of "no degradation" of groundwater resources, employing both voluntary adoption and regulatory requirements of BMPs to achieve that goal.[69] The commissioner of agriculture is responsible for identifying BMPs, promoting their voluntary adoption, and where necessary, promulgating regulatory requirements for agricultural polluters. For all other sources of groundwater pollution, these responsibilities rest with the MPCA.[70]

WELL PERMITS AND REGULATIONS

The Minnesota Department of Health has major responsibility for protecting the quality of drinking water, especially from groundwater. It sets standards and issues permits for the drilling, construction, modification, repair, and sealing of wells; examines and licenses well contractors; and has authority to inspect and test all wells in the state "at all reasonable times."[71] In some cases, the state Department of Health has delegated some or all of its inspection, permitting, and enforcement duties to local boards of health.[72]

The Department of Health also performs source water assessments for public water supply systems to ensure the safety of their water supplies and administers the state's Wellhead Protection Program as required by the federal Safe Drinking Water Act.[73] In addition, the Department of Health establishes health risk limits for groundwater contaminants; these in turn are incorporated into groundwater protection regulations promulgated by the MPCA or, in agricultural areas, the Department of Agriculture to address significant occurrences of groundwater pollution.[74]

CONCLUSIONS

As the foregoing survey suggests, Minnesota's water law is exceedingly complex, a composite of residual common-law principles with layer after layer of incremental legislative enactments intended to protect the state's critically important water resources. In many areas, Minnesota has been a legal pioneer and innovator, and some of its innovations stand as national models, widely admired by those who would do more to protect their own water resources.

Yet a balanced view would also find much to criticize. First is the sheer complexity of it: even if each of the parts works reasonably well, they arguably are not woven together into a coherent, seamless whole. The confusing welter of rules and administrative bodies makes it difficult for ordinary citizens to understand what their rights and legal responsibilities are, possibly to the detriment of compliance.

Second, such a complex set of arrangements may have costly inefficiencies and redundancies. In the area of wetlands protection, for example, the multiple layers of bureaucracy that an applicant must go through to secure permission to develop a wetlands site surely operate at the margins as a costly deterrent to development. Yet the substantive standards at each stage of that bureaucratic morass are sufficiently flexible—or weak, depending on one's point of view—that many wetlands development projects ultimately do go forward, provided the developer is able to pony up enough cash to purchase wetlands restoration credits. Consequently, it may be that Minnesota's complex bureaucratic approach to wetlands protection simultaneously creates both over- and under-deterrence, thwarting some relatively harmless projects because of the sheer cost and complexity of the administrative process, while lacking the regulatory "teeth" to stop big-ticket projects that may be truly damaging to wetlands values.

Third, despite the multiplicity of laws, regulations, and programs, large gaps may exist. The present legal structure is jerry-built out of a lengthy series of piecemeal legislative reforms, most of them relatively narrow in scope and designed to address particular problems. Little thought has gone into how the pieces fit together as a whole, with the result that some substantial problems may go unaddressed. Indeed, if projections of increased stress on Minnesota's water resources as a result of climate change and growing demand are correct, there is little reason for confidence that the current system, predicated on a history of water abundance, is well designed to handle those changing realities.

Some sentiment exists in the legislature in favor of streamlining and rationalizing Minnesota's water law—simplifying, coordinating, eliminating redundancies and unnecessary complexities, and filling gaps. No one has yet produced even a preliminary sketch, much less a blueprint, of what such a major overhaul might look like, but that discussion is beginning. If successful, this effort could once again propel Minnesota into the forefront as a leader in crafting a water law for the 21st century.

NOTES

1. See also *Pratt v. State*, 309 N.W.2d 767, 771 (Minn. 1981).
2. *Bartell v. State*, 284 N.W. 2d 834, 838 (Minn. 1979).
3. Minn. Stat. § 103G.265, subd. 1.
4. Minn. Stat. § 103G.711.
5. *State v. Longyear Holding Co.*, 224 Minn. 451 (Minn. 1947) (quoting *United States v. Holt State Bank*, 270 U.S. 56 (1926)).
6. See also *Schmidt v. Marschel*, 211 Minn. 539 (Minn. 1942).
7. Minn. Stat. § 103G.201.
8. Minn. Stat. § 103G.245.
9. Minn. Stat. § 103G.205, subd. 15.
10. Minn. Stat. § 103G.205, subd. 15a.
11. Minn. Stat. § 103G.401(b).
12. Minn. Stat. § 103G.245.
13. Minn. Stat. § 103G.315, subd. 3.
14. Minn. Admin. Rules 6115.0210.
15. Minn. Stat. § 103G.271; Minn. Admin. Rules 6115.0620.
16. Minn. Stat. § 103G.315.
17. Minn. Stat. § 103G.261.
18. Minn. Stat. § 103G.301.
19. Minn. Stat. § 103G.281.
20. Minn. Stat. § 103G.315.
21. Minn. Stat. § 103G.285.
22. *Nelson v. De Long*, 7 N.W.2d 342 (Minn. 1942).
23. *State by Head v. Slotness*, 185 N.W.2d 530 (Minn. 1971).
24. *Sanborn v. People's Ice Co.*, 82 Minn. 43 (Minn. 1900).

25. *Petraborg v. Zontelli*, 217 Minn. 536 (Minn. 1944); *Johnson v. Seifert*, 257 Minn. 159 (Minn. 1960).

26. *Bartell v. State*, 284 N.W.2d 834, 838 (Minn. 1979).

27. *Bartlett v. Stalker Lake Sportsmen's Club*, 168 N.W.2d 356 (Minn. 1969).

28. *Flynn v. Beisel*, 257 Minn. 531 (Minn. 1960); *Barth v. Stenwick*, 761 N.W.2d 502 (Minn. App. 2009).

29. Minn. Stat. § 103G.205.

30. Minn. Stat. § 103A.201.

31. Minn. Stat. § 103G222.

32. *Id.*

33. Minn. Stat. § 103G.245.

34. 33.U.S.C. § 1344.

35. 15.U.S.C. § 1381.

36. Minn. Stat. § 103E.

37. Minn. Stat. § 103E.015.

38. Minn. Stat. § 103E.401.

39. Minn. Stat. § 103E.011, § 103E.255, § 103E.301.

40. Minn. Stat. § 103D.355.

41. *Minch v. Buffalo–Red River Watershed District*, 723 N.W.2d 483 (Minn. App. 2006).

42. Minn. Stat. § 103D.101.

43. Minn. Stat. § 103B.231.

44. *Id.*; Minn. Admin. Rules 8410.0100.

45. Minn. Admin. Rules 6264.0050.

46. Minn. Stat. § 97A.095.

47. Minn. Stat. § 97A.101.

48. Minn. Stat. § 97A.085.

49. Minn. Stat. § 97A.095, § 97A.083, § 97A.137.

50. Minn. Stat. § 103F.211, § 103F.215, § 103F.221.

51. Minn. Stat. § 103F.227.

52. Minn. Admin. Rules 6120.5400.

53. Minn. Admin. Rules 6120.5700, 6120.5800.

54. Minn. Admin. Rules 6120.5800.

55. Minn. Stat. § 103F.315.

56. Minn. Stat. § 103F.311.

57. Minn. Stat. § 103F.335.

58. Minn. Stat. § 103F.331.

59. Codified at 28 U.S.C. § 1271 et seq.

60. Pub. L. 92-560, 1972.

61. Minn. Stat. § 103F.369.

62. *Id.*

63. Minn. Stat. § 103F.371.

64. Minn. Stat. § 103G.801; S. J. Res. 45, 110th Cong., 2008.

65. CWA § 301(a), 33 U.S.C. § 1311(a).

66. CWA § 402(a), 16 U.S.C. § 1342(a).

67. CWA § 303(d), 33 U.S.C. § 1313(d).

68. 40.C.F.R. § 130.

69. Minn. Stat. § 103H.001, § 103H.101.

70. Minn. Stat. § 103H.151.

71. Minn. Stat. § 103I.101.

72. Minn. Stat. § 103I.111.

73. Minn. Admin. Rules 4720.5100 et seq.; 42 U.S.C. § 300h-7.

74. Minn. Stat. § 103H.201, § 103H.275.

REFERENCES

Carlyle, William J. 1985. Water in the Red River Valley of the North. *Geographical Review* 74 (3):331–358.

Dellapenna, Joseph W. 2010. Global Climate Disruption and Water Law Reform. *Widener Law Review* 15:409–445.

DNR (Minnesota Department of Natural Resources). 2000. *Minnesota's Water Supply: Natural Conditions and Human Impacts.* St. Paul: DNR.

———. 2003. *DNR Trout Stream Easements: Answers to Questions Landowners Often Ask.* St. Paul: DNR.

———. 2009a. *History of Water Protection: Origin of State Authority over Public Waters.* St. Paul: DNR.

———. 2009b. *Water Year Data Summary 2007–08.* St. Paul: DNR.

———. 2010a. *Minnesota's Designated Wildlife Lakes.* St. Paul: DNR.

———. 2010b. *Water Laws in Minnesota: Questions and Answers about Minnesota Water Laws.* St. Paul: DNR.

Hearne, Robert R. 2007. Evolving Water Management Institutions in the Red River Basin. *Environmental Management* 40:842–852.

Klein, Christine A., Mary Jane Angelo, and Richard Hamann. 2009. Modernizing Water Law: The Example of Florida. *Florida Law Review* 61:403–474.

MAWD (Minnesota Association of Watershed Districts). 2006. *Annual Report 2005–2006.* St. Paul: MAWD.

McCorvie, Mary R., and Christopher L. Lant. 1993. Drainage District Formation and the Loss of Midwestern Wetlands, 1850–1930. *Agricultural History* 67 (4):13–39.

MHB (Mississippi Headwaters Board). 2002. *Mississippi Headwaters Management Plan.* Backus, MN: MHB.

Moehlman, E.H. 1935. The Red River of the North. *Geographical Review* 25 (1):79–91.

MPCA (Minnesota Pollution Control Agency). 2002. *NPDES/SDS Permits: Permitting Process for Surface Water Dischargers.* St. Paul: MPCA.

———. 2007. *Lake Pepin Watershed TMDL: Eutrophication and Turbidity Impairments Project Overview.* St. Paul: MPCA.

———. 2008. *Minnesota's Nonpoint Source Management Program Plan.* St. Paul: MPCA.

———. 2009. *Draft MPCA 2010 303(d) List.* St. Paul: MPCA.

NPS (National Park Service). 2001. *Lower St. Croix River National Scenic Riverway Record of Decision, 66 Fed. Reg. 56848–56852 (2001) and Final Cooperative Management Plan/Final Environmental Impact Statement: Lower St. Croix National Scenic Riverway.* Washington, DC: NPS.

USACE (U.S. Army Corps of Engineers). 2008. *Review Plan: Red River Basin Watershed Feasibility Study.* Washington, DC: USACE.

PART III

MANAGING THE WATERS AS BIOPHYSICAL RESOURCES

CHAPTER 6

Local-Scale Institutions: Lake Management Associations and Watershed Districts

Louis N. Smith and John C. Kolb

*W*hereas the desire to protect and improve our lakes may arise from the purest, simplest motivation to conserve nature's creation, the work of lake stewardship quickly encounters complexity. We are blessed with the fact that all levels of government have expressed their concern about water resources, but we are also cursed with an amazingly complex regulatory structure that few mortals can negotiate successfully. Our present circumstances are indeed the product of many different approaches to managing water resources.

Nationally, water laws are highly fragmented, with different federal statutes governing safe drinking-water standards, the Great Lakes, wetlands, coastal zones, hydropower development, Wild and Scenic Rivers, waterways on federal lands, and others. States further regulate wetland, floodplain, and lakeshore development and nonpoint-source pollution, and can implement stricter effluent standards than required by the Clean Water Act (Goldfarb 1994). Our state government has, over the past century, enacted laws to facilitate drainage for agriculture, protect wetlands, regulate impacts to public waters, promote shoreline protection, and protect water quality. Local municipalities are usually responsible for water supply and use issues, as well as land use regulation.

As a result, water resource problems seldom fall under the jurisdiction of a single agency. Water resource management responsibility lies in the hands of multiple federal, interstate, state, county, and municipal government bodies, most with overlapping jurisdictions. Comprehensive and integrated water resource management is often hindered by institutional rivalries, disagreements, and inconsistencies, as well as by the divergent demands of different constituencies with different economic and environmental needs (Goldfarb 1994).

Our traditional organization of government authority based on political, rather than hydrologic, boundaries has often led to disparate jurisdictions and local governments often pushing water problems to the next community downstream.

The result is that each jurisdiction pursues solutions to its development problems without consideration of the effects they may have on neighboring or downstream jurisdictions in a sort of "tragedy of the commons" (Goldfarb 1994). This multilevel, jurisdictional web separates water resource management decisions from land use decisions, which are mostly made at the local level. From the perspective of local government, the maze of authorities responsible for water management decisions can be overwhelming.

Despite this complexity, Minnesota has developed a range of innovative institutional structures for management at the local watershed level. At the local scale, institutions such as watershed districts, soil and water conservation districts, and lake improvement districts can be influential and effective in leading water management. Nongovernmental organizations, including lake associations, can also play an important role. In some locations within the state, a combination of local government and private initiative is effectively managing issues such as runoff from homeowners' property and septic systems, use of water-conserving technology like porous paving and rain gardens, and lake management issues like boat speeds and aquatic plants. In other situations, the collection of institutions seems highly ineffective at protecting human value and ecosystem services. This chapter reviews various local water management institutions, with a special focus on the watershed scale, asking which suite of institutions is most effective and why.

As we pursue this question, several basic principles seem helpful to keep in mind. First, the state of our lakes and other water resources is greatly impacted by what we do on the land that drains into them. Second, much of our best conservation work is done locally. As human beings, we may be capable of deep concern for faraway places, but we tend to care best for that which is close by. And third, as a result of the first two principles, we will do our best lake and stream conservation work when local people can effectively integrate decisions about land use and water resources in a comprehensive fashion.

LOCAL WATER MANAGEMENT IN MINNESOTA

Minnesota has created several types of local institutions with important jurisdiction or interest in water resources. Our analysis of their effectiveness best begins with an examination of their legal origins.

Watershed Districts and Water Management Organizations

Water knows no political boundaries. The idea that we should manage water resources by watershed or along hydrologic boundaries is currently quite popular, but it is not new. In fact, one of the greatest proponents of watershed management was John Wesley Powell, the 19th-century explorer of the Colorado River and the American West. Powell thought our political boundaries should be drawn hydrologically and envisioned all of the western states organized into "natural hydrographic districts, each one to be a commonwealth within itself" (Powell 1890).

The United States did not heed Powell's suggestion, but today, more than 100 years later, the notion of managing water resources by local watershed is nearly a national consensus. Following an era of expansive environmental legislation from 1965 to 1986, the pendulum seemed to swing away from concentration of federal authority in 1987, when Congress enacted amendments to the Clean Water Act. Grounded in the concern that the country had failed to make adequate progress in addressing nonpoint-source pollution, the amendments made adequate nonpoint-source control a national goal but the responsibility of state and local governments. The U.S. Environmental Protection Agency (EPA) began to promote a local watershed approach as the means of implementing this nonpoint-source program. The "watershed approach" is now the national model and new hope for effective management of water resources. "The intimacy of the smallest watersheds may be a key to their restoration: At that level every individual can have an effect. 'It's almost impossible to address water quality on the main stem of a river,' says James Fisher of the National Watershed Coalition. 'If you do it one small watershed at a time, you still have public support. Small size is the advantage. This replaces Big Brother with Joe down the creek'" (Parfit 1993).

Local watershed councils have thus emerged along many streams in the United States, typically the product of government agency initiatives to address nonpoint-source pollution. Using hydrologic boundaries as their jurisdictional parameters, these councils promote broad stakeholder engagement to identify voluntary actions to improve water quality (Sabatier et al. 2005). These efforts typically take the form of citizen groups, ad hoc coalitions, interagency partnerships, or nonprofit organizations (Leach and Pelkey 2001; Genskow and Born 2006; Moore and Koontz 2003).

Few states have created local watershed entities with the powers of local government. Minnesota has pioneered this concept, adopting the Minnesota Watershed District Act in 1955. At the time, managing flooding impacts on downstream communities or water resources from new development or drainage projects was becoming a serious concern. Raymond Haik, one of the key drafters of the Watershed Act, has explained that one important goal of the act was to provide for a special-purpose local unit of government that could protect wetlands and other water resources in parallel to local drainage authorities (Haik 2009).

The Minnesota Watershed District Act, now codified in Minnesota Statutes Chapter 103D, provides for establishment of watershed districts "to conserve the natural resources of the state by land use planning, flood control, and other conservation projects ... using sound scientific principles for the protection of the public health and welfare and provident use of the natural resources." The act recognizes several fundamental concepts in the effective management of water resources:

• *Managing with a hydrologic focus.* The law recognizes that water does not adhere to political boundaries and, thus, allows for establishment of watershed districts as local government units bounded by hydrologic divides as opposed to political borders. As a result, lakes and streams, and the land draining into them, are regulated by one local entity with a central comprehensive vision for managing the entire water resource.

- *Supplementing land use regulation.* The law recognizes that regulation of land use within a watershed is an essential component in protecting and preserving water resources within the watershed. Watershed districts supplement municipal land use regulation with an exclusive focus on water quality and flood control in a manner designed to avoid the problem of pushing detrimental effects of development downstream.
- *Financing water resources improvements.* Watershed districts provide a workable and rational means of financing improvements for water resources. Typically, local municipal jurisdictions lack the necessary resources to fund critical improvements designed to restore water quality or provide flood control for local lakes and streams. Assessing costs across the entire watershed that contributes drainage to these lakes and streams provides a more equitable and effective approach.
- *Engaging citizen ownership.* Watershed districts, as local entities with boards composed of local citizens, provide an effective means of engaging citizen ownership and management of valued local water resources. Watershed district board members, called "managers," are appointed by county boards of commissioners, rather than elected. These citizen managers are often intimately familiar with local land use issues and are charged with using governmental powers of regulation, taxation, and construction of public improvements to achieve water resource goals.

Watershed districts are created only through a voluntary petition process governed by the state Board of Soil and Water Resources. A petition to create a watershed district must be signed by one or more of the following groups:

- half or more of the counties within the proposed watershed district;
- counties having 50% or more of their area within the proposed watershed district;
- a majority of the cities within the proposed watershed district; or
- 50 or more resident owners residing in the proposed watershed district, excluding resident owners within the corporate limits of a city if the city has signed the petition.[1]

Watershed districts are thus created through local initiative and typically with the support of local units of government. There are currently 46 watershed districts in Minnesota; these 46 districts cover less than half of the state.

The Metropolitan Surface Water Management Act of 1982 added several important powers to watershed districts in the seven-county metropolitan area. Based in part on a concern that watershed district capital projects should be the product of more intensive long-term planning, this act mandated that each watershed area within the metropolitan region develop and implement a 10-year plan. The act also required that every "minor watershed unit" in the metropolitan area be subject to this planning process, to be performed by a watershed district, a joint powers water management organization, or the county.[2] By mandating that the entire metropolitan area have a watershed plan, the legislature recognized that many local units of government, principally cities, would not support the creation of a watershed district within their communities. As a result, the legislature provided an option for a joint powers water management organization to develop the watershed plan.

After more than two decades of watershed planning experience, many observers have come to consider the joint powers organizations (JPOs) to be ineffective. These organizations generally lack taxing authority and rely on the financial contributions of each member municipality to fund watershed planning and implementation. They rarely have adequate funding to undertake significant activities, and controversies among members frequently paralyze the development of any water resource project. Although some JPOs are willing to review and comment on development projects, most do not exercise meaningful regulatory power. In several cases, the JPO has successfully obtained legislative approval for its own tax levies. In other cases, however, JPOs have been deemed by the Board of Water and Soil Resources and counties to be failures, and watershed districts have been created in their place.

A fully functioning watershed district that utilizes its statutory powers, has an adequate tax base, cooperates effectively with other local units of government, and engages citizens in watershed stewardship can be tremendously successful. These watershed districts are able to use their intensive monitoring and assessment abilities to diagnose water quality problems and their causes, then plan, design, and implement projects and programs to address them.

Soil and Water Conservation Districts

Soil and water conservation districts (SWCDs) in Minnesota trace their origin to 1937 legislation aimed at addressing the impacts of wind and water erosion. This followed a national effort in the wake of the Dust Bowl droughts to encourage local conservation practices. Today, as their authority is codified in statutes,[3] SWCDs are local units of government that manage and direct natural resource management programs at the local (typically county) level. Their scope has moved beyond agricultural practices to also provide assistance in forested lands, lakes, and urban areas.

Minnesota has 91 SWCDs, which collectively cover all of the state. Although encouraged in statute to cooperate with adjacent government, plan, and organize to implement conservation programs on a subwatershed basis, SWCDs are organized along county boundaries. Each district is governed by a locally elected board that establishes the policies and objectives of the district. In both urban and rural settings, SWCDs and their staff coordinate with other units of government and landowners to implement programs to develop and conserve soil, water, and related resources.

The legislative policy for soil and water conservation is articulated in Minnesota Statutes Sections 103A.206 and 103C.005:

Maintaining and enhancing the quality of soil and water for the environmental and economic benefits they produce, preventing degradation, and restoring degraded soil and water resources of this state contribute greatly to the health, safety, economic well-being, and general welfare of this state and its citizens. Land occupiers have the responsibility to implement practices that conserve the soil and water resources of the state. Soil and water conservation measures implemented on private lands in this state provide benefits to the general public by reducing erosion, sedimentation, siltation, water pollution, and damages caused by floods. The soil and water conservation policy of the state is to encourage land occupiers to conserve soil, water, and the natural resources they support through the implementation of practices that:

(1) control or prevent erosion, sedimentation, siltation, and related pollution in order to preserve natural resources;
(2) ensure continued soil productivity;
(3) protect water quality;
(4) prevent impairment of dams and reservoirs;
(5) reduce damages caused by floods;
(6) preserve wildlife;
(7) protect the tax base; and
(8) protect public lands and waters.

Despite this strong legislative policy, SWCDs are the only unit of local government that is present in every county of the state yet has no ability to generate operating revenues from tax sources. Each SWCD must annually prepare and present a budget to its county board or boards for approval and appropriation. Thus, in many counties, an SWCD is viewed as an extension of the county government and competes, like other departments, for priorities within the county's budget.

Because of their "unit of government" status and special purposes and authorities, however, SWCDs are uniquely suited to cooperate with state and federal agencies, as well as with private organizations, to enhance the implementation of public and private conservation initiatives within their jurisdictions. Throughout Minnesota, SWCDs are implementing clean water, nonpoint-source pollution reduction, habitat, and other conversation programs in cooperation with the U.S. Department of Agriculture (USDA), EPA, Minnesota Pollution Control Agency, Pheasants Forever, Ducks Unlimited, Trout Unlimited, U.S. Fish and Wildlife Service, and the state's Board of Water and Soil Resources. Other agencies, including the Minnesota Department of Agriculture and Department of Natural Resources, work closely with SWCDs to ensure consistent application of environmental and agriculture policy. Additionally, many SWCDs enjoy cooperative relationships with local watershed districts and have established cooperative funding and technical assistance agreements.

Because of their limited revenue-generating capabilities, partnerships are essential for SWCDs. An example of healthy partnering is found in the USDA's recent implementation of the Mississippi River Basin Initiative (MRBI). Targeted at select watersheds within the Mississippi River Basin, the MRBI seeks to provide funding for practices to improve water quality, the hypoxic zone in the Gulf of

Mexico being a major driver for the initiative. In spring 2010, the Sauk River Watershed District (SRWD) was notified that its watershed was one of three in Minnesota selected under the initiative. Because the Sauk River watershed covers parts of five counties, the SRWD plays a coordinating role for the various SWCDs to develop priorities for funding applications under the MRBI. Priorities developed by the SWCDs include practices to reduce nutrient runoff, improve wildlife habitat, and maintain agricultural productivity. The primary focus is to address and manage nitrogen and phosphorus within fields to minimize runoff and reduce downstream nutrient pollution. As with most SWCD-implemented programs, funding will be provided to cooperating private landowners for voluntary implementation of the practices. Through this cooperative effort, the SWCDs within the Sauk River watershed have been able to secure $12 million in funding over a five-year period to be paid to cooperating landowners for projects within the watershed.

SWCDs are annually responsible for the delivery of an estimated $38 million of state and federal funds to individuals and organizations for conservation, protection, preservation, and enhancement of the state's soil, water, air, plant, and wildlife resources (BWSR 2008). Progressive counties leverage the administrative and technical services of SWCDs to focus and manage county water planning and implementation efforts. Thus, although they do not exercise jurisdiction over a watershed, SWCDs do serve to coordinate and synchronize water resource management efforts on a subwatershed level within a county and, if cooperating with adjacent units of government, across county boundaries.

Minnesota statute encourages SWCDs to work with landowners to advance conservation practices and projects. Although most of these practices are individual and voluntary in nature, a process detailed in statute allows a more programmatic approach to resource management. SWCDs may undertake works of improvement on request by affected landowners or by directive from a county board. Like the SWCD budget, all works of improvement are subject to county board approval and funding decisions.

SWCD programs are structured as improvement works units, each of which contains a program for works of improvement. A county or joint county board may, on its own motion or on receipt of a petition, adopt a resolution directing a SWCD to establish such a unit. Alternatively, an affected landowner may request that the SWCD board establish a unit and a program of works. Units are established when the board determines that the works are feasible; will be of public utility and benefit; will promote public health, safety, and welfare; and will further the authorized purposes and best interests of the district.

Despite the statutory opportunity for public involvement in establishing projects and programs within an SWCD, the "works of improvement" provisions are seldom used and, where attempted, have seldom succeeded. In one example, a lake preservation effort in Hubbard County failed because the county board declined to approve the improvement works unit. This is not to say that SWCDs are not effective and beneficial units of government. In the web of organizations seeking to implement water policy, however, they may not be the most efficient organizations for affecting watershed-level management.

Lake Improvement Districts

The Minnesota legislature first authorized establishment of lake improvement districts (LIDs) in 1973. The Lake Improvement District Law is now codified in Minnesota Statutes Sections 103B.501 to 103B.581 for the purpose of preserving and protecting the state's lakes and increasing and enhancing their use and enjoyment.

The legislature found it in the public interest to establish a statewide lake improvement program to preserve the natural character of lakes and their shoreland environment, where feasible and practical; improve the quality of water in lakes; provide for reasonable assurance of water quantity in lakes, where feasible and practicable; and assure protection of lakes from detrimental effects of human activities and certain natural processes. Minnesota currently has 32 active LIDs. The three primary reasons for their establishment are to manage water quality (8 LIDs), water level control (8), and aquatic vegetation (15). The state's lake improvement program is administered by the Minnesota Department of Natural Resources (DNR).

LIDs may be established for one or multiple lakes based on hydrologic boundaries of the lake or lake system, and in multiple counties, they operate under a joint powers agreement. Lake improvement is limited to bodies of water in which the public has access to some portion of the shoreline. A LID may be established in one of three ways: by resolution, petition, or DNR commissioner's order. A county board may initiate establishment of a lake improvement district in a portion of the county by adopting a resolution declaring its intent to establish such a district. After a hearing, the board may establish a lake improvement district by order if it determines that the proposed district is necessary or public welfare will be promoted by the establishment of the district, property to be included in the district will be benefited by establishing the district, and formation of the district will not cause or contribute to long-range environmental pollution.

Upon receipt of a petition signed by a majority of property owners within the proposed district, the county board must hold a hearing and determine whether the conditions for establishment of a lake improvement district exist. Within 30 days of the hearing, the board must enter an order establishing or denying establishment of the proposed district. If the county denies establishment of the proposed district, landowners may petition the DNR commissioner to issue an order establishing the district. A LID has never been approved by the DNR after being denied by local government. All establishment orders are subject to a referendum vote if requested by at least 26% of property owners within the district.

A LID is a form of government that is subordinate to the county or counties that established it, has no taxing powers of its own, and is limited to just those authorities that the county, counties, or other government entities give to it. County boards, joint county authorities, statutory and home rule cities, and towns may, by order, delegate certain powers to the board of directors of a LID to be exercised within the district. These powers include acquisition and management of water control works, alteration of watercourses and water bodies, implementing programs and projects to improve water quality, construction and management of wastewater systems, and regulation of the surface use of waters.

Historically, local governments were hesitant to approve petitions for LIDs because of the added responsibility and cost involved in oversight of their funding, complexities of program administration, and public and political acceptance. This may be why a state with tens of thousands of lakes has only 32 active LIDs. This hesitancy may be changing, however. In areas of the state with actively engaged lake associations or coalitions of lake associations, and where the Heritage Foundation has provided assistance through its Healthy Lakes Program, there is greater acceptance. Most successful LIDs are formed around a lake association that is already well organized and active in various aspects of lake management.

Crookneck LID in Morrison County is a good example of a successful LID program. Although control of exotic species was the primary reason for its establishment, the LID also stresses the importance of education and other methods to manage water quality impacts. Morrison County is unique in that it is now willing to initiate approval of LIDs through county resolution rather than through petition. In Morrison County, Crookneck and Sullivan LIDs were approved by resolution in 2005; Alexander LID was approved in 2006. All of these had support of the Healthy Lakes Program and good participation by local DNR fisheries and ecological services staff.

LIDs provide a great opportunity for local management on a lake watershed level. The greatest challenge to their effectiveness is the willingness of county and other governments to delegate authority and commit necessary resources.

Lake Associations

In Minnesota, lake associations can play an important role in promoting watershed stewardship. With varying degrees of formality, these groups of primarily lakeshore property owners have formed to advocate for water quality, promote fish and wildlife habitat, and express concern about aquatic invasive species. The advantage of lake associations is that their members are typically motivated by their lakeshore property ownership to promote a healthy lake environment. Initial motivation to form an association may be a local problem (e.g., high or low water levels, or invasive species such as Eurasian watermilfoil) or a proposed new development or nearby industrial facility.

Once organized, lake associations may mobilize members to participate in water quality monitoring. Some associations find grants and partnerships with local agencies to develop lake management plans. Others may undertake programs to educate owners about shoreline protection and restoration or work with the DNR to stock fish. As a lake association develops specialized knowledge about the lake environment based on water quality data and a well-informed vision of the lake's potential through a management plan, the association can become an important player in local government. Tracking local government land use decisions in reviewing proposed development, updating ordinances, and keeping a watchful eye on enforcement are all valuable roles for a lake association.

The power of most lake associations is the clear interest of its members: most lakeshore property owners know that a healthy lake supports strong property values, and an unhealthy lake can quickly erode one's lifetime investment.

There are, however, limitations on the effectiveness of lake associations, the most significant of which goes back to the concept of the watershed. Often, all the best practices in the world by lakeshore owners may not be enough to improve an unhealthy lake, because the causes of most lakes' problems are in the land use practices upstream in the watershed. Lakeshore owners can find like-minded allies in other neighboring lake associations. Minnesota has a number of county coalitions of lake associations (COLAs) that can provide valuable settings for these alliances.

Lake associations present one of the best and most effective means for engaging citizens in water resource stewardship. As a few watershed districts have discovered, aligning citizen engagement with the legal and policy expertise of watershed districts represents a potent combination.

PARTNERSHIP OPPORTUNITIES: WATERSHED DISTRICTS AND LAKE ASSOCIATIONS

As unique and frequently less visible local units of government, watershed districts have come to recognize that active citizen support for their mission can be an essential asset in negotiating the waves of local politics and land use controversies that can characterize watershed management. Lake associations can provide a key source of watershed allies. Similarly, lakeshore owners often confront the limits of their association's financial wherewithal and legal authority. A local watershed district may be just the right source of necessary funding or regulatory perspective to realize the goals of an association's lake management plan. Here are the makings of powerful partnerships! Two cases of watershed district–lake association alliances illustrate this potential.

Minnehaha Creek Watershed District's Watershed Association Initiative

The Minnehaha Creek Watershed District (MCWD) in Hennepin and Carver Counties has elected to invest heavily in promoting the formation of nonprofit organizations to encourage lake and stream stewardship. The MCWD's 10-year water resource management plan recognizes that citizen education and communication programs play an important role in furthering water quality and protection (MCWD 2007, 73). Coupled with this recognition of the value of civic engagement is the reality that watershed districts historically have not been the most visible form of local government. Many citizens are not aware of the existence of their local watershed district, if there is one. The Watershed Act does require that each watershed district appoint a citizens advisory committee, which can be a helpful source of civic engagement. Often, however, the citizens who do participate have a specific grievance and may lack the broader context for their concern.

In 2007, the MCWD decided that it wanted to augment its civic engagement work and thus turned to Minnesota Waters, a statewide organization dedicated to promoting responsible stewardship of water resources by engaging citizens, local and state policymakers, and other partners in the protection and restoration of Minnesota's lakes and rivers. Minnesota Waters' model of locally led conservation follows the "data to information to action" pathway. Its work focuses on helping citizens gather water quality data, convert these data into meaningful information, and act on what they learn, often by contacting the local government officials who make land use decisions.

Recognizing that Minnesota Waters, the product of a merger of the former Minnesota Lakes Association and Rivers Council of Minnesota, had unique experience and independent credibility with citizens, the MCWD hired the organization to recruit and train new lake associations. The MCWD's Watershed Association Initiative has several objectives:

- empower citizens to help meet local water quality goals by strengthening stewardship practices;
- educate property owners on ways to form healthy, effective, and sustainable citizen-led, water-focused organizations;
- present workshops for newly forming and established water protection groups to help them create action plans to meet fund-raising, organizational, and communication goals;
- strengthen and encourage collaboration among local businesses, government, nonprofit, and citizen-led lake, stream, or watershed associations;
- promote a network of peer-to-peer learning and opportunities for citizen groups; and
- provide funding through the Mini-Grant Program to help establish groups.

Through Minnesota Waters, the MCWD has helped form nine new lake or stream associations and provides ongoing support to several others. In October 2009, the MCWD and Minnesota Waters hosted the third annual lake and river leaders' summit for citizen leaders. While the MCWD has a strong technical and educational program staff to inform these lake association initiatives, Minnesota Waters offers critical and complementary capabilities to reach citizens and engage them in watershed stewardship activities. As individual associations become more organized, Minnesota Waters has supported members in making connections to local government, organizing citizen outreach to local officials to brief them on lake issues.

This arrangement is not without its complexities. The MCWD was warned before launching this initiative that to date, the only termination of a watershed district in Minnesota was the product of a local lake association's intense opposition to its existence. It has been important for MCWD to be clear up front that local lake associations are not beholden to the district and are free to develop their own agendas. The MCWD seeks to build local citizen capacity to protect water resources through these new lake associations, but it is not seeking to control outcomes of their work, trusting that the watershed will be better off with more engaged local citizens. Using Minnesota Waters as a third-party vendor to provide citizen engagement services helps posture this relationship more effectively.

The citizens who have been trained through this Watershed Association Initiative continue to care about their local lakes, but also have gained a broader watershed perspective through workshops with other local lake association members. If they did initially come with a specific grievance, they now have more tools with which to pursue their concerns more effectively. The MCWD has also found that these citizens have diverse and rich expertise to offer from their own professional lives and are willing to volunteer their talents to pursue grants or support watershed projects. In only three years, the MCWD's partnership with Minnesota Waters has yielded dozens of local lakeshore leaders and watershed activists who have developed skills and are committed to lake and watershed stewardship.

Rice Creek Watershed District's Bald Eagle Lake Initiative

Bald Eagle Lake lies in the Rice Creek watershed and in three counties, Ramsey, Washington, and Anoka. The lake suffers from an infestation of curlyleaf pondweed, an invasive aquatic species that contributes to phosphorus impairment of lake water quality. The Bald Eagle Lake Association recognized the seriousness of the problem but lacked the formal structure and funding to pursue solutions. Association leaders grew weary of knocking on doors to collect contributions for lake improvement activities and concluded that individual solicitation was not an equitable or sustainable funding system.

Bald Eagle Lake Association members circulated a survey to lakeshore property owners and, through a "block captain" approach, received responses from 85% of property owners. More than 80% of the respondents expressed support for paying some form of tax or fee increase of $150 per year to contribute to the improvement of the lake. Aware of the Rice Creek Watershed District (RCWD) as a local unit of government devoted to water quality and resource management, the association called on the district for help.

The RCWD recognized that partnering with the lake association could be mutually beneficial. It also recognized that while its watershed district-wide, ad valorem property tax levy could provide some revenue for lake improvement projects, it was also appropriate to consider more localized funding for lake programs, especially when the lake association requested such a measure.

Watershed districts have several funding mechanisms available; in this case, the RCWD chose to respond to the lake association's petition by creating a stormwater utility. A stormwater utility fee is the single clearly viable means of collecting revenue on a subwatershed basis available under current Minnesota law. The funding raised can be applied to a "water management district" or districts created under the authority given in Minnesota Statutes Chapter 103D. To date, only a handful of watershed districts have implemented subwatershed funding through the stormwater utility mechanism. The Pelican River and South Washington Watershed Districts have successfully used this funding system for several years. Others are actively exploring the possibility.

Minnesota Statutes Section 444.075 authorizes watershed districts to establish a stormwater utility, thus enabling a charge to be assessed against all properties, including nonprofit- and government-owned properties not subject to property taxes. A benefit of a stormwater utility fee over an ad valorem tax is that properties are charged based on their contribution to the problem to be addressed (e.g., pollution from stormwater runoff) rather than their economic value. The statute recognizes that assessing such contributions involves consideration of a number of more or less measurable variables, and it provides that charges may be established:

(1) by reference to the square footage of the property charged, adjusted for a reasonable calculation of the storm water runoff; or
(2) by reference to a reasonable classification of the types of premises to which service is furnished; or
(3) by reference to the quantity, pollution qualities, and difficulty of disposal of storm water runoff produced; or
(4) on any other equitable basis, including any combination of equitable bases referred to in clauses (1) to (3), but specifically excluding use of [water use] and otherwise without limit.[4]

In other words, land use and imperviousness are clearly stated bases for setting the fee rate. The stormwater utility statute also empowers the process of determining reasonable charges to include "all costs of the establishment, operation, maintenance, depreciation and necessary replacements of the [stormwater management] system, and of improvements, enlargements and extensions necessary ... including ... the costs of obtaining and complying with permits required by law."[5]

On its face, the stormwater utility law limits the use of moneys collected to stormwater facilities constructed under the authority granted in the statutes,[6] but the term "stormwater facilities" has been interpreted broadly to encompass many watershed projects that manage runoff and related programs.

A watershed district establishes a water management district by amendment to its watershed plan. The amendment must describe the following "with particularity":

• the territory or the area to be included in the water management district;
• the amount of the necessary charges;
• the methods used to determine charges; and
• the length of time the water management district will remain in force.[7]

After the amendment is approved by the Board of Water and Soil Resources and adopted by the watershed district board of managers, the amendment must be filed with the county auditor and county recorder.[8] Notice to other local government units within the relevant water management district is also required when a specific project is initiated.[9]

A watershed district's stormwater utility fee can be "billed and collected in a manner the district shall determine, including certification to the counties with territory in the district for collection by the counties."[10] The county, in turn, has discretion to determine how it will collect the fee for the district if asked to do so, or the county and district can follow the process provided by statute, whereby the fee is collected in the manner of a special assessment.[11] Charges are collected along with property taxes and are due in January of each year. A watershed district working with a county to collect a fee using this system will be obligated to provide the county with the list of properties from which the fee should be collected and the amount of the fee for each before October 15. If, for any reason, the county collection method is not feasible, watershed districts can consider other collection mechanisms such as partnering with electrical or water utility billing systems.

The RCWD completed this stormwater utility process and now collects approximately $35,000 annually. It has committed to using its districtwide ad valorem tax levy to match this amount, creating a $70,000 dedicated annual funding source to finance implementation of a Bald Eagle Lake management plan. Association members have been able to devote their energies to informing the development of this plan, as well as lakeshore owner outreach and education—activities they find highly preferable to fund-raising. Most of this localized funding and the watershed match will focus on in-lake issues, especially treatment for curlyleaf pondweed. Meanwhile, the RCWD is leading the process to address the external loading, or watershed issues that impact Bald Eagle Lake, and have furthered its placement on the state's impaired waters list. The district recognizes that these larger water quality issues arise throughout the watershed, and addressing them will most likely continue to be funded through the districtwide ad valorem tax levy.

For the watershed district, this partnership with the Bald Eagle Lake Association has brought a much stronger connection between the district and local citizens. And with this new dedicated funding source created by the RCWD, the lake association has a much more sustainable and useful role to play in protecting and improving Bald Eagle Lake.

CONCLUSIONS

At the outset of this chapter, we contrasted the blessing of governmental concern for water resources with the curse of complex regulatory structures. A similar contrast could be made between the abundance of statutory entities authorized to manage, protect, and rehabilitate water resources and the institutional friction that prevents their efficient or effective function. Minnesota probably does not need more new agencies to manage water, and despite perennial studies that propose to streamline water management, it is unlikely that any of the existing agencies will disappear. What would help most is partnership, especially at the local level.

This chapter has examined but a few of the institutions available to implement local water management. The diversity of institutional structures and characteristics can allow for a best-fit management solution, depending on the needs and priorities of the community of interest. However, one of the objectives of water resource management should be to achieve prioritized, synchronized, and coordinated management practices on a meaningful scale. This objective is embodied in the watershed approach. Indeed, partnerships between watershed districts or other local units of government, such as SWCDs or LIDs, and interested nongovernmental organizations, such as farm coalitions for drainage and waterways, homeowners associations, or business organizations for stormwater issues, offer great promise. It is through such strategic local partnerships that our work can best address our relationships with land and water—and hence protect and restore our lakes and streams.

The watershed approach, augmented by strong community involvement, represents the most effective institutional model for water resource management. Unlike the other institutions examined, watershed districts have the governmental tools of regulation and taxation with which to connect land and water stewardship effectively. Promising new partnerships between watershed districts and lake associations offer great potential to combine local public governance with strong civic engagement. Where lake association members find themselves without a watershed district in their area, they may find it worthwhile to establish one.

NOTES

1. Minn. Stat. § 103D.205, subd. 3.
2. Minnesota Laws 1982, Ch. 509, § 21.
3. Minn. Stat. Ch. 103C.
4. Minn. Stat. § 444.075, subd. 3b.
5. Minn. Stat. § 444.075, subd. 3g.
6. "Any watershed district may build, construct, reconstruct, repair, enlarge, improve, or in any other manner obtain storm water systems, including mains, holding areas and ponds, and other appurtenances and related facilities for the collection and disposal of storm water, maintain and operate the facilities, and acquire by gift, purchase, lease, condemnation, or otherwise any and all land and easements required for that purpose" (Minn. Stat. § 103D.730(a)).
7. Minn. Stat. § 103D.729, subd. 2.
8. *Id.*
9. Minn. Stat. § 103D.729, subd. 3.
10. Minn. Stat. § 444.075, subd. 2a(a).
11. Minn. Stat. § 444.075, subd. 2a(b).

REFERENCES

BWSR (Board of Water and Soil Resources). 2008. Statement of Revenues and Expenditures. www.bwsr.state.mn.us/finance/2008_Revenue_and_Expenditures.pdf (accessed December 9, 2010).

Genskow, Kenneth D., and Stephen M. Born. 2006. Organizational Dynamics of Watershed Partnerships: A Key to Integrated Water Resources Management. *Journal of Contemporary Water Research & Education* 135:56–64.

Goldfarb, William. 1994. Watershed Management: Slogan or Solution? *Boston College Environmental Affairs Law Review* 21 (3):483–509.

Haik, Raymond. 2009. Remarks at the 50th Anniversary of Nine Mile Creek Watershed District. September 2009, Bloomington, MN.

Leach, William D., and Neil W. Pelkey. 2001. Making Watershed Partnerships Work: A Review of the Empirical Literature. *Journal of Water Resources Planning and Management* 127 (6):378–385.

MCWD (Minnehaha Creek Watershed District). 2007. Comprehensive Water Resources Management Plan, 2007–2017, Plan 6.3. www.minnehahacreek.org/MCWD_comprehensive_plan.php#download (accessed January 18, 2011).

Moore, Elizabeth A., and Tomas M. Koontz. 2003. A Typology of Collaborative Watershed Groups: Citizen-Based, Agency-Based, and Mixed Partnerships. *Society and Natural Resources* 16 (5):451–460.

Parfit, Michael. 1993. New Ideas, New Understanding, New Hope. In *Water: The Power, Promise and Turmoil of North America's Fresh Water.* pp. 113–114. Washington, DC: National Geographic Society.

Powell, John Wesley. 1890. Institutions for the Arid Lands. *Century* 40 (May–October):111–116.

Sabatier, Paul, Will Focht, Mark Lubell, Zev Trachtenberg, Arnold Vedlitz, and Mary Matlock. 2005. Collaborative Approaches to Watershed Management. In *Swimming Upstream: Collaborative Approaches to Watershed Management.* pp. 3–21. Edited by Paul Sabatier, Will Focht, Mark Lubell, Zev Trachtenberg, Arnold Vedlitz, and Marty Matlock. Cambridge, MA: MIT Press.

CHAPTER 7

Policy Decisions and the Changing Face of Wetlands

Jay A. Leitch and Gyles Randall

P rior to settlement and into the early 20th century, Minnesota's landscape was dominated by forests and prairies, highly punctuated by wet areas. Glaciers that influenced all but the southeastern corner of the state over 12,000 years ago left it with a landscape full of potholes, including Minnesota's more than 12,000 lakes and hundreds of thousands of wetlands. (Wet areas smaller than lakes have variously been referred as sloughs, potholes, swamps, bogs, wetlands, and other monikers. For consistency, we use the term "wetland" for these features throughout this chapter.)

The National Wetlands Inventory (NWI), which has been called "the best and most current statewide data available for existing wetlands" (DNR 1997, 9), reports that about 10.6 million acres of wetlands exist in Minnesota. Other inventories have come up with a little less (Anderson and Craig 1984) or about the same (DNR 1997). The ballpark consensus for how many wetlands existed before European settlement of the state is around 20 million acres. Minnesota has converted half of its wetlands; however, inventorying wetlands is difficult because of seasonal variations, problems defining wetlands, and the sheer numbers of them in the state.

Settlement began in earnest in Minnesota toward the end of the 19th century, around 1875, following the Indian wars and the coming of the railroad. The earliest settlers avoided areas too wet to farm. As land became more precious and farming technology improved, farmers began to find ways to convert these wet areas to more productive cropland (Leitch 1989). Overall, original wetlands— those existing at statehood in 1858—have all but disappeared in some of Minnesota's 87 counties in the agricultural region and in urban or built-up areas, but nearly all the original wetlands are intact in some counties in the northeastern coniferous forest region (Figure 7.1) (Weirens et al. 2005).

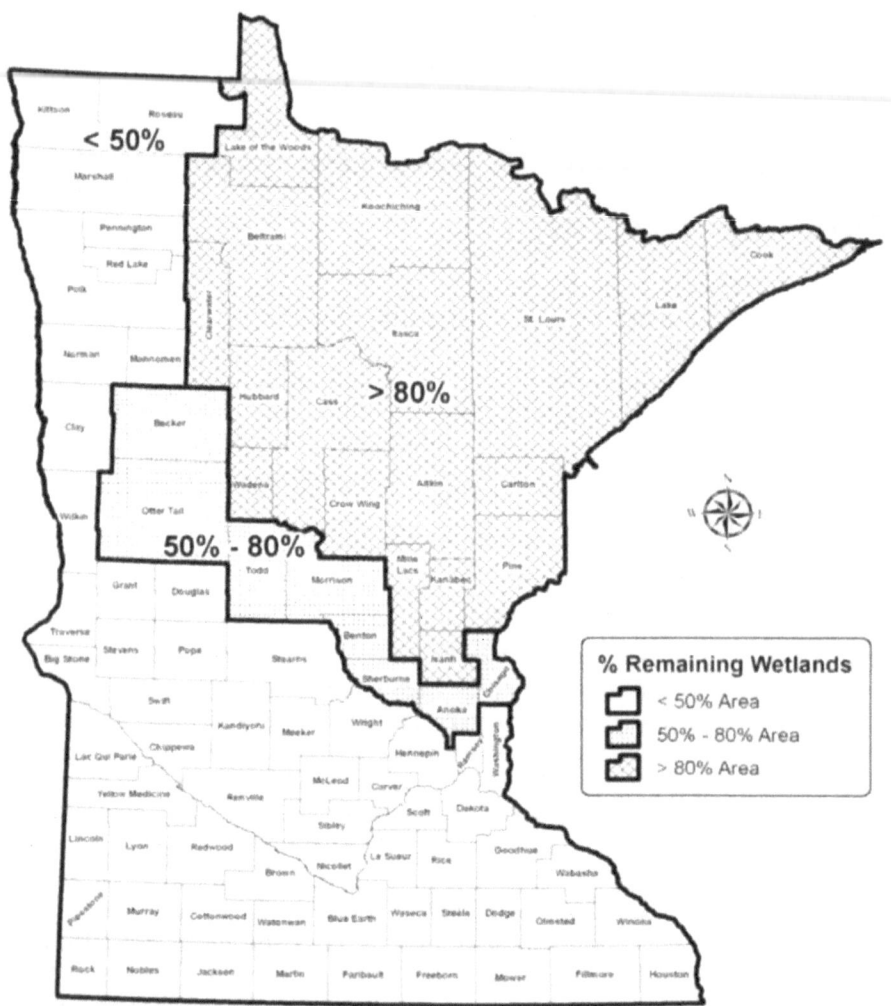

Figure 7.1 *Pre-statehood wetland areas*

Minnesota has three general types of natural vegetation landscapes. The northern conifer forest occupies the northeastern 40% of the state, where more than 80% of wetlands are intact. A diagonal band from northwest to southeast represents the deciduous forest transitional region, where 50% to 80% of wetlands are intact. The prairie/grassland agricultural region of the southwest has less than 50% of wetlands intact. Soil-forming factors and soil properties largely explain the development of these three distinct landscapes and characterize their nature. Cool climates with about five months of frozen conditions coupled with a glacial remnant of many medium- to coarse-textured soils over rather shallow bedrock have resulted in forest domination, primarily confers, in northeastern Minnesota. Even though soils high in organic matter do exist in many of the wetland areas, the climate is too cool and the growing season too short for most cultivated annual crops; thus subsurface tile drainage has not been used in these wetlands to increase agricultural productivity.

In the deciduous forest transition region, extending from Becker County in the northwest to Chisago County in the east-central part of the state, the landscape is marked by large areas of sandy, coarse-textured soils and thousands of lakes. Many of these soils require irrigation to be agriculturally productive. Even with artificial subsurface drainage, few areas of poorly drained soils are suitable for intense crop production, primarily because of the short, cool growing season.

The southern and western areas of Minnesota were originally dominated by prairie grasslands and wetlands and are markedly different from the other two regions. The soils are medium- to fine-textured (silt loam to clay), deep (many support crop root growth of 5 feet), and range from well drained to very poorly drained, depending on the clay content. Coupled with the higher temperatures, longer growing season, and ample water, these soils support intense and highly productive agriculture. Many of the poorly and very poorly drained wetland soils were obstacles to settlers attempting to farm these soils in the late 1800s. As a result, streams were channelized and ditches dug to improve surface drainage. Later, subsurface drainage was accomplished by installing tile lines greater than 3 feet deep that connected to drainage ditches and streams. During these activities, many wetlands were drained, essentially converting them from areas of reed and sedge to fields of corn and soybeans. Today these areas tend to have the darkest (indicating high organic matter), most fertile, and highly productive soils in the U.S. Corn Belt.

Two types of drainage systems are used to improve the internal drainage of poorly drained soils. Surface drainage consists of ditches that quickly remove excess water from the soil surface to larger ditches, streams, and rivers. Subsurface drainage consists of tile lines—formerly clay or concrete pipes, but now mostly corrugated plastic tubing—installed 3 to 4 feet deep in the soil spaced 40 to 100 feet apart. These subsurface tiles drain excess water from the soil profile to the depth of the tile, delivering the water to ditches and streams. This improves the aeration of the profile, providing an excellent environment for crop root growth.

These two types of drainage systems roughly correlate with two different drainage situations. The first and earliest is drainage of wetlands, or ponded water, and is the subject of this chapter. The second, more recent situation is drainage of wet lands, or saturated soils. Surface ditches have been the usual method to drain wetlands, whereas subsurface drainage has been used to drain wet lands. We focus on wetlands, not wet lands, in this chapter.

Both types of drainage systems affect landscape hydrology by increasing connectivity and reducing groundwater recharge. Most unaltered landscapes in the wetland areas of Minnesota consist of numerous individual depressional spots, or potholes, that are not connected. As precipitation falls or snow melts, each of these potholes acts as a storage basin, increasing the residence time of water temporarily stored throughout the landscape. When drainage systems are installed, water in each of the individual basins becomes connected, by ditch or tile line, as the water is shunted off to receiving streams and rivers. Because increased connectivity decreases residence time, water is transported more quickly to rivers.

Consequently, drainage has been identified as a major factor causing flooding in some highly drained areas following heavy thunderstorms. However, that effect is highly dependent on antecedent conditions: whether the soil is saturated and the depressional areas are full. Also, with fewer storage basins in the landscape, and as a greater proportion of precipitation is transported from the terrestrial landscape to rivers, groundwater recharge is likely reduced.

The human activities and public policies that have shaped the face of Minnesota's wetland landscape evolved over three time periods. The first was the era of wetland conversion, which lasted from settlement to early into the 20th century. The second was the era of wetland awareness, from the Dust Bowl years to the 1970s. The current period, the era of preservation and restoration, continues today.

ERA OF WETLAND CONVERSION

A variety of public policy, market, and cultural forces have resulted in the conversion of about 50% of Minnesota's presettlement wetlands to other uses. Government policies provided strong incentives to drain wetlands. Demand for commodities and the increasing price of cropland were the primary market influences for conversion. Farming culture, a general lack of appreciation for "nonproductive" land, and strong desires to "develop" the landscape spurred conversions. Each of these forces played an important role in the state's early wetland policy.

The Role of Public Policy

State and federal policies (see Box 7.1) in the early to mid-20th century encouraged wetland conversion mainly to promote economic growth and development. The federal government was the principal overt player, with the state and local governments playing more subtle roles.

The federal Swamp Land Act of 1849 granted Louisiana control of all swamplands for the purpose of reclaiming wetlands (Mitsch and Gosselink 1993). Reclaiming meant draining and converting them to more productive uses. Minnesota was added to the act in 1860, marking the start of federal involvement in conversion of the state's wetlands.

Until as recently as 1977, the U.S. Department of Agriculture (USDA), through the Agricultural Stabilization and Conservation Service (ASCS), administered programs to subsidize on-farm wetland drainage. In 1962, the USDA subsidies for draining types 3, 4, and 5 wetlands, which were larger wetlands with open water during much or most of the growing season (Shaw and Fredine 1971), were discontinued, but it was not until 15 years later, in 1977, that all federal agriculture programs encouraging wetland drainage were ended.

State and local governments encouraged cropland drainage indirectly by providing access roads to rural areas and directly by establishing local drain boards. Construction of the rural road system also contributed directly to wetland losses.

Box 7.1. *Minnesota Wetland Milestones*

1858 Minnesota becomes a state, law allows organizations to be formed for draining lands

1919 State law establishes the Department of Drainage and Waters

1934 Federal Duck Stamp Act provides money for wetland conservation

1947 State law defines public waters as "waters providing substantial public use ..."

1951 Save the Wetlands program enacted

1962 Most federal drainage subsidies discontinued

1972 Federal Water Pollution Control Act regulates fill in navigable waterways, including wetlands (Section 404)

1976 Inventory of public waters starts

1977 Executive Order 11990: Protection of Wetlands (42 Fed. Reg. 26961) issued by President Carter

1985 Federal Food Security Act, or Farm Bill, includes "Swampbuster" provisions

1986 Reinvest in Minnesota (RIM) begins

1991 Governor Carlson signs Executive Order 91-3, wetland "no net loss" policy for state departments and agencies

1992 Wetland Conservation Act enacted, creates a "no net loss" policy

2009 Wetland Conservation Act overhauled

Roads in Minnesota generally follow the rectangular township-range grid system, with a road at least every mile, both east-west and north-south. These roads in the prairie region look like a checkerboard or waffle when viewed from above. The system of roads, road ditches, and culverts provides an incentive to drain adjacent cropland. In addition, road builders went in straight lines whenever possible, going through and across countless wetlands (see Figure 7.2).

A state–local partnership of drain boards established a network of judicial ditches providing the collection artery for adjacent and nearby field drainage. This network remains today and is administered by local watershed management boards, under the purview of Minnesota's Board of Water and Soil Resources (BWSR).

A small but nonetheless culturally important impact on wetlands was their use as local dumps or landfill sites. From settlement well into the 20th century, wetlands provided a convenient place to dispose of all kinds of household and municipal solid waste, reinforcing the mindset that wetlands were wastelands. This practice has all but stopped, largely as a result of pollution control regulations rather than concern for wetland conservation.

Figure 7.2 *Wetland bisected by roads originally built almost 100 years ago*

Source: Google Earth image from 2003.

In summary, from the beginning, all levels of government encouraged private landowners to convert wetlands to more productive uses, either directly or indirectly. Additionally, the public sector was responsible for wetland losses through urban sprawl, landfills, and road building across the prairie.

The Market's Influence

The market's influence on wetland drainage in Minnesota is both transparently straightforward and subtly complex. Market incentives to produce agricultural commodities and to provide space for suburban development and for built-up uses in rural areas are straightforward. Forces outside the conventional market system, or externalities, work to create incentives for wetland preservation and are often difficult to measure.

The free market provided monetary incentives to convert wetlands to other market-based uses. These included turning what was viewed as "idle wasteland" into productive farmland, developing land for commercial and residential uses, and using wetlands for public infrastructure purposes, such as landfills and roads. The market provided monetary incentives that benefited the owner if a wetland were drained.

The market provided incentives to convert wetlands partly because many of the public benefits of wetlands are not captured in the market. In a market-driven economy, benefits that accrue outside of the market, known as external benefits or externalities, usually are not considered explicitly during decisionmaking. Most of the public functions and values of wetlands, such as biodiversity maintenance, aesthetics, and primary production, are externalities. As broadly demonstrated elsewhere (Mitsch and Gosselink 2007), the nonmarket benefits of many wetlands vastly exceed the market or private benefits. If these external benefits are somehow internalized, such as through a property rights or regulatory system, or are recognized and valued through public policy, wetlands will be more efficiently allocated to benefit all of society.

Cultural Forces

Cultural forces also are responsible for conversion of Minnesota's wetlands. Farmers have traditionally believed—and have been told by everyone from neighbors to bankers to government leaders—that they should develop their lands to their highest agricultural potential. Farming around a wetland was not seen as being progressive and left unsightly voids in otherwise flowing, long, straight rows in grain fields. The advent of bigger farm machinery, largely driven by market forces, also contributed to the culture of clean, level, well-drained fields that were free of all "nuisance" wetlands.

Farmers also realized that if drainage systems to remove excess soil water were installed, these wet land areas, including wetlands, had the potential to produce very high yields on a consistent basis. In addition to being deep, high in organic matter, and flat (and thus not vulnerable to water erosion), these soils usually have a very high available water content. Therefore, in years when growing-season precipitation is marginal and limits yields on many soils, sufficient available water exists in these soils to sustain the crops, enabling farmers to achieve very high and profitable yields on a consistent basis.

Nonfarmers were likewise acculturated to converting wetlands to what were seen as better uses. Whether because of the relative abundance of wetlands, the lack of knowledge about wetland values, or a general preoccupation with matters more important to them as individuals, the nonfarming public was largely oblivious to the declining numbers of wetlands.

Agriculture's Impact on Wetlands

Agriculture had a profound effect on Minnesota's wetlands, primarily through installation of surface and subsurface drainage systems. Similarly to other cause-and-effect relationships, there often are unintended downsides or consequences that accompany the upsides of wet land and wetland drainage. The upside is very tangible, evidenced by consistently greater crop yields and profitability, resulting in overall greater agricultural productivity. The downside is much less tangible, especially to the farmer. It consists primarily of hydrologic effects on the landscape, such as downstream flooding, as well as increased losses of nutrients, pesticides, pathogens, and antibiotics via drainage water.

Because drainage removes excess water from the soil, the profile becomes more aerobic, creating an environment conducive for microbes to mineralize soil organic matter. In the process of mineralization, substantial quantities of nitrate are produced, especially in previously wetland soils with high organic matter content. Numerous research studies in the Corn Belt and England have identified the significant impacts of drainage-induced soil mineralization on the loss of nitrate to groundwater and surface waters.

The type of cropping system has the largest influence of any management factor on drainage volume and nitrate loss in subsurface tile drainage. Studies in the United States, Canada, and Europe generally indicate that the largest drainage volumes and greatest nitrate concentrations in water emanate from fertilized annual row crops, particularly corn; intermediate volumes and losses are associated with annual cereal crops; and the smallest volumes and losses are seen with perennial grass and alfalfa crops. In a four-year drainage study at Lamberton, Minnesota, drainage volumes were reduced by 25% in a Conservation Reserve Program (CRP) grass and legume system and 50% in alfalfa, compared with annual corn and soybean row crops (Randall et al. 1997). Nitrate-N concentrations in drainage water were greatest for continuous corn (28 milligrams per liter [mg/L]), slightly less for a corn–soybean rotation (22 mg/L), and considerably less for alfalfa (1.6 mg/L) and the CRP system (0.7 mg/L). (The nitrate-N limit for drinking water set by the U.S. Public Health Service is 10 mg/L.) Nitrate-N loadings (drainage volume times nitrate-N concentration) from corn and soybeans were 30 to 50 times larger than from perennial alfalfa or CRP crops. Because many converted wetlands in Minnesota are planted to corn and soybeans, one can readily see the impact of this cropping system on surface waters. The hypoxic conditions in the Gulf of Mexico each summer are considered to be due in large part to nitrate losses in subsurface drainage from these annual row crops.

Artificial wetland drainage has greatly increased agricultural productivity of over a million acres in Minnesota, but the negative consequences on water quality are significant. This is particularly true when annual corn and soybean row crops, the backbone of Minnesota's crop agriculture, are planted on these inherently very poorly drained soils. Best management practices such as reducing the depth of tile installation, reducing drainage intensity, using the recommended rate of nitrogen rather than augmenting with an additional "insurance" rate, using a spring preplant fertilizer application, applying the proper fertilizer at the right time, and using a nitrification inhibitor when appropriate all will help reduce nitrate losses to drainage water. Converting from annual row crops to perennial crops would result in the greatest improvement in drainage water quality, although that also would have the greatest cost in economic productivity.

Policy implementation to support, accelerate, and subsidize drainage of wetland soils in the mid-20th century was clearly ahead of science, which has since identified and quantified the associated negative consequences. If the present science-based knowledge had been available to the decisionmakers of that time, the face of Minnesota wetlands could look quite different from what presently exists. Wetlands strategically located in the agricultural portions of the state may provide multiple benefits, including greater land use diversity.

ERA OF AWARENESS AND RESEARCH: DUST BOWL TO THE 1970s

Environmental awareness and subsequent pressure from the public, primarily special-interest groups, in the mid- to late 20th century paved the way for a wide range of policies to curtail, or even reverse, wetland losses. The cultural momentum behind encouraging farmers to farm wetlands, which lasted decades and several generations of Minnesota farm families, was difficult to stanch. What was to become a major change in attitudes started with recognition of the public values of wetlands, followed by public policies reflecting these values, including incentives to farmers.

Public Values of Wetlands

Once wetlands came to the attention of society, ecologists, nature lovers, wetland scientists, and special interests wasted no time in enumerating the multiple functions (what they do) and values (what the functions are worth) of wetlands. The functions and values of wetlands are well documented in the literature (e.g., Mitsch and Gosselink 2007; Niering 1988). They include the following (based on BWSR 2009a; CAST 1994; NRC 1995):

- floodwater retention;
- groundwater recharge;
- water quality enhancement;
- fish and wildlife habitat provision;
- flora and fauna diversity;
- primary production;
- biodiversity maintenance;
- shoreline protection;
- aesthetics/recreation/education;
- commercial uses; and
- other direct and indirect functions

Not all wetlands provide all functions, but overwhelming evidence demonstrates that most have values well beyond those that can be captured by individual landowners.

What might be the first academic research on the economic value of wetlands was an analysis of wetland problems in Minnesota (Goldstein 1967). That was followed shortly by another study of wetland values in the Prairie Pothole Region, of which Minnesota is a part (Hammack and Brown 1974).

While it has been widely demonstrated that wetlands provide benefits to the public, most wetlands are on property owned by individuals, and individual owners also derive benefits from wetlands. A major policy issue has become how to maintain public values of wetlands that are under private ownership. Some wetland preservation policies were seen by landowners as "takings" by the government and became legal issues. Recently proposed revisions of the federal Clean Water Act (CWA), such as HR 2421, the Clean Water Restoration Act, have brought property rights and federal land use control issues back to the fore. Revisions of the laws would remove the word "navigable" from their description of waters of the United States, greatly extending the reach of the CWA. Yet it is apparent that the property rights issues related to government policies about private wetlands remain highly controversial.

Policies Followed Value Perceptions

As the list of wetland functions and values grew and was better quantified, policies were modified to fit emerging public attitudes about wetlands. President Jimmy Carter's Executive Order 11990, Protection of Wetlands, was perhaps the turning point in policy. Not long after, federal agencies, followed closely by Minnesota lawmakers and agencies, began developing policies to protect, preserve, and restore the state's wetland resources. At the same time, nongovernmental organizations such as Ducks Unlimited, The Nature Conservancy, and Sierra Club became engaged in the effort to conserve wetlands.

As is often the case in policymaking, the devil was in the details. The era of conservation spawned a plethora of scientific research about wetlands, as effective government policy requires knowing what a wetland is, what wetland functions are worth to society, and how to write effective laws and regulations. Minnesota has been a leader among the states in these areas, all of which have required extensive research and deliberation.

As a result, the technical and scientific aspects of federal and state wetland policies have been largely ironed out. There are still some areas of contention, such as property rights, and academics are still doing research in many physical and natural science areas concerning wetlands. However, the era of awareness has since ushered in the current era of preservation and restoration.

ONGOING ERA OF WETLAND PRESERVATION AND RESTORATION

Although Minnesota has a "no net loss" wetland policy in the form of the Wetlands Conservation Act (WCA), wetland protection policies within the Reinvest in Minnesota (RIM) program (BWSR 2009b), and a duck recovery plan with a goal of restoring 2 million acres of habitat (30% wetland), it appears as if the state's wetland base is still shrinking (DNR 2006). In addition to losing wetlands as CRP acres are returned to cropland, wetland losses continue through road building, housing developments, channelization, shoreline alteration, aquatic vegetation removal, and increased watercraft use (DNR 2006).

The most recent changes to the WCA (BWSR 2009c), an optimistic goal for wetland restoration (Minnesota leads the nation in spending for this purpose), the country's best wetland inventory data, and citizen support through a dedicated sales tax[1] are all indications that the face of the state's wetlands is still changing.

Minnesota's wetland policy is more comprehensive than ever and can be summarized in five areas: definition and delineation; assessment; mitigation, banking, and restoration; regulation; and cooperation.[2]

Definition and Delineation

Defining "wetland" and outlining a wetland on the landscape for regulatory purposes, known as delineation, were major issues during the era of awareness and remain somewhat an issue today, especially on a case-by-case basis. Minnesota statute[3] says, "Wetland boundaries must be determined using the methodologies in the US ACE *Wetlands Delineation Manual* (January 1987), including subsequent updates and supplements, and guidance provided by the board" (BWSR 2009c). Delineation has been the subject of much discussion and scientific inquiry; however, the debate as far as Minnesota's WCA is concerned has been largely stilled. The WCA defines "wetlands" as follows:

> "Wetlands" means lands transitional between terrestrial and aquatic systems where the water table is usually at or near the surface or the land is covered by shallow water. For purposes of this subpart, wetlands must:
>
> (1) have a predominance of hydric soils;
> (2) be inundated or saturated by surface water or groundwater at a frequency and duration sufficient to support a prevalence of hydrophytic vegetation typically adapted for life in saturated soil conditions; and
> (3) under normal circumstances, support a prevalence of hydrophytic vegetation.[4]

Assessment

In order to justify, write, implement, and operationalize regulations, decision-makers and regulators need to know the functions and values of the state's wetlands in general and of individual wetlands in particular. The BWSR has approved two functional assessment methods for use in Minnesota. Shortly after the WCA was passed in 1992, MnRAM, a wetland assessment method specifically for Minnesota, was developed. A number of other states have assessment methods for this and other purposes, but Minnesota's MnRAM, which started as a qualitative method, has matured through several major revisions to include a quantitative component as well. The primary purpose of MnRAM is to serve as a tool for local authorities to make sound wetland management decisions.

The second BWSR-approved method is the Hydrogeomorphic (HGM) Approach, as detailed in Gilbert et al. (2006). HGM was designed for the CWA's Section 404 Regulatory Program but has been modified to serve other uses, such as assessing functions of prairie potholes and aiding in developing assessment models. HGM uses indices of the physical and natural functions of wetlands to assess site-specific wetlands. It does not consider the socioeconomic setting of specific wetlands.

Mitigation, Banking, and Restoration

Under the state's "no net loss" policy, entities that convert wetlands to other uses are required to mitigate those losses, often at a ratio greater than one-to-one to achieve functional equivalency. As an alternative to contemporary mitigation, entities are allowed to purchase wetland credits from the Minnesota Wetland Bank. Credits are deposited in the bank by others who have restored or created wetlands somewhere else in the state. The price of credits varies widely, from $1,000 per acre in rural areas to $20,000 or more in the Twin Cities metro area (BWSR 2005). The bank serves more to facilitate mitigation than it does to encourage or promote restoration.

The goal of the Minnesota Department of Natural Resources (DNR) is to restore 2 million acres of waterfowl habitat in the state by 2056. This goal represents almost 10% of all cropland in the state, which may be overly optimistic. Restoration of converted wetlands and creation of new wetlands are the only means to offset, or mitigate, losses due to development or increase the amount of wetland in the state. However, debate is ongoing about whether the value of restored wetlands should be measured in acres or functional equivalency, as not all wetlands are created equal.

Regulation

While the definition of a regulated wetland and the scope of the public's interest in private land management remain somewhat contentious, the fate of the state's remaining wetlands is more certain than ever. Minnesota's wetland policy is both the most grassroots and the most comprehensive, by far, among the 50 states. Whereas several other states have assumed control over Section 404, none have taken it to the extent that Minnesota has. Minnesota's current Wetland Conservation Act (BWSR 2009c) was adopted following an extensive two-year review-and-revise period. The policy revision process started with a memorandum of understanding between the U.S. Army Corps of Engineers, which is the administrator of Section 404 of the Clean Water Act and Section 10 of the Rivers and Harbors Act, and the Minnesota BWSR, which has oversight responsibilities for the Wetland Conservation Act (Kramer and Pfenning 2007).

Cooperation

Public wetland management in Minnesota is a model of grassroots and interagency cooperation and coordination. Wetland policy is grassroots in that, according to the WCA, "local government units are responsible for making decisions on applications made under [Chapter 8420]" (BWSR 2009c). One example of interagency cooperation and coordination is the Drainage Work Group (DWG), established in 2006 as a stakeholder group to advise the BWSR on a specific study. On completion of the study, the DWG agreed to continue to meet to provide input to the BWSR. The DWG includes stakeholders from drainage authorities, farm groups, environmental groups, state agencies, the Minnesota legislature, and several other state-level associations.

Another group, the Drainage Management Team (DMT), consists of individuals from different agencies at both the state and federal levels (BWSR, DNR, Minnesota Pollution Control Agency, USDA, and Natural Resources Conservation Service), as well as the University of Minnesota. The team meets to pool information regarding drainage issues and provide information and education to drainage authorities throughout Minnesota (Peterson 2009). Together, the DWG and DMT bring representation from various wetland and drainage perspectives together to influence the state's wetland policy.

CONCLUSIONS

From an all-out campaign to convert wetlands to "beneficial" uses in the era of conversion to a contemporary, nearly unanimous move to preserve, conserve, and restore them, policy regarding Minnesota's wetlands has shifted 180 degrees in the past 100 years. Agriculture had a tremendous impact on the face of wetlands, but it was not alone in its impact. Urban sprawl, road building, landfills and development-oriented public policies also contributed to the conversion of wetlands.

Although the era of wetland conversion resulted in the loss of many wetlands in the southeastern agricultural region of the state, other parts experienced much less change. The northern conifer forest, an area that occupies 40% of the state, still retains 80% of its wetlands; the deciduous forest transitional region retains 50% to 80% of its wetlands. However, it is important to be aware of the fact that this analysis focuses on actual wetlands, not saturated lands (i.e., not wet lands). Wetland ecosystems are now recognized as producing a wide range of services, including groundwater recharge, floodwater retention, wildlife habitat, recreation, and biodiversity. In contrast, wet or saturated lands, although providing benefits such as groundwater recharge and flood storage, are nearly ignored by Minnesota water policy.

Despite the fact that some technical and scientific issues remain around the edges, unless there are major changes in public policy or funding, the fate of the state's wetlands is clear. Unlike during the early years, numerous special interest groups are now actively promoting wetland preservation and supporting public wetland protection policies. Minnesota's Wetland Conservation Act has been quite comprehensive in specifying regulatory provisions leading up to and including the current rules. Property rights challenges, property-specific delineation questions, mitigation issues, and research to fine-tune the scientific foundations for policies will likely continue in the years to come.

NOTES

1. Minnesota voters approved the Outdoor Heritage Fund in 2008 (Minn. Stat. 2008, 97A.056), increasing the state general sales tax rate by three-eighths of 1%. A majority of the money must be spent restoring, protecting, and enhancing wetlands, prairies, forests, and habitat for fish, game, and wildlife.

2. Most of the literature and documents about wetland policy in Minnesota during this era are agency reports, legislation, or gray literature. The BWSR maintains an excellent index at www.bwsr.state.mn.us/wetlands/index.html.

3. Minn. Rules 8420.0405, subp. 1.

4. Minn. Rules Ch. 8420, Part 8420.1100 Definitions, subp. 71.

REFERENCES

Anderson, J.P., and W.J. Craig. 1984. *Growing Energy Crops on Minnesota's Wetlands: The Land Use Perspective*. Minneapolis: Center for Urban and Regional Affairs, University of Minnesota.

BWSR (Minnesota Board of Water and Soil Resources). 2005. Wetland Banking Fact Sheet. www.bwsr.state.mn.us/wetlands/wetlandbanking/fact_sheet.pdf (accessed September 29, 2009).

———. 2009a. Wetland Functional Assessment: MnRAM. www.bwsr.state.mn.us/wetlands/mnram/index.html (accessed September 3, 2009).

———. 2009b. RIM [Reinvest in Minnesota] Reserve Fact Sheet. www.bwsr.state.mn.us/easements/rim/factsheet.html (accessed December 3, 2009).

———. 2009c. Wetland Conservation Act: Minnesota Rules Chapter 8420. www.bwsr.state.mn.us/wetlands/wca/WCA_Rule_1-15-09_Draft-clean.pdf (accessed September 1, 2009).

CAST (Council for Agricultural Sciences and Technology). 1994. *Wetland Policy Issues*. Comments from CAST No. CC1994-1. Ames, IA: CAST.

DNR (Minnesota Department of Natural Resources). 1997. Minnesota Wetlands Conservation Plan, Version 1.0. http://files.dnr.state.mn.us/eco/wetlands/wetland.pdf (accessed September 10, 2009).

DNR (Minnesota Department of Natural Resources). 2006. Long Range Duck Recovery Plan. http://files.dnr.state.mn.us/recreation/hunting/waterfowl/duckplan_042106.pdf (accessed September 12, 2009).

Gilbert, Michael C., P. Michael Whited, Ellis J. Clairain Jr, and R. Daniel Smith. 2006. *A Regional Guidebook for Applying the Hydrogeomorphic Approach to Assessing Wetland Functions of Prairie Potholes*. ERDC/EL TR-06-5. Washington, DC: U.S. Army Corps of Engineers.

Goldstein, Jon H. 1967. An Economic Analysis of Wetland Problems in Minnesota. PhD diss., University of Minnesota. (Also published as Jon H. Goldstein. 1971. *Competition for Wetlands in the Midwest: An Economic Analysis*. Washington, DC: Resources for the Future).

Hammack, Judd, and Gardner Mallard Brown Jr. 1974. *Waterfowl and Wetlands: Toward Bioeconomic Analysis*. Baltimore: Johns Hopkins University Press for Resources for the Future.

Kramer, Randy, and Michael Pfenning. 2007. Interagency Memorandum of Understanding: Wetland Mitigation Guidelines. www.bwsr.state.mn.us/wetlands/BWSR-COEmemo.pdf (accessed September 1, 2010).

Leitch, Jay A. 1989. Politicoeconomic Overview of Prairie Potholes. In *Northern Prairie Wetlands*. Edited by Arnold van der Valk. Ames: Iowa State University Press.

Mitsch, William J., and James G. Gosselink. 1993. *Wetlands*. 2nd ed. New York: Van Nostrand Reinhold.

———. 2007. *Wetlands*. 4th ed. New York: John Wiley & Sons.

Niering, William A. 1988. *The Audubon Society Nature Guides: Wetlands*. New York: Chanticleer Press.

NRC (National Research Council). 1995. *Wetlands: Characteristics and Boundaries*. Washington, DC: National Academy Press.

Peterson, Joel. 2009. Drainage Management Team. Minnesota Board of Water and Soil Resources. www.bwsr.state.mn.us/drainage/management.html (accessed September 17, 2009).

Randall, G.W., D.R. Huggins, M.P. Russell, D.J. Fuchs, W.W. Nelson, and J.L. Anderson. 1997. Nitrate Losses through Subsurface Tile Drainage in CRP, Alfalfa, and Row Crop Systems. *Journal of Environmental Quality* 26:1240–1247.

Shaw, Samuel P., and C. Gordon Fredine. 1971. *Wetlands of the United States: Their Extent and their Value to Waterfowl and Other Wildlife*. Circular 39. Washington, DC: Fish and Wildlife Service, U.S. Department of the Interior.

Weirens, David, and BWSR staff. 2005. 2001–2003 Minnesota Wetland Report. St. Paul: Minnesota Board of Water and Soil Resources. www.bwsr.state.mn.us/wetlands/publications/wetlandreport.pdf (accessed December 2, 2010).

CHAPTER 8

Groundwater Policy at State and Local Levels: The Science–Policy Linkage

Bob Tipping, Scott Alexander, and
E. Calvin Alexander, Jr.

Groundwater is a vital resource in Minnesota, providing drinking water for at least 75% of its population (DNR 2009a). In addition, groundwater is essential to industry and agriculture in the state. Perhaps even more significantly, groundwater has essential ecosystem functions that do not just support a diversity of biological life, but also are the foundation of Minnesota's recreation and tourism, key components in our quality of life. These ecosystem functions have not been considered until recently in groundwater policy decisions.

Groundwater resources in Minnesota are not uniformly distributed in terms of either quantity or quality (DNR 2009b; Kanivetsky 1986). Highly productive bedrock aquifers that provide water to the cities of southeastern and east-central Minnesota are absent in the rest of the state; bedrock aquifers in northeastern Minnesota yield only a small fraction of the water supplied by the southeastern aquifers. In the majority of the state, groundwater is withdrawn from both buried and surficial sand and gravels deposited by the advance and retreat of glaciers during the last ice age. These aquifers are highly variable in extent and, as with bedrock aquifers, have natural water quality ranging from good to undrinkable, largely related to the type of pathway the water has followed and length of time it has been underground.

Groundwater managers make decisions that hinge on questions regarding the groundwater quality in their area, whether the groundwater is getting worse or better, and what the consequences would be if use of the groundwater were increased. These questions are challenging on several levels. Groundwater is not a single thing. Groundwaters are diverse, heterogeneous resources. Groundwater is not easily seen. It can be difficult to manage a resource that is out of sight and has long time constants over which changes are noticed. This is further compounded by the assumption of an almost magical or mythic ability of contaminants to be filtered out of groundwater. Finally, most decisionmakers' knowledge of these resources is incomplete and is in the midst of a paradigm shift. The emerging paradigm emphasizes the heterogeneous nature of the resources, based largely on our increasing ability to measure groundwater systems over a broader range of spatial and temporal scales.

This chapter addresses the current state of groundwater research and how it affects policy decisions at the state and local levels. Traditionally, groundwater policy has been shaped by the assumption of long, slow, uniform flow paths. Quarterly, or perhaps annual, sampling was viewed as sufficient to define changes in water usage and quality. This has created an environment where shortage or contamination events are almost always surprising, leading to reactive or crisis approaches to changing groundwater conditions. The goal of current scientific research, recognized at both state and federal levels, is to improve our understanding of groundwater cause-and-effect relationships, which in turn can lead to proactive policy decisions (SWAQ 2007).

Current hydrologic research emphasizes obtaining high-frequency, multi-parameter information from a high density of sampling points (Hondzo 2009). Our knowledge of the range of temporal and spatial scales for groundwater flow systems continues to broaden as our ability to measure them grows. Specifically, research is focusing on information about heterogeneities in the groundwater resources. Information on how long it takes for water to move from the land surface to discharge from a well, wetland, river, stream, or lake (termed its "residence time") and what routes it takes to get there (its "pathways") is markedly improving. Equally important, the tools to transfer this knowledge to decisionmakers in a tangible way are improving. Geographic information systems (GIS) now can display three- and four-dimensional distributions of groundwater flow paths and hydrochemical compositions in ways that are more intuitive to a nontechnical audience. The information presented in this chapter builds on the themes of addressing hydrologic systems as a whole and the impacts of urbanization on groundwater flow systems, which are examined elsewhere in the book (see Chapters 3, 9, and 16)

CURRENT GROUNDWATER RESEARCH

Groundwater research initially focused on water supply questions. In the broadest sense, initial paradigms used for modeling groundwater flow systems simplified hydrogeologic settings into three components: recharge, discharge, and hydraulic gradient. Geologic controls on groundwater movement were incorporated into models by extrapolating point measurements of porosity and permeability across the model domain and adjusting their values until recharge, discharge, and hydraulic gradient were balanced. Residence times and pathways of groundwater flow systems were calculated for any point within the modeled domain based on this water balance in an assumed single continuum. Recent research has increasingly focused on increasing the density of temporal and spatial measurements of aquifer properties, quantifying residence times and pathways of groundwater flows and contaminant transport, and improving numerical models such that now they incorporate better data and conceptual models. From society's perspective, this shift is being driven by the inability of single-continuum models to accurately predict contaminant transport. Contaminants have repeatedly moved faster, farther, and at higher concentrations than models predicted.

Our understanding of groundwater flow systems has benefited greatly from research on porosity and permeability of geologic materials. The most significant change has been recognition of secondary porosity and permeability as dominant hydraulic features in sedimentary rocks (Runkel et al. 2003). Fracture and conduit flow, typically associated with karst conditions in carbonate rocks, have also been found to be pervasive in sandstones, fine-grained siltstones, and shales. New approaches used to identify enhanced permeability due to fracture and conduit flow include borehole hydrogeophysics, outcrop investigations, and discrete interval hydraulic testing.

A complementary approach to understanding pathways and residence times is provided by chemical, isotopic, and anthropogenic tracer data. As water moves from the land surface, through an aquifer to a well, the water develops a geochemical fingerprint that records the water's path and residence time. These data can then be used to establish baseline chemical compositions and hydrogeochemical facies unique to particular flow paths.

The term "environmental tracers" describes a broad range of substances that are incorporated into the water cycle at the land surface. Measurable concentrations of such tracers can persist in the subsurface at levels that distinguish them from natural background concentrations. Tracers that have been used to track groundwater flow include agricultural by-products such as pesticides, herbicides, and their breakdown products; wastewater effluent compounds such as caffeine, pharmaceuticals, and insect repellents; industrial wastes such as solvents, gases, and specialized compounds like perfluorochemicals (PFCs); and isotopic tracers related to atmospheric testing of nuclear weapons. The most direct and intuitive example of the use of tracers is introduction of a unique artificial tracer at one place and its detection somewhere else. Detection of the tracer indicates a hydraulic connection between the two points, regardless of what a model predicts (Kendall 2005). Including groundwater flow path and residence time data in a GIS has proved to be an effective tool to help managers as well as constituents understand the limits of current numeric models. Flow path and residence time data provide dynamic, flow-based targets that reduce uncertainty in models that were previously calibrated only on how well they matched static measurements of hydraulic head.

Within the Twin Cities metropolitan area, contaminant plumes are inadvertent tracers that provide critical information on patterns and timing of groundwater flow not revealed in groundwater models. In 2005, PFCs were discovered in municipal well water from the city of Oakdale, in Washington County (ATSDR 2008). PFCs are conservative in groundwater; they are relatively inert and travel freely through the subsurface. Disposal of these chemicals at several landfill sites within the county in the 1950s to early 1970s provided a start time from which to calculate approximate residence times of groundwaters where these chemicals were detected.

Subsequent testing of domestic wells showed a much more extensive plume, in length, breadth, and depth, than predicted by groundwater modeling in the area (Figure 8.1). Southern Washington County, the area where PFCs were detected, extends over 100 square miles (MDH 2008). Distribution of these chemicals revealed stratified flow within rocks that had been considered a single unstratified aquifer. The PFC crisis generated a density of data and information that would not have been available without the public perception of a threat to human health. In this case, managers gathered sufficient data to define and understand the crisis before trying to remediate it. Their response caused a serendipitous reevaluation of how groundwater models are calibrated.

Beyond the use of tracers, continuous data logging can further define the interplay of rapid and slow responses in the ways hydrologic systems respond to changes in the landscape. Continuous monitoring of temperature, conductivity, and water level help refine and reshape conceptual hydrogeologic frameworks used to construct groundwater models. In Minnesota, this is most clearly recognized in continuous monitoring of springs in the southeastern part of the state. As is illustrated in Figure 8.2, continuous monitoring of spring temperature, stage levels, and chemical composition often shows spring flow and chemistry to be closely linked to local hydrologic changes such as storm events and pumping cycles. The initial response of springs to these events is often on the order of hours to days, and lingering effects can be measured for weeks or longer. These short-term causes and effects cannot be revealed by quarterly or seasonal sampling.

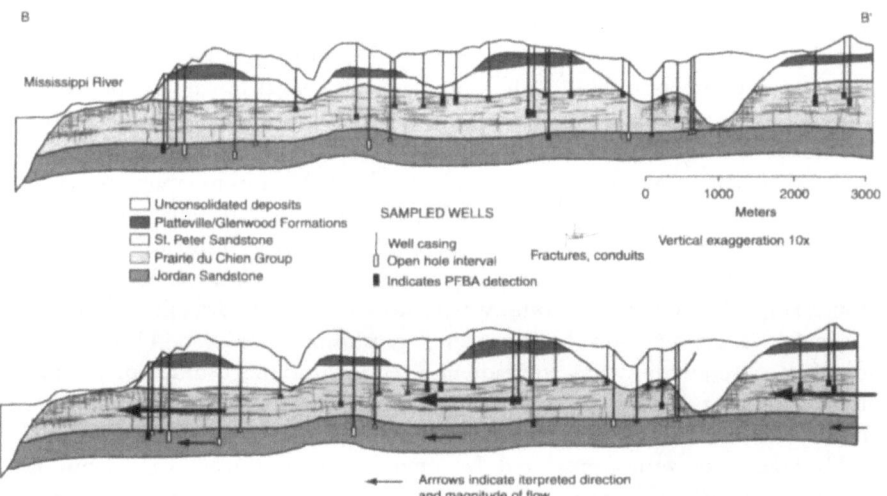

Figure 8.1 *Cross section of PFC contamination in Southern Washington County*

Notes: The majority of contaminated wells (dark rectangles) are found within the upper half of the Prairie du Chien Group and not in the underlying Jordan Sandstone, indicating different flow paths and residence times with depth. These two bedrock units traditionally have been treated as a single aquifer system in both conceptual and numeric models. Cross-section location is shown as B–B' in Figure 8.3.

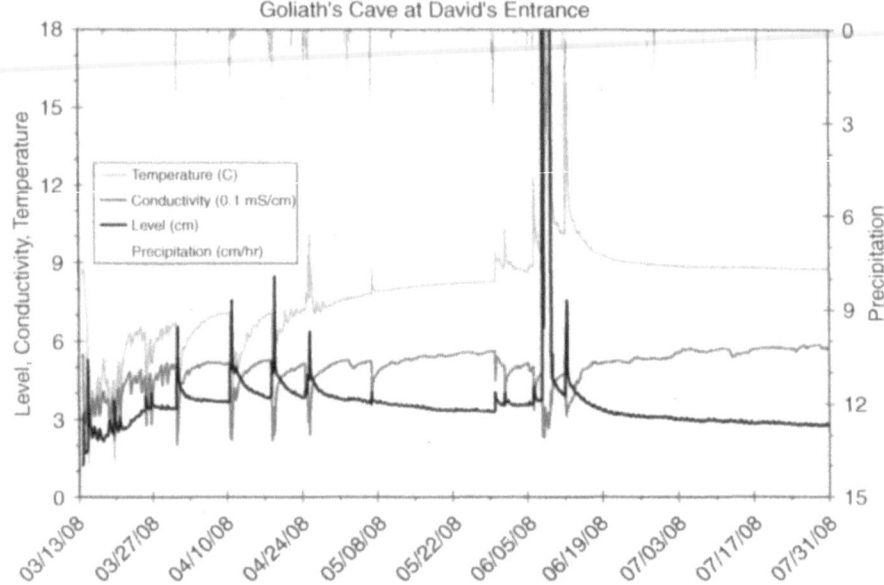

Figure 8.2 *Continuous surface rainfall and underground stream water level, conductivity, and temperature in Goliath's Cave in Fillmore County, March 13–August 13, 2008*

Note: This record illustrates very rapid major changes of water temperature, conductivity, and level in response to surface precipitation and snowmelt events.

In urban settings, continuous water-level monitoring over longer terms is changing our understanding of the temporal and spatial impact of high-capacity pumping on groundwater flow directions. As part of ongoing groundwater quality investigations in the Edina, Minnesota, area, state and local authorities collected water-level data from three monitoring wells at 30-minute intervals over the course of nine months. The effect of high-capacity wells in the area on the regional flow regime was striking. Previous mapping based on historical data showed regional flow in the area consistently moving east-southeast. Data from continuous monitoring showed that groundwater flow changed direction hourly and seasonally, depending on high-capacity pumping schedules. Heavy seasonal pumping caused wide swings in gradient direction and magnitude during summer months, whereas lower demand in the winter months resulted in more stable gradients (Rzepecki and Robertson 2009).

The range of times measured by environmental tracers runs from hours to millennia. Whereas short time scales are often important in contaminant transport, longer time scales are often significant in groundwater production. A lag often occurs in system response to changes in groundwater recharge or withdrawal rates. Changes in climate, alteration of surficial flow paths, and variation in pumping conditions all can produce long-term effects. For such groundwater systems, data collection over one or more decades is required to recognize trends (Taylor and Alley 2001). The ability to collect and display spatial and temporal data from long-term monitoring of both surface water and groundwater systems in a GIS makes this type of data collection a critical component of any sustainable management of groundwater resources. Yet long-term groundwater monitoring has been vulnerable in a crisis-driven, public-funded system of water planning.

STATE GROUNDWATER POLICY

Groundwater resources are largely in the realm of "out of sight, out of mind," and traditionally, decisionmakers have not been inclined to act until there was a tangible problem, such as a contaminated public supply well, drought, or conflicts resulting from competing uses. A major challenge in attempts to create comprehensive, statewide water policy in Minnesota has been the historic division of water management across many state agencies with conflicting goals and priorities. This is true of groundwater policy as well. In Minnesota, issues related to supply are handled by the Department of Natural Resources (DNR); nonagricultural contamination and ambient groundwater quality are the jurisdiction of the Minnesota Pollution Control Agency (MPCA); impacts on groundwater quality related to agricultural practices are handled by the Minnesota Department of Agriculture (MDA) and Board of Water and Soil Resources (BWSR); and drinking water protection, both wellhead and source-water protection for surface water resources, is handled by the Minnesota Department of Health (MDH). Guidelines for regional policy coordination at the statewide scale come from the Environmental Quality Board (EQB). For the Twin Cities metropolitan area (TCMA), regional policy coordination comes from the Metropolitan Council.

In February 2004, the three state agencies charged with monitoring groundwater quality, the MPCA, MDA, and MDH, signed a letter of agreement to address redundancies in monitoring efforts and consolidate planning activities for new monitoring (MPCA 2004). At present, efforts are under way to reevaluate water policy practices and develop plans for sustainable groundwater use (EQB 2008; UMWRC 2009). As a result of these discussions as well as related legislative directives, changes within agencies include a revival of the MPCA's ambient groundwater-monitoring program, a shift within the DNR to evaluate water supply on a regional basis (Reeves 2009), and development of plans to establish long-term groundwater monitoring in the 11-county TCMA (DNR 2009a) and long-term protection of groundwater and surface water resources.[1] These efforts coincide with efforts by the U.S. Geological Survey (USGS) to evaluate the condition of aquifers nationwide and coordinate a national framework for groundwater monitoring (USGS 2009a, 2009b).

One of the main difficulties in managing groundwater resources from a state perspective is definition of a management area. Aquifers and aquitards usually extend beyond local political boundaries. The Mount Simon Aquifer and overlying Eau Claire aquitard extend over much, but not all, of the TCMA

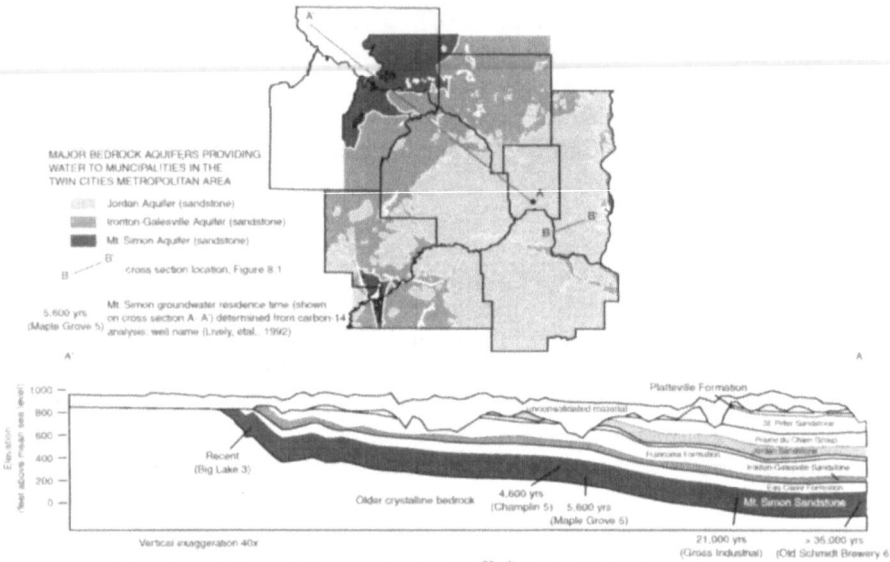

Figure 8.3 *Map of major bedrock aquifers providing water to municipalities in the Twin Cities metropolitan area*

Notes: Line B–B' locates the cross section in Figure 8.1. Cross section A–A' includes selected Mount Simon groundwater residence times based on carbon-14 data. Residence times within the subcrop area of Mount Simon are considerably younger than Mount Simon water in the downtown area (right side of cross section).

ᵃLively et al. (1992).

(Figure 8.3). Recharge to the Mount Simon Aquifer is considered to occur primarily in areas where the Eau Claire is absent. Residence times of water within the Mount Simon Aquifer (shown in cross section in Figure 8.3) clearly illustrate a temporal disconnect among waters within the aquifer, recharge, and the time scales typically addressed in land use and water policy. During the drought of 1988, groundwater use in the TCMA was carefully reevaluated as water levels in the Mississippi River dropped near water supply intake pipes for the city of Minneapolis (DNR 1989). The practice of using water from wells completed in the Mount Simon Aquifer for once-through cooling of buildings in the metro area counties was prohibited, and the legislature terminated existing withdrawal permits at the end of 1992.[2] In addition, elevated natural levels of radium restrict the use of water from the Mount Simon in some areas. The legislature also prohibited new once-through cooling use of Minnesota groundwater from all aquifers and required existing permits to be terminated by 2010.[3] In addition, it restricted use of Mount Simon Aquifer groundwater to potable water supplies, and then only if there were "no feasible or practical alternatives to this source, and a water conservation plan is incorporated with the permit."[4] St. Bonifacius is one of the few communities that has met these requirements and increased its use of groundwater from the Mount Simon Aquifer.

The Mount Simon example illustrates what has been characteristic of groundwater management in Minnesota: a crisis results in political action and a change in policy. In order to be proactive rather than reactive, decisionmakers and their constituents need to conceptualize groundwater flow systems in tangible ways. Decisionmakers and their constituents often find the information provided by the new tools described above more understandable than traditional models. Measurements of residence times within the Mount Simon Aquifer showed that water being harvested from it was properly considered groundwater mining; as a result, the managers changed policies.

The division of groundwater quality and quantity jurisdiction among state agencies further compounds the problem of defining groundwater management areas. Water quality concerns over natural contaminants such as arsenic and radium have primarily been the focus of the MDH; concerns over contamination from agricultural chemicals have been the focus of the MDA; and concerns over other anthropogenic contaminants have been the focus of the MPCA. Programmatic compartmentalization of water chemistry data by these agencies limits use of the data as a tool to help clarify groundwater flow paths and residence times. As the DNR, whose previous water quantity focus was on a permit-by-permit basis, moves toward regional management of groundwater resources, strengthening links between quantity and quality information will be necessary in order to develop comprehensive groundwater policy.

LOCAL GROUNDWATER POLICY

The Clean Water Partnership (CWP), established by the MPCA in 1987, provides local units of government with state and federal resources to protect waters. The focus of the program is control of nonpoint-source pollution through watershed management, to protect and improve surface water and groundwater in Minnesota. Additional financial support for local water planning comes from the BWSR. BWSR programs include local government assistance, comprehensive local water planning, and wetland management. The board also provides a link between state and local programs.

The emphasis of both the CWP and BWSR programs has been on surface water quality, with funds directed toward groundwater issues only through the impetus of local water planners. Typically, that impetus has been driven by contaminated groundwater concerns, but in some cases, it has been driven by recognition of groundwater-dependent resources or concerns over future water supply. In this way, local water planning has shown itself to be flexible and proactive about groundwater resources, developing innovative approaches that in turn have helped guide water policy at the state level.

Local water planners are often faced with groundwater problems whose causes are outside their jurisdictional control. As an example, a local unit of government may be located outside the recharge area for the aquifer that supplies its water; in this setting, land use changes outside the local jurisdiction affect the community's resources. Additionally, contaminant plumes that come from somewhere else can harm water quality. The city of New Brighton, whose municipal water supply was contaminated by a trichloroethylene (TCE) plume from the Twin Cities army ammunition plant, is an example of this condition.

Innovative local policy is most evident in areas where groundwater flow systems, from recharge to discharge, occur within the local jurisdiction. This is the case in Washington County, where groundwater flow recharge to the most productive aquifers occurs in the central portion of the county, and groundwater flow directions are southwest to the Mississippi River and east-southeast to the St. Croix River. The current Washington County water plan includes a policy to protect groundwater-dependent resources, such as fens, other wetland seepage communities, and trout streams. It also recognizes links between groundwater and surface water lake and stream water quantity and quality (Washington County 2009b, 2009c). The county, through its water consortium of county officials, state agency representatives, water management organizations, and research institutions, is currently in the process of developing rules that can be incorporated into existing and newly developed watershed district rules. The changes address groundwater appropriations, stormwater volume control, groundwater quality, and groundwater-dependent resources (Washington County 2009a).

In July 1999, Dakota County obtained a CWP grant to quantify and map elevated nitrate in groundwater in the city of Hastings and surrounding townships. County staff applied for the grant after noticing increasing nitrate levels in Hastings city wells, as well as a number of private wells. The program was expanded into an ambient groundwater study, which included testing for a number of constituents besides nitrate, in addition to residence time estimates based on tritium concentrations. The result is a comprehensive water chemistry dataset combined with a well network that allows the county to establish spatial and temporal trends in water quality (Dakota County 2006). The policy impact of this study has been greater at the state than the local level, because it is to date the most comprehensive evaluation of agricultural impacts on shallow and deep aquifers at the county scale in Minnesota, and because it demonstrates the need to monitor groundwater chemistry on a regional basis.

The city of Rochester, in Olmsted County, has a long association with the USGS, contracting with the agency to do some of the first regional-scale groundwater modeling in Minnesota outside the TCMA (Delin 1991). Results of that modeling revealed that the city is receiving a large percentage of recharge to its water supply aquifers, the Prairie du Chien Group and Jordan Sandstone, through a narrow strip defined by the eroded edges of the Decorah Formation and the St. Peter Sandstone. This geographic area, known as the Decorah Edge, surrounds the city on three sides. Recharge through this zone comes from the Galena Group above the Decorah Formation and contains groundwater that is high in nitrates. Concern over possible nitrate contamination of the city's water supply prompted the city and county planning offices to expand wetland administrative rules and zoning policies to include the Decorah Edge. The goal was to minimize the potential for additional contaminant loading, while providing the opportunity for bioremediation via nitrate uptake by plants as groundwater passes through the Decorah Edge. The zoning changes allow higher-density development on land parcels that are not on the Decorah Edge and extend the setback distances from wetlands (Olmsted County 2008). Fillmore County adopted parallel changes in its zoning ordinances to protect the Decorah Edge. The U.S. Department of Agriculture expanded definition of the set-aside program to include the Decorah Edge in the Conservation Reserve Program. These changes represent federal government recognition of the Decorah Edge as a definitive water resource control point.

In this case, as in many others, much of the proactive groundwater work accomplished by local planning has influenced regional planning at the state level. Local programs and policies offer creative solutions to groundwater problems that often succeed because they have grassroots support.

CONCLUSIONS

Society is fundamentally interested in the current and future condition of its water resources. Because decisionmakers have little direct experience with groundwater, unlike their often extensive experience with surface waters such as lakes, rivers, and streams, they rely on the expertise of the hydrogeologic community to provide analyses of groundwater flow systems, asking experts to predict future conditions. Numeric modeling has been and will continue to be the leading tool to make these predictions. However, standard groundwater models based on assumptions of isotopic, porous media do not always work reliably for estimates of yield and do not provide reliable estimates of contaminant transport, often erring by several orders of magnitude.

At the local scale, examples from Washington and Dakota Counties and the city of Rochester illustrate how the threat of contamination has motivated local planners to work toward proactive groundwater policy. At the regional scale, distribution of groundwater residence times in the Mount Simon Aquifer has prompted state water planners to recognize groundwater in the Mount Simon Aquifer, at least in part, as a nonrenewable resource. From both local and state perspectives, distribution of contaminants in groundwater has raised public awareness about vulnerability of groundwater resources. Rapid water movement and distribution of geochemical composition in groundwater systems, documented through use of borehole geophysics, tracers, and continuous data logging, is becoming more available but is not yet a typical part of model calibration. By including these additional calibration targets in future modeling efforts, the accuracy of groundwater models can be increased.

Combining the spatial and temporal distribution of geochemical composition in groundwater systems with results of numeric modeling is changing local and state approaches to groundwater policy. Decisionmakers now have the means to view groundwater flow systems and understand elusive concepts of groundwater residence times and pathways. Policy should, by nature, be iterative and flexible; these attributes of groundwater policy will increase as our understanding of groundwater flow systems grows. As the potential impacts of decisions about land and water use on groundwater systems become better understood, more informed decisions will be made.

NOTES

1. Minn. Law, 86th Leg. sess., 2009–2010, Ch. 37, HF No. 2123, § 4, subd. 3.
2. Minn. Stat. § 103G.271, subd. 4a(b).
3. Minn. Stat. § 103G.271, subd. 5(a).
4. Minn. Stat. § 103G.271, subd. 4a(a).

REFERENCES

ATSDR (Agency for Toxic Substances and Disease Registry). 2008. Public Health Assessment for Perfluorochemical Contamination in Lake Elmo and Oakdale, Washington County, Minnesota. www.health.state.mn.us/divs/eh/hazardous/sites/washington/lakeelmo/phaelmooakdale.pdf (accessed January 5, 2010).

Dakota County Environmental Services. 2006. Dakota County Ambient Groundwater Quality Study, 1999–2003 Report. www.co.dakota.mn.us/EnvironmentRoads/Reports/Water/Ambient.htm (accessed December 31, 2009).

Delin, G.N. 1991. *Hydrogeology and Simulation of Ground-water Flow in the Rochester Area, Minnesota.* Water-Resources Investigation 90-4081. Washington, DC: U.S. Geological Survey.

DNR (Minnesota Department of Natural Resources). 1989. Drought of 1988. http://climate.umn.edu/pdf/drought88.pdf (accessed December 29, 2009).

———. 2009a. Groundwater: Plan to Develop a Groundwater Level Monitoring Network for the 11-County Metropolitan Area. http://files.dnr.state.mn.us/publications/waters/groundwater_level_monitoring_report_october_2009.pdf (accessed December 29, 2009).

———. 2009b. Groundwater Provinces. www.dnr.state.mn.us/groundwater/provinces/index.html (accessed January 6, 2010).

EQB (Environmental Quality Board). 2008. Managing for Water Sustainability: Report of the EQB Water Availability Project. www.eqb.state.mn.us/documents/Managing_for_Water_Sustainability_12-08.pdf (accessed October 9, 2009).

Hondzo, M. 2009. WATERS Test Bed Site: Minnehaha Creek Watershed. www.watersnet.org/wtbs/wtbs06/index.html (accessed October 9, 2009).

Kanivetsky, R. 1986. Major Constituent Chemistry of Selected Phanerazoic Aquifers in Minnesota. *Minnesota Geological Survey Miscellaneous Map Series M-61* St. Paul: Minnesota Geological Survey.

Kendall, C. 2005. Isotope Hydrology Workshop. Materials presented at Minnesota Ground Water Association Fall Workshop. November 2005, St. Paul.

Lively, R.S., R. Jameson, E.C. Alexander, Jr, and G.B. Morey. 1992. *Radium in the Mt. Simon–Hinckley Aquifer, East-Central and Southeastern Minnesota.* Minnesota Geological Survey Information Circular IC-36.

MDH (Minnesota Department of Health). 2008. PFBA in the Groundwater of the Southeast Metro Area. www.health.state.mn.us/divs/eh/hazardous/topics/pfbasemetro.html (accessed December 29, 2009).

MPCA (Minnesota Pollution Control Agency). 2004. Minnesota's Water Quality Monitoring Strategy, 2004 to 2014. www.pca.state.mn.us/water/pubs/wqms-report.html (accessed December 31, 2009).

Olmsted County. 2008. County Zoning Ordinance Amendments. Article IX: Overlay Districts. Section 9.20: Decorah Edge Overlay District. Rochester-Olmsted Planning Department. www.co.olmsted.mn.us/departments/docs/Planning/countydecorahedgezoning20060829.pdf (accessed January 6, 2010).

Reeves, L. 2009. Buffalo Aquifer and Bonanza Valley Studies Initiated. *Minnesota Ground Water Association December Newsletter* 28 (4):1.

Runkel, A.C., R.G. Tipping, E.C. Alexander Jr, J.A. Green, J.H. Mossler, and S.C. Alexander. 2003. *Hydrogeology of the Paleozoic Bedrock in Southeastern Minnesota*, Minnesota Geological Survey Report of Investigations 61.

Rzepecki, P., and S. Robertson. 2009. Hydraulic Gradient Fluctuations: Prairie du Chien–Jordan Aquifer System, East-Central Hennepin County, June 2007–March 2008. *Minnesota Ground Water Association June Newsletter* 28 (2):5.

SWAQ (Subcommittee on Water Availability and Quality). 2007. A Strategy for Federal Science and Technology to Support Water Availability and Quality in the United States. Report of the National Science and Technology Council, Committee on Environment and Natural Resources, Subcommittee on Water Availability and Quality. www.ostp.gov/galleries/NSTC/Fed%20ST%20Strategy%20for%20Water%209-07%20FINAL.pdf (accessed October 9, 2009; site discontinued).

Taylor, C.J., and W.M. Alley. 2001. Ground-Water-Level Monitoring and the Importance of Long-Term Water Level Data. U.S. Geological Survey Circular 1217. pubs.usgs.gov/circ/circ1217/html/pdf.html (accessed December 31, 2009).

UMWRC (University of Minnesota Water Resources Center). 2009. Minnesota Water Sustainability Framework. http://wrc.umn.edu/watersustainabilityframework (accessed December 29, 2009).

USGS (U.S. Geological Survey). 2009a. National Water-Quality Assessment (NAWQA) Program. http://water.usgs.gov/nawqa (accessed December 31, 2009).

———. 2009b. A National Framework for Ground Water Monitoring in the United States. Advisory Committee on Water Information, Technical Reports of the Subcommittee on Ground Water. http://acwi.gov/sogw/pubs/tr/index.html (accessed December 29, 2009).

Washington County. 2009a. Groundwater/Surface Water Management in Washington County. Department of Public Health and Environment. www.co.washington.mn.us/info_for_residents/environment/water_resources/groundsurface_water_management (accessed January 22, 2010).

———. 2009b. Washington County Groundwater Plan: 2003-2013. Department of Public Health and Environment. www.co.washington.mn.us/client_files/documents/phe/ENV/ENV-2003GroundwaterPlan.pdf (accessed January 5, 2010).

———. 2009c. Washington County Groundwater Workplan 2009. Department of Public Health and Environment. www.co.washington.mn.us/client_files/documents/phe/ENV/ENV-GW-2009WorkPlan.pdf (accessed January 5, 2010).

PART IV

MANAGING COMPETING WATER USES

CHAPTER 9

Urbanization and Water Resources: Policy Implementation in an Era of Austerity

Greg Lindsey

*I*n early May 2009, the Minnesota legislature was embroiled in controversy over the largest budget deficit—$4.6 billion—in the state's 151-year history. In Minnesota, however, opening day for walleye season is big news, and for a day or two, news of fishing overshadowed the financial crisis. Local television stations featured extensive coverage of the governor's choice of fishing locations, reporting that for the first time in history, a governor chose to open the season in the Twin Cities metropolitan area. The reports also noted that water levels in suburban White Bear Lake, the governor's choice, were near record low levels because of a seasonal drought. In news reports, Governor Tim Pawlenty explicitly linked recreational use of the state's water resources to the health of the Minnesota economy: "'Tourism in Minnesota is an enormous part of our economy,' the Republican governor said. ... 'It's licenses and fuel and boats, tackle and equipment, restaurants and lodging.'" (Northland Outdoors 2009).

This vignette illustrates three points that provide context for this chapter: First, urban Minnesotans want and demand use of high-quality water resources. Second, the availability of urban water resources is not assured; use can be disrupted by hydrologic and human factors. Third, economic issues inherently and inextricably influence Minnesota's governance of water resources. Sustainable stewardship of the state's water resources requires political leadership and depends on adequate investment.

In Minnesota, as in other states, the last century of population change can be characterized succinctly as one of increasing urbanization. Urbanization has well-known effects on hydrologic systems, and urbanized populations place great demands on these systems for water supply, wastewater disposal, energy and industrial production, water-related recreation, and ecological services. Although the effects of urbanization on water resources generally are known, they are not completely understood, and research to increase our understanding of them is ongoing. In contrast, strategies for mitigating the impacts of urbanization are not well developed, and effective management practices are only beginning to be implemented. Federal, state, and local officials and multiple stakeholders are struggling—in turbulent natural, political, and financial environments—to adopt policies and implement programs to sustain urban water resources.

This chapter builds on preceding ones by exploring the implications of urbanization for water resources policy and management. It presents a broad conceptual framework for thinking about the evolution of environmental and water resources policy; describes historical and projected patterns of urbanization in Minnesota; examines the effects of urbanization on hydrologic systems and strategies for mitigation; looks at demand for water, focusing on metropolitan areas; explores financial challenges in achieving urban water policy objectives; and outlines challenges to and strategies for implementation of sustainable urban water policy that complements the five core principles introduced in earlier chapters: interconnectedness, informed decisionmaking, precaution, transparency, and accountability.

TRENDS IN WATER RESOURCE POLICY AND ADMINISTRATION

Scholars have documented broad trends and transformations in national environmental policy and governance. These trends include a move away from top-down, command-and-control approaches to regulation to more collaborative processes to engage stakeholders in long-term, adaptive, and sustainable management (Durant et al. 2004; Kettl 2002; Vig and Kraft 2006). Mazmanian and Kraft (2009) characterize three overlapping epochs in this evolutionary process:

- 1970s–1990: regulation for environmental protection
- 1980–2000s: emergence of efficiency-based regulatory reform and flexibility
- 1990–present: beginnings of a transition to a focus on sustainable communities

Perspectives on epochs of water resource policy include Kraft's (2009) assessment of Wisconsin's move to collaborative decisionmaking in the Fox–Wolf Basin and studies by Lubell et al. (2009) of the effectiveness of collaborative watershed partnerships. Kraft found that successful partnerships were established in the Fox–Wolf Basin, and that these partnerships led to "effective working relations among business officials, state and local policymakers, scientists, environmentalists, and community leaders." However, "cooperation and voluntary cleanup programs, especially those directed at contaminated sediments, could only go so far in light of the high costs of remediation, ill-defined goals, and limited public understanding of the issues. ... The major constraints that so limited this process seemed to be a reluctance to pay the high cost of cleanup and continuing doubts that the anticipated ecological and health benefits could justify doing so" (2009, 135). Kraft concluded that collaborative processes are more likely to be successful in reaching agreement on general goals than on specific implementation strategies, and that selective use of the "best" regulatory approaches likely will continue to be necessary in the new epoch of sustainability.

Drawing on studies of the National Estuary Program and the Watershed Partnership Project, Lubell et al. (2009) reached similar conclusions. Arguing that form should follow function, they presented a sequential, conceptual framework for understanding the effectiveness of watershed partnerships. This framework includes the following:

- first-order outputs, such as development of human and social capital;
- second-order outputs, such as consensus agreements;
- third-order outputs, including more specific plans for implementation, monitoring, and cooperation; and
- outcomes that include indicators of both actual and perceived effectiveness.

Lubell and his colleagues concluded that the most important factors affecting the success of partnerships include "perceived fairness, level of trust, density of social networks, the quality of science, presence of neutral facilitators, intensiveness of deliberation, and the availability of financial resources." Noting, however, the lack of evidence linking the efficacy of partnerships to water quality outcomes, they also pointed out that "in the case of watershed management in the United States, it is more accurate to say that the epoch of sustainability has not supplanted environmental regulation *at all*" (2009, 282, 284).

The research literature has yet to document the efficacy of collaborative processes heralded as central to sustainability. These collaborative processes appear to have limits, especially in the face of high costs that must be borne by one or more stakeholders. There is little evidence that these new collaborative processes will supplant older approaches in the near future, especially when meaningful progress requires significant financial investment. Continued reliance on core regulatory strategies therefore is likely to be essential to achieve water quality policy objectives.

URBANIZATION IN MINNESOTA

Like other states, Minnesota likely will become progressively more urban throughout the 21st century. Table 9.1 presents estimates of Minnesota's total and metropolitan populations since 1970, including population projections through 2030. The state's population in 2010 was estimated at about 5.4 million people, approximately 4 million (73%) of whom reside in seven metropolitan statistical areas (MSAs).[1] Between 1970 and 2000, Minnesota's population increased by 29%. The growth in metropolitan population was more than 47%, while the population in nonmetropolitan Minnesota increased by only 0.2%. Among the seven MSAs (Minnesota portions only), growth was uneven: two lost population, while five grew. The Twin Cities and Rochester MSAs grew at the greatest rates: 58%

Table 9.1 Historical and projected populations in Minnesota and its metropolitan statistical areas (MSAs)

MSA (Minnesota portions of population only)	1970	1980	1990	2000	2010 projected	2020 projected	2030 projected	Population change, 1970–2000 (%)	Population change, 2000–2030 projected	Population change, 2000–2030 projected (%)	Percent of state population change, 2000–2030
Duluth–Superior, MN–I MSA	220,693	222,229	198,213	200,528	235,000	242,400	247,300	−9.1%	46,772	23.3%	3.4%
Fargo–Moorhead, ND–MN MSA	46,585	49,327	50,422	51,229	57,100	63,000	66,900	10.0%	15,671	30.6%	1.1%
Grand Forks, ND–MN MSA	34,435	34,844	32,498	31,369	31,900	33,400	34,300	−8.9%	2,931	9.3%	0.2%
La Crosse, WI–MN MSA	17,556	18,382	18,497	19,718	20,400	21,300	22,100	12.3%	2,382	12.1%	0.2%
Minneapolis–St. Paul–Bloomington, MN–WI MSA	1,813,647	2,070,271	2,413,873	2,868,847	3,248,400	3,583,200	3,828,500	58.2%	959,653	33.5%	69.6%
Rochester, MN MSA	84,104	92,006	106,470	124,277	192,700	217,900	236,200	47.8%	111,923	90.1%	8.1%
St. Cloud, MN MSA	134,585	163,256	190,921	167,392	198,000	225,000	245,700	24.4%	78,308	46.8%	5.7%
State of Minnesota	3,804,971	4,075,970	4,375,099	4,919,479	5,446,530	5,943,240	6,297,950	29.3%	1,378,471	28.0%	100.0%
Nonmetropolitan Minnesota	1,453,366	1,425,655	1,364,205	1,456,119	1,463,030	1,557,040	1,616,950	0.2%	160,831	11.0%	11.7%
Metropolitan Minnesota	2,351,605	2,650,315	3,010,894	3,463,360	3,983,500	4,386,00	4,681,000	47.3%	1,217,640	35.2%	88.3%

Source: U.S. Census Bureau (2005).

and 48%, respectively. Among Minnesota's 87 counties, the fastest-growing all were in MSAs.

The population of Minnesota is expected to grow to 6.3 million by 2030, an increase of 28% over the 2000 population. During this 30-year period, the rate of growth in the seven MSAs (31%) is projected to be more than 14 times that in nonmetropolitan Minnesota. Two MSAs, Twin Cities and Rochester, will account for nearly 78% of the absolute growth in the state's population (U.S. Census Bureau 2005). The Minneapolis–St. Paul MSA will add nearly 1 million people, more than eight and a half times the absolute increase in population projected for the Rochester MSA, and roughly six times the projected increase in all of nonmetropolitan Minnesota.

Minnesota's metropolitan areas also differ from the balance of the state with respect to population density and income. Population density is an indicator of consumption of land, demand on water resources, and the potential complexity of managing water resource problems. Table 9.2 includes estimates and projections of population densities in Minnesota MSAs for 1970 through 2030. Population density varies greatly between metropolitan and nonmetropolitan areas and among and within MSAs. These estimates of density illustrate the relationship between population change and use of land.

Perhaps the most striking information in Table 9.2 is the decrease in population density in the Twin Cities MSA, a change that reflects the addition of more counties to the MSA as a result of suburbanization. Thus, while the population of the Twin Cities MSA grew by more than 1 million people between 1970 and 2000 (more than 58%), its population density decreased by nearly 49%, from 1,177 to approximately 602 persons per square mile.

Another important observation from these data is that none of the other six MSAs is nearly as densely populated as the Twin Cities MSA. In 2000, after the density of the Twin Cities essentially was halved by the expansion of its boundaries, its population density remained three times greater than that of the next most densely populated MSA, Rochester. Assuming no additional changes in the boundaries of MSAs through 2030, population density in the Twin Cities MSA will remain more than twice as great as the Rochester MSA, an outcome reflecting the historic concentrations of populations in Minneapolis and St. Paul. Another indicator of the large differences in density between the Twin Cities area and other MSAs is that population density for the state as a whole is greater than the density of four MSAs and likely will remain so for at least the next three decades. This fact is attributable to the concentrations of population in the Twin Cities, Rochester, and St. Cloud MSAs.

These changes in population density reflect well-documented trends toward increasing land consumption per person and household—trends popularly called "urban sprawl." Marshall, for example, has shown that newcomers to urban areas in the United States between 1950 and 2000 generally occupied "about twice the land area per capita of existing residents." He also noted that few other measures of urban form have remained "so constant" over time (2007, 1889). Absent significant regulatory interventions into local land development processes

Table 9.2 Historic and projected population densities in MSAs

Metro area (Minnesota portion of MSAs only)	1970	1980	1990	2000	2010 projected	2020 projected	2030 projected	Percent change, 1970–2000	Projected percent change, 2000–2030	Projected percent change, 1970–2030
Duluth–Superior, MN–WI MSA	35.4	35.7	31.8	32.2	37.7	38.9	39.7	−9.1%	23.3%	12.1%
Fargo–Moorhead, ND–MN MSA	44.6	47.2	48.2	49.0	54.6	60.3	64.0	10.0%	30.6%	43.6%
Grand Forks, ND–MN MSA	17.5	17.7	16.5	15.9	16.2	17.0	17.4	−8.9%	9.3%	−0.4%
La Crosse, WI–MN MSA	31.4	32.9	33.1	35.3	36.5	38.1	39.6	12.3%	12.1%	25.9%
Minneapolis–St Paul–Bloomington, MN–WI MSA	1176.8	532.2	557.6	602.0	681.6	751.8	803.3	−48.8%	33.5%	−31.7%
Rochester, MN MSA	128.8	140.9	163.0	190.3	295.1	333.7	361.7	47.8%	90.1%	180.8%
St. Cloud, MN MSA	61.5	74.6	87.2	95.5	113.0	128.4	140.2	55.4%	46.8%	128.0%
Minnesota	47.8	51.2	55.0	61.8	68.4	74.6	79.1	29.3%	28.0%	65.5%

Note: Population density is persons per square mile.
Source: U.S. Census Bureau (2005).

by regional, county, and municipal governments, these patterns of urbanization are likely to continue.

Median household income, an indicator of both demand and ability to pay for use of water resources, also varies greatly between urban and rural areas. In 2007, the estimated median household income in Minnesota was $55,664, which was approximately 10% higher than the median household income in the United States (U.S. Census Bureau 2008). This positive indicator masks considerable variability across the state. Among Minnesota counties in 2007, for example, estimated median household incomes ranged from $34,503 in rural Clearwater County to $80,038 in urban Carver County.

Median household incomes in MSAs areas ranged from $42,615 in the Duluth–Superior MSA to $63,898 in the Minneapolis–St. Paul–Bloomington MSA. While the median household income in the Twin Cities MSA was 15% higher than the state's overall, the median household incomes in five MSAs were lower than the state's median, again reflecting the dominance of the Twin Cities. The median household income in rural, non-MSA counties, $44,025 in 2007, was comparable to incomes in the three poorest MSAs. Just as population is centralized in the Twin Cities area, so too are the households with the highest incomes. Holding other factors equal, these differences indicate that both demand and ability to pay for management of water resources will vary across the state, with both greatest in the Twin Cities MSA.

The key points here are that urbanization increases variation in population and is associated with changes in distribution of incomes. Urbanization consumes land and increases local demands on fixed water resources for all uses. This concentration of demand increases the probability that demand will outstrip supply, thus increasing the potential for conflict and the complexity of management. The ability of local residents to pay for programs to achieve water policy objectives also varies geographically, complicating political processes and the adoption of policies required for sustainable stewardship. While urbanization will have the greatest impact on water resources in the Twin Cities MSA, this area also will have the highest income and thus the greatest capacity to manage impacts.

EFFECTS OF URBANIZATION ON WATER RESOURCES

For more than 100 years of Minnesota's 150-year history, urbanization occurred with little regard for impacts on water resources. Given the widespread availability, it was assumed water always would be available, and water was withdrawn from surface water and groundwater systems with little understanding of their capacity to sustain use. Industrial, municipal, and household wastes were discharged indiscriminately into rivers and lakes, with little concern for their effects on water quality, aquatic resources, or other uses. The effects of development on hydrologic systems—of paving land with impervious surfaces, filling wetlands, and destroying headwaters—were not understood, and no effort was made to control them.

The impacts of urbanization on water resources now are better understood. Processes of development, for example, adversely affect hydrologic systems and local water budgets by altering the structure of the physical landscape. Grading and denuding land during development destroys ephemeral and first-order streams, reducing habitat and increasing sediment and other pollutant loads in receiving waters. Construction of buildings and paving land for roads and parking lots permanently decreases infiltration, increases volume, and increases the velocity of runoff, resulting in both higher flood flows and lower base flows. These changes simultaneously increase the probabilities of floods and seasonal droughts and alter physical dimensions of stream and river channels, further decreasing water quality, habitat, and species diversity. Populations of rough fish such as carp that can tolerate lower-quality habitat, with lower volumes of water, greater turbidity, and higher temperatures, replace more sensitive species like walleye. Without mitigative measures, urbanization inevitably reduces the availability of water for withdrawal for water supply and other uses, the capacity of water resources to absorb discharges of waste, and the attractiveness of streams for fishing and other forms of recreation.

Based on improved understanding of these aspects, planners and engineers have developed a number of mitigative strategies:

- careful site design, including clustering of structures on development sites, minimizing encroachment on floodplains and wetlands, and preserving first-order streams and critical habitat;
- erosion and sediment control during development, including implementation of practices to minimize and sequence disturbances of earth surfaces, minimize exposure of nonvegetated surfaces to wind and precipitation, and maintain temporary sediment capture devices during construction;
- construction of stormwater management and flood control facilities to reduce permanent increases in flow volumes and velocities and to maximize pollutant settling and removal;
- permanent maintenance of stormwater management and flood control facilities, with periodic sediment removal and rehabilitation of structures;
- continuing education of citizens and industry to adopt stormwater "best management practices," including pollution prevention through behavioral change (e.g., proper disposal of waste oil);
- development of local regulatory and institutional capacity to ensure and enforce good site design, effective erosion and sediment control, and long-term stormwater management; and
- identification and dedication of permanent funding sources for program management, including maintenance of facilities.

In Chapter 4, Rob Johansson and Faye Sleeper noted that the federal Clean Water Act first targeted gross, visible pollutants from pipes discharged by politically palatable targets—industrial dischargers and municipal wastewater treatment facilities—using command-and-control approaches, technological requirements, and for municipal dischargers, subsidies. Although experts knew that sediments from agricultural and urban runoff were among the largest pollutants by volume, no attempts to regulate or manage them were made initially. Processes of land development and urbanization, if regulated at all, were regulated by local rather than federal or state governments.

In the 1987 amendments to the Clean Water Act, however, Congress directed the U.S. Environmental Protection Agency (EPA) to extend the National Pollutant Discharge Elimination System (NDPES) permit system to stormwater discharges from industries and both large and smaller municipalities. The regulations governing municipalities, called Municipal Separate Storm Sewer System (MS4) regulations, are most relevant here because they govern most suburban municipalities where urbanization is occurring. Because of the complexity, cost, and feasibility of establishing effluent limitations for individual stormwater discharges, EPA moved to a new system of programmatic regulation that requires municipalities to implement best management practices (BMPs). Specifically, the MS4 regulations require more than 220 Minnesota municipalities, watershed districts, and counties to implement stormwater pollution prevention plans (SWPPPs) that include six control measures:

1. public education and outreach;
2. public participation and involvement;
3. illicit discharge detection and elimination;
4. construction site runoff control;
5. postconstruction runoff control; and
6. pollution prevention/good housekeeping (e.g., street sweeping).

Two of these measures, construction site and postconstruction runoff control, effectively require intervention in development processes and thus constitute federal and state intrusion into local land use regulation. However, the core rationale for this programmatic regulation—that implementation of BMPs will reduce pollutant loadings and improve water quality—has not been proved over long time periods. Consideration of one requirement, construction site erosion and sediment control, illustrates the challenges of achieving the new policy objectives. For Minnesota's erosion and sediment control (ESC) programs to achieve the intended water quality outcomes, state and local administrators and operators of construction sites must achieve essentially 100% effectiveness in implementation of five distinct components of construction site regulation (Lindsey 1995):

1. *Coverage.* Each of Minnesota's regulated jurisdictions must adopt ESC ordinances (or collaborate with other jurisdictions) to ensure that all developers and operators of construction sites know they are regulated, apply for permits, and comply with regulatory requirements.
2. *Plan review.* Local engineers or soil conservation specialists must review and certify that ESC plans for construction sites include BMPs that will effectively reduce or eliminate the release of sediments during the construction process. Examples of these BMPs include sequencing of construction to minimize

areas of disturbance and exposed soil, placement of erosion controls (e.g., vegetative mats) on temporarily denuded surfaces, installation of silt fences, and construction of temporary sediment traps for settling of sediments prior to discharge of storm flows to receiving waters.

3. *Installation*. Construction site workers must install all devices in accordance with plans.
4. *Maintenance*. Construction site workers must maintain all BMPs in working order throughout the construction process, which may continue for several years.
5. *Monitoring and enforcement*. Local regulatory officials, typically inspectors from planning or public works departments or soil and water conservation districts, must inspect operations and use enforcement tools (e.g., notices of violation, stop work orders, civil and criminal penalties) as required to ensure compliance.

Successful implementation of this single program aimed mainly at a single pollutant—sediment—involves adoption of new regulations by hundreds of jurisdictions, creation of new regulatory structures, and new educational programs to change the behaviors of hundreds of administrators and construction site operators. It is clear that local jurisdictions and developers will bear new costs. It also is clear that achievement of these regulatory objectives will require political commitment, including the willingness to empower local inspectors to temporarily close a construction site or fine construction site operators for failure to maintain BMPs during construction. Finally, it is clear that water policy objectives will not be achieved if these programs are not implemented successfully. It is unclear, however, that the converse is true. Given the uncertainties associated with this type of programmatic regulation and the pragmatically necessary absence of requirements for monitoring sediment discharges in runoff, it is unclear that implementation of ESC programs will achieve water quality objectives. All other factors being equal, this uncertainty will be greatest in those areas where rates of urbanization are highest, specifically the Twin Cities and other growing MSAs.

The challenges appear more daunting when the full scope of programs needed to mitigate the effects of urbanization is considered. For example, the same framework for implementation—coverage, plan review, installation, maintenance, and monitoring and enforcement—can be applied to postconstruction runoff controls, such as stormwater retention and detention facilities or artificial constructed wetlands. Municipalities must adopt ordinances that prescribe design objectives, implement plan review procedures to ensure that plans include proper design, work with construction site operators to ensure that facilities are built as designed, educate homeowners associations and other property owners about their responsibilities as owners of facilities, and use monitoring and enforcement tools to ensure proper construction and perpetual maintenance by owners. This maintenance includes inspection of stormwater facilities throughout their 30- to 50-year lives and rehabilitation of them as required. Along with elimination of illicit connections to separate storm sewers, ESC and postconstruction runoff control are the best-understood requirements of the MS4 permit programs.

Regulators have little real understanding of the water quality improvements to be expected with the remaining MS4 requirements.

DEMAND FOR WATER RESOURCES IN MINNESOTA

The demands for use of water resources in Minnesota include municipal water supply, waste disposal, industrial consumptive use, irrigation and other agricultural uses, energy production, ecological services, and recreation. This discussion focuses on demand for water supply, recognizing that this is only one category of demand.

Minnesota law requires the Environmental Quality Board (EQB) of the Department of Natural Resources "to coordinate a biennial assessment of the availability of water to meet the state's long range needs" (Van Buren and Wells 2007, i). The most recent EQB report indicates that total annual demand for water in Minnesota grew 18% between 1995 and 2005, from slightly more than 1.2 quadrillion to more than 1.4 quadrillion gallons (Van Buren and Wells 2007). Per capita use grew 6% during this same time. In other words, the increase in demand for water occurred both because the state's population increased and because people on average consumed more individually. Total demand for 2030 is expected to increase in proportion to changes in the state's population, which is expected to increase by 26%. Planners optimistically estimate that per capita consumption will remain flat because of the introduction of new conservation measures.

With respect to net water use, heavy demand (consumptive use relative to available supply) occurs principally in the urban areas, specifically in the Twin Cities MSA. In 2005, seven counties used 39% or more of their renewable resource; each was in the Twin Cities MSA (Figure 9.1). Seven other counties used between 10% and 22% of their renewable resources, while 73 of Minnesota's 87 counties used less than 10%. By 2030, the patterns of use are predicted to remain comparable, but more counties will use higher proportions of their renewable resources. Only one county (Ramsey) used more than 100% of its renewable resource in 2005, but four counties are expected to use 99% or more of their renewable resources in 2030. The number of counties using more than 10% of their renewable resources is predicted to increase from 14 to 19. Permitted uses—volumes authorized for withdrawal in permits issued by the DNR—exceed actual withdrawals; hence if permitted volumes actually are withdrawn, these estimates of availability will worsen.

These facts led the EQB to conclude that "the label of Minnesota as water rich does not fit as well as once thought. The growth corridor from south of the Twin Cities to St. Cloud already makes significant demands on its renewable water resources, making water supply management a special concern. In the remainder of the state, care also must be taken by local and state officials in planning to meet the demands for and allocations of water" (Van Buren and Wells 2007, i). Stated another way, meeting the growing demand for water in Minnesota will be increasingly difficult, especially in the rapidly urbanizing metropolitan areas.

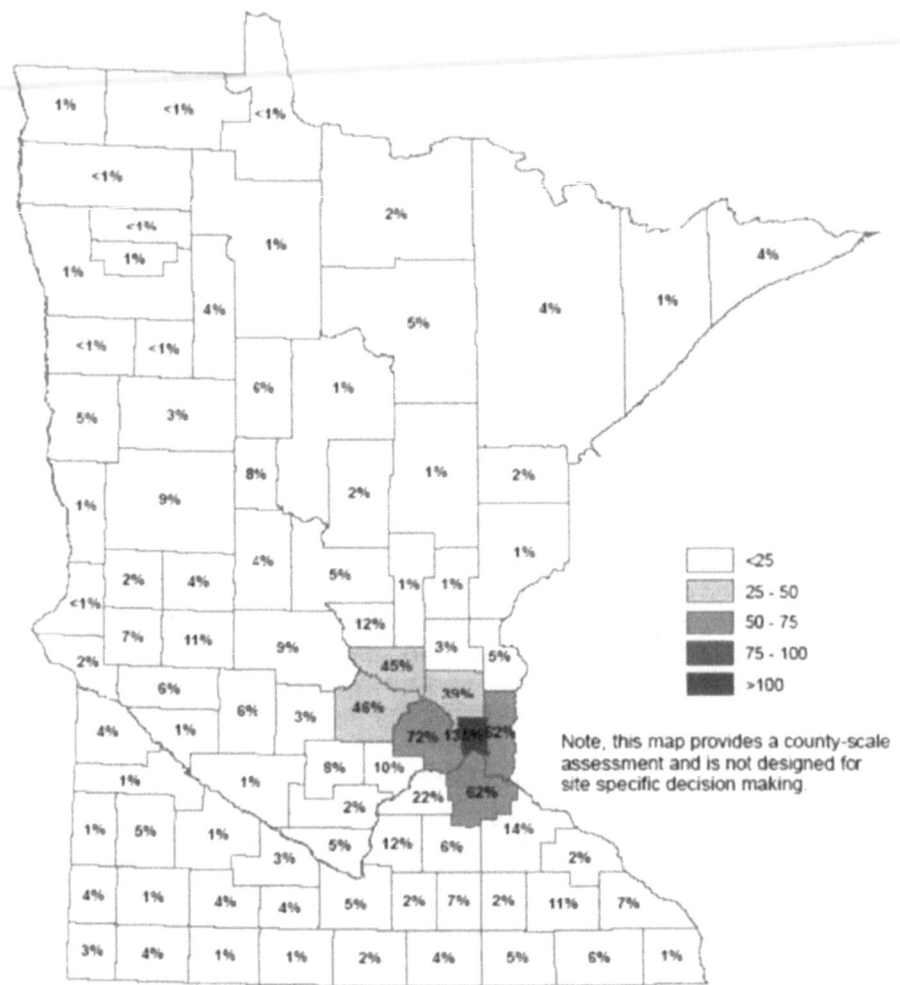

Figure 9.1 *Estimated 2005 net water use as a percent of the renewable resource*

Source: Van Buren and Wells (2007, 5).

FINANCIAL CHALLENGES IN URBAN WATER MANAGEMENT

Protection of water resources under pressure from urbanization requires investments to mitigate the effects of landscape change, manage the concentration and discharge of municipal and industrial wastes, and provide clean water for drinking and other purposes. No comprehensive estimates of the costs of water resource management in Minnesota exist, but in collaboration with EPA, the Minnesota Pollution Control Agency (MPCA) and Minnesota Department of Health (MDH) have developed partial estimates of the costs of wastewater, stormwater, and water supply infrastructure required to comply with federal regulatory requirements. In 2008, the MPCA published the Wastewater Infrastructure Needs Survey (WINS) and a companion document, the Storm Water Infrastructure Needs Survey (SWINS). In 2007, the MDH published the Drinking Water Needs Survey. When added together, these estimates of the costs of known needed investments in Minnesota water infrastructure exceed $11.6 billion. Most of these costs will be incurred in urban areas. The estimates for wastewater and drinking-water infrastructure both are substantially higher than previous estimates, indicating that costs have risen as populations have increased impacts on water resources, use of water resources has increased, and technical understanding of the magnitude and complexity of the problems has grown.

The 2008 WINS Survey

Recognizing these financial challenges, the Minnesota legislature has directed the MPCA to report periodically on future wastewater infrastructure needs and capital costs, cost increases to residential users, the affordability of costs, and the likely impacts of implementation of the total maximum daily load (TMDL) regulatory process.[2] The most recent estimate, based on a survey of Minnesota municipalities in 2007, is $4.53 billion for more than 1,100 projects planned through 2027 (MPCA 2008). This estimate includes costs for publicly owned wastewater treatment systems and public costs to provide sewers and treatment in unsewered areas; 32% of the total costs are for new interceptor sewers, and an additional 30% is for secondary wastewater treatment, using biological processes to remove organic matter.

The $4.53 billion is only a partial estimate of total costs. It does not include costs for permitted industrial facilities that do not discharge to public facilities, privately owned subsurface sewage treatment systems (e.g., the septic tanks and fields that serve 25% of Minnesota's population), or very small public systems (MPCA 2008). It also does not include the costs of new municipal programs required as a condition of MS4 permits.

Capital costs for wastewater projects will be concentrated in the Twin Cities metropolitan area. More than half (52%) will be incurred by Metropolitan Council Environmental Services (MCES), a regional authority that serves 7 of the 11 counties in the MSA. An additional 10% will be incurred by the 180 municipalities in the MCES service area. Capital costs for wastewater projects in "greater Minnesota," 80 of the state's 87 counties, will account for just 38% of the state's total costs.

The MPCA concludes that costs will be least affordable for residents in the smallest (generally rural) communities located throughout greater Minnesota. The U.S. Department of Agriculture considers projects that will cost less than 1.5% of a community's median household income to be affordable. Across the 10 economic regions in greater Minnesota, the percentage of communities with needs that will exceed this guideline ranges from more than 25% to nearly 70%. In the MCES service area, because of economies of scale and the greater income and wealth, communities will be more likely to finance projects without imposing significant hardships on the majority of households. Although costs are projected to exceed this guideline in fewer than 10% of communities in the MCES service area, given the greater populations of municipalities and the variability of household incomes within them, large numbers of households in the MCES area nonetheless could face hardships.

The 2008 SWINS Survey

In 2007, to obtain information about the costs of stormwater programs required under the MS4 permits, the MPCA initiated its first Storm Water Infrastructure Needs Survey (SWINS). Although only 25% of the jurisdictions responded, and only half of these estimated costs of future capital projects, the responses provided useful information about the scope and costs of programs, a basis for estimating statewide costs, and insights into the challenges associated with the new programs. Acknowledging the uncertainty in the data, the MPCA extrapolated to "suggest the probable overall statewide costs of storm water infrastructure needs" (n.d., 5).

The estimated costs for stormwater infrastructure needs were $1.1 billion by 2027, 51% for projects planned within 5 years and the balance for projects between 5 and 20 years into the future. Among respondents, nearly 60% had implemented some type of stormwater utility fee for costs of implementation; the median annual fee was $38 per household, more than double the reported median annual expenditure on stormwater infrastructure operations and maintenance ($16), exclusive of capital costs.

Jurisdictions also reported the types of permanent, postdevelopment storm-water management structures being installed and who would bear responsibility for their management. The three most commonly reported types of facilities were sump manholes, wet retention basins, and water quality basins. Other types of facilities included constructed wetlands, dry detention basins, infiltration basins, grassed swales, and vegetative filter strips. Each of these facilities, if properly designed, installed, and maintained, has the potential to prevent some pollutants from entering receiving waters, but none is completely effective across the array of pollutants in runoff. Furthermore, the extent to which facilities are effective at a large scale in removing nitrogen, oils, and toxics remains unknown.

Most facilities are being built to mitigate effects of new development, but many jurisdictions are assessing how to retrofit land developed prior to adoption of the MS4 regulations. In most cases, when new facilities are constructed as a condition of receiving development approvals, the facilities remain under private ownership, either homeowners' associations in the case of residential development or industrial or commercial owners in the case of nonresidential development. Respondents to the SWINS survey, for example, indicated that approximately 90% of residential and commercial facilities and 79% of industrial facilities were financed and constructed by private developers.

Policies that require private financing, ownership, and maintenance are consistent with a "polluter pays" principle, but they raise other issues in implementation, such as how best to help homeowners associations understand, plan for, and pay the costs of maintaining facilities in perpetuity. These policies also potentially place jurisdictions in the role of enforcing maintenance requirements against these same homeowners associations—a new area of local regulation. In contrast, when jurisdictions retrofit existing systems, the costs of design, construction, and maintenance almost always are public, because they involve modifications of existing infrastructure that has been owned by a local government, in some cases for 150 years or more. Retrofitting existing stormwater conveyance systems such as ditches and storm sewers with facilities to control the volume and quality of water is more complex and costly than installing facilities during development, when engineers are unconstrained by existing structures, can take advantage of natural features, and integrate facilities into overall development plans.

These factors underscore the MPCA's conclusion that the current $1.1 billion cost estimate is uncertain and not reflective of the scope of programs that will be required to control and mitigate pollutants in stormwater. Most jurisdictions have not developed comprehensive plans or cost estimates for retrofitting existing areas of development. Although local governments now typically require developers to install facilities to control stormwater as a condition of development, in most cases new facilities will be maintained by private owners with little understanding of how facilities must be maintained, and most local governments have not committed resources to enforce requirements for maintenance. As state and local governments confront these challenges, the costs of stormwater pollution prevention certainly will rise.

The 2007 Drinking Water Infrastructure Needs Survey and Assessment

In an effort analogous to the Clean Watersheds Needs Survey, EPA submits to Congress a quadrennial Drinking Water Infrastructure Needs Survey and Assessment (DWINS). The DWINS includes estimates of the costs of capital improvements in public water system infrastructure needed to ensure that people have safe drinking water. The 2007 DWINS covers projects eligible for funding under state revolving loan programs that are planned between January 1, 2007, and December 31, 2026.

Minnesota has drinking-water needs approaching $6 billion (Table 9.3). Nearly half (47%) of the projected need is for transmission and distribution—getting water to people—and a third is for capital projects related to treatment. The remaining 20% is for storage facilities, development of supplies, and other projects. While the 2007 needs are approximately $0.5 billion less than needs estimated in 2003 ($6.5 billion), they are more than $1.9 billion higher than needs estimated in 1999 ($4.1 billion). These differences reflect both implementation of projects and refinement of methods for estimating needs.

Estimated capital needs are greatest for the 158 medium-size public drinking-water systems that serve populations between 3,301 and 100,000. These systems account for nearly 61% of total needs. Large systems that serve more than 100,000 people, of which there are only two in Minnesota, account for 11.2% of total needs. The smallest systems, which account for less than 25% of total needs, serve smaller municipalities, mainly in nonmetropolitan Minnesota. Not-for-profit community water systems, such as those for schools, churches, and camps, also face significant challenges.

Table 9.3 *Estimated financial needs for drinking water*

Category of need	Estimated need	Percent of total need
Transmission/distribution	$2,819,300,000	47.1%
Source	$372,000,000	6.2%
Treatment	$1,982,900,000	33.1%
Storage	$770,300,000	12.9%
Other	$43,900,000	0.7%
Total	$5,988,400,000	100.0%
Taype of system	**Estimated need**	**Percent of total need**
Large	$672,000,000	11.2%
Medium	$3,631,700,000	60.6%
Small	$1,416,500,000	23.7%
Non-for-profit, noncommunity	$268,300,000	4.5%
Total	$5,988,500,000	100.0%

Source: EPA (2009).

EPA and the states have established state revolving loan funds (SRFs) that provide below-market-rate loans to assist owners of drinking-water systems with financing of projects. In Minnesota, the Department of Health estimates that approximately $50 million is available each year through its SRF (MDH 2009). To place this amount in perspective, if Minnesota's total needs were annualized over a 20-year period, the annual needs would be approximately $299 million—roughly six times the amount available annually through the fund. Alternatively, if the revolving loan fund were the sole source of funds available, and no new capital needs were identified in the future, roughly 119 years would be required to finance current needs. Thus, although the revolving loan fund is an important financing mechanism, it falls far short of total demand. Most systems must identify other sources of financing.

The Magnitude of the Financial Challenge

Together, the costs for wastewater, stormwater, and drinking-water infrastructure approach $12 billion. Most must be paid by local jurisdictions and households, especially in the Twin Cities MSA. This estimate of costs certainly is an underestimate of the total costs to achieve water management policy objectives. The authors of these estimates acknowledge that the costs are incomplete. Historically, as studies for wastewater and drinking water have been updated, estimates of future costs typically have increased, even though projects have been completed. The principal reasons include increased understanding of the complexity and severity of problems and the need for more comprehensive and far-reaching solutions.

CONCLUSIONS

This chapter began with a vignette to illustrate three points: Minnesotans place significant demands on urban water resources, their use of these resources can be disrupted, and economic factors are important in political discussions of water resource policy. These points frame the challenges of the future.

From a broad policy perspective, we are in the epoch of sustainability, an epoch characterized by attempts to complement or supplant historic command-and-control approaches to regulation with collaborative processes aimed at reaching consensus on policies and management strategies. Although these processes have succeeded in reaching agreements about new policy directions and goals, they have foundered when confronted with the politics of paying for cleanup and other programs.

Minnesota's population will continue to urbanize for at least the near future, placing additional demands and stresses on water resources. To mitigate the effects of lower-density sprawl on water resources, administrators are implementing new programs to control pollution from construction sites and manage stormwater runoff following construction. The long-term effectiveness of the BMPs being put in place, however, has not been documented.

As these newer programs have been implemented and demand for water supply in metropolitan regions has increased, estimated costs for infrastructure to meet water quality and drinking-water needs have risen. The numbers of regulated facilities and systems have increased, while federal and state budgets have remained constant, and lagging enforcement efforts have fallen further behind. Financial needs in Minnesota approach $12 billion, and these estimates do not include all needs, such as the costs of enforcement.

The legacy of the financial crises of 2008–2009 with respect to the environment and water quality will not be known for many years, but the prognosis is not good. For the 2010–2011 biennium, Minnesota leaders faced a deficit of $4.6 billion but were able to mitigate impacts on state agencies somewhat through the use of federal stimulus funding. No stimulus funds are expected to be available for the 2012–2013 biennium, however, when state deficits are expected be larger.

These facts do not augur well for Minnesota. The epochs that reflect the evolution of environmental policy are being layered with a new epoch of financial austerity. While the historic and the newer collaborative processes to manage Minnesota's water resources will continue, it is difficult not to conclude that Minnesota's principal water resource challenge for the next decade will be that of financing needed investments and programs, not expanding regulatory or policy frameworks.

Three unrelated news articles from September 2009 illustrate the challenges of this new epoch of austerity. First, on September 13, 2009, the *New York Times* detailed the nation's failure to enforce existing water pollution laws and regulations. EPA administrator Lisa P. Jackson acknowledged that "enforcement of water laws is unacceptably low." Minnesota congressman James L. Oberstar, chairman of the House Transportation and Infrastructure Committee, which has jurisdiction over many water quality policies, stated: "I don't think anyone realized how bad things have become. ... The E.P.A. and states have completely dropped the ball." The *Times* continued: "State officials, for their part, attribute rising pollution rates to increased workloads and dwindling resources. ... Though the number of regulated facilities has more than doubled in the last 10 years, many state enforcement budgets have remained essentially flat when adjusted for inflation" (Duhigg 2009).

Two days later, the *Minneapolis Star Tribune* reported on controversy over the degradation of Round Pond in suburban Bloomington, Minnesota, since homeowners agreed to a city proposal to discharge stormwater runoff directly to it. Noting that "algae-choked storm water ponds across the metro are a real sore spot for shoreline homeowners," the *Star Tribune* recognized the paradox of the problem: "Storm water ponds clogged with plants may be doing exactly what they were intended to do: hold back and filter water before it flows into larger lakes, creeks, and rivers. But residents who live around these ponds usually want water that has a reflective surface just like a lake" (Smetanka 2009). In this specific case, the city and homeowners have agreed to share costs of treatment to control watermeal, a tiny water plant that thrives on phosphorus in runoff and coats the pond's surface. The costs for treatment, which in 2009 were approximately $60 per household, were for "aesthetics only," and are in addition to taxes and fees homeowners already pay for water, wastewater, and stormwater service. Adjacent homeowners also had paid for installation of vegetated buffers around Round Pond in their struggle to maintain it. Meanwhile, the senior civil engineer responsible for water resources in Bloomington acknowledged that keeping urban water like Round Pond clear of algae "is to some extent a losing battle." It is worth noting that the fees paid by Round Pond homeowners were more than three and a half times the median annual expenditure on stormwater infrastructure operations and maintenance ($16) reported by municipalities in response to the SWINS survey.

Finally, on September 28, 2009, while Minnesota Management and Budget (2009) reported that estimates of the state's projected deficit for the 2012–2013 biennium ranged from $4.4 billion to $7.2 billion, an amount equivalent to between 15% and 20% of the state's annual budget, Minnesota Public Radio (MPR) reported that Democrats in the Minnesota legislature had appointed a new subcommittee to make recommendations for solving the budget crisis. Noting that options on the table included consolidating state agencies and reorganizing county government, MPR also reported that the standoff over taxes between Democrats and the Republican governor was continuing.

Stated more simply, we have failed to enforce provisions of existing water policies even as we have enacted more regulations, expanding the number of regulated entities that financially strapped administrative agencies must oversee. Some solutions to the water resource problems exacerbated by urbanization paradoxically may lead to new problems, further challenging administrative agencies and contributing to additional increases in costs. The financial crisis has increased political divisiveness and inevitably will leave as one of its legacies reduced funding for water resource programs and regulation. When state and local leaders are faced with the imperative of making significant budget cuts, it is doubtful any will argue that inspectors to enforce construction site erosion control regulations are more important than police officers. Recognition of the inevitability of reduced funding for water resources will place additional stress on the newer collaborative processes, particularly those in watershed management, for it is precisely over the problems of financing and project implementation that these processes have foundered.

How can the many diverse interests engaged in the implementation of water resource policy in Minnesota make lemonade from the budgetary lemons that will be harvested in the coming years? There are no simple recipes, but elaborating on some of the principles outlined in this book with more specific strategies seems relevant and appropriate. Consistent with the principle of informed decisionmaking, one strategy is to explicitly recognize the economic importance of Minnesota's water resources. Governor Pawlenty did this early in 2009, when he asserted that the fishing industry alone adds $5 billion annually to the Minnesota economy. As the financial stakes grow higher, policy issues increasingly will be influenced by financial considerations. Policymakers and advocates interested in sustainable water resource management will be more effective if they can make economic arguments for investing in water infrastructure and programs.

A second strategy is transparency, being explicit about the magnitude of the financial challenge. Without transparency about costs, state and local leaders cannot make informed decisions, and citizens cannot hold them accountable. Reports like the Clean Watersheds and Drinking Water Infrastructure Needs Surveys often do not draw significant attention, yet constituencies are emerging to focus on these needs. The City Engineers Association of Minnesota, Minnesota County Engineers Association, Minnesota Public Works Association, and League of Minnesota Cities are collaborating to call attention to the condition of the state's infrastructure (Eggum 2009). By engaging other advocates for water quality, these organizations can increase the effectiveness of communications with officials, thus helping ensure that investments in water infrastructure and resources remain as effective as possible.

A third strategy is to continue to plan, which is the first step in implementing a precautionary principle. Minnesota has shown itself to be a leader in producing information to inform policy decisions. The Metropolitan Council's planning processes can be used to guide development and minimize impacts on water resources. Reports such as the EQB report on sustainable water use are illustrative of the types of information elected officials and administrators need to develop plans that recognize the interconnectedness among components of water resource management systems, establish priorities, and inform inevitable trade-offs among conflicting needs. These plans need to address challenges such as the 2009 drought, as well as emerging problems such as increasing evidence of the public health risk posed by endocrine disruptors and other exotic pollutants in our waters.

A fourth strategy, as suggested elsewhere in this volume, is to pursue policies that increase efficiency of infrastructure investments and programs. This strategy involves seeking opportunities to use markets and pricing to allocate resources as efficiently as possible, creating and providing economic incentives for mitigating the impacts of urbanization, and conducting the research necessary to determine the relative cost-effectiveness of different investments. Examples include, respectively, moving from average to marginal cost pricing for drinking water, granting density bonuses for developments that incorporate conservation design and protect water resources, and increased use of cost–benefit and cost-effectiveness analysis to prioritize alternative projects for watershed restoration. These emphases on efficiency must be balanced with considerations for equity and economic hardship. Although household incomes generally are highest in metropolitan areas, where urbanization will place the greatest stresses on water resources, the costs of achieving water policy objectives will be most burdensome in rural communities with declining populations where legacies of contamination remain.

A fifth strategy is to continue to engage politically in policy processes. Although the challenge of sustainable water management may be daunting, Minnesotans historically have demonstrated their commitment to stewardship of their water resources. Most recently, their commitment emerged in the form of the Clean Water, Land, and Legacy Amendment, a voter-approved increase in sales tax to fund environmental restoration. One-third of the funds generated by the tax increase—originally estimated to be about $80 million to $90 million—will be spent to "protect, enhance, and restore water quality in lakes, rivers, streams, and groundwater, with at least 5% of the fund spent to protect drinking water sources" (DNR 2009). This infusion of capital will be insufficient to meet all the state's water resource needs, and because the financial crisis has reduced household expenditures, revenues will be lower than originally projected. Nonetheless, the spirit exemplified by passage of the Legacy Amendment is the type of spirit that will be necessary in the future to meet the state's needs. Continuation of this spirit will be essential to sustain urban water management in the new epoch of fiscal austerity.

NOTES

1. The U.S. Office of Management and Budget (OMB) defines an MSA as a geographic entity with "a core urban area of 50,000 or more population" that "consists of one or more counties" including "adjacent counties that have a high degree of social and economic integration (as measured by commuting to work) with the urban core" (www.census.gov/population/www/metroareas/metroarea.html).
2. Minn. Stat. § 115.03, subd. 9.

REFERENCES

DNR (Minnesota Department of Natural Resources). 2009. Clean Water, Land, and Legacy Amendment. www.dnr.state.mn.us/news/features/amendment.html (accessed October 17, 2009).

Duhigg, Charles. 2009. Clean Water Laws Are Neglected, at a Cost in Suffering. *New York Times* Sept. 13. www.nytimes.com/2009/09/13/us/13water.html?fta=y (accessed September 13, 2009).

Durant, R.F., D.J. Fiorino, and R. O'Leary. 2004. *Environmental Governance Reconsidered: Challenges, Choices, and Opportunities.* Cambridge MA: MIT Press.

Eggum, Thomas. 2009. State of MN Infrastructure. *Minnesota Cities* (August):4–5.

EPA (U.S. Environmental Protection Agency). 2009. Drinking Water Infrastructure Needs Survey and Assessment. Fourth Report to Congress. Office of Water, Office of Ground Water and Drinking Water, Drinking Water Protection Division, Washington, DC. www.epa.gov/safewater/needssurvey/pdfs/2007/report_needssurvey_2007.pdf (accessed January 7, 2011).

Kettl, Donald. 2002. *Environmental Governance. A Report on the Next Generation of Environmental Policy.* Washington, DC: Brookings Institution Press.

Kraft, Michael. 2009. Cleaning Wisconsin's Waters: From Command and Control to Collaborative Decision Making. In *Toward Sustainable Communities: Transition and Transformations in Environmental Policy,* 2nd ed. pp. 115–140. Edited by D. Mazmanian and M. Kraft. Cambridge, MA: MIT Press.

Lindsey, G. 1995. Managing Implementation of Environmental Programs: The Case of Erosion and Sediment Control. *Public Productivity and Management Review* 18 (3):247–261.

Lubell, Mark, William D. Leach, and Paul A. Sabatier. 2009. Collaborative Watershed Partnerships in the Epoch of Sustainability. In *Toward Sustainable Communities: Transition and Transformations in Environmental Policy,* 2nd ed. pp. 255–288. Edited by D. Mazmanian and M. Kraft. Cambridge, MA: MIT Press.

Marshall, Julian. 2007. Urban Land Area and Population Growth: A New Scaling Relationship for Metropolitan Expansion. *Urban Studies* 44 (10):1889–1904.

Mazmanian, Daniel A., and Michael E. Kraft. 2009. *Toward Sustainable Communities: Transition and Transformations in Environmental Policy.* Cambridge, MA: MIT Press.

MDH (Minnesota Department of Health). 2009. Drinking Water Revolving Loan Fund General Information. www.health.state.mn.us/divs/eh/water/dwrf/index.html (accessed October 8, 2009).

Minnesota Management and Budget. 2009. *General Fund, Fund Balance Analysis, End of 2009 Legislative Session,* June 11, St. Paul.

MPCA (Minnesota Pollution Control Agency). 2008. *Future Wastewater Infrastructure Needs and Capital Costs: Report to the Legislature.* St. Paul: MPCA.

———. No date. *2008 Stormwater Infrastructure Needs & Costs. Preliminary Results and Findings of the Minnesota Pollution Control Agency (MPCA) 2007 Stormwater Infrastructure Needs Survey.* St. Paul: MPCA.

Northland Outdoors. 2009. www.northlandoutdoors.com/index_articles.cfm?id=23680& property_id=3 (accessed September 5, 2009; site discontinued).

Smetanka, Mary Jane. 2009. Sublime to Slime. *Minneapolis Star Tribune* Sept. 15, B1, B8.

U.S. Census Bureau. 2005. Table A1: Interim Projections of the Total Population for the United States and States: April 1, 2000 to July 1, 2030. www.census.gov/population/projections/SummaryTabA1.pdf (accessed January 7, 2011).

———. 2008. 2007 Data Source: U.S. Census Bureau, Small Area Estimates Branch. Washington, DC: U.S. Census Bureau.

Van Buren, Princesa, and John Wells. 2007. *Use of Minnesota's Renewable Water Resources: Moving toward Sustainability.* Minneapolis: Minnesota Environmental Quality Board. www.eqb.state.mn.us (accessed September 11, 2009).

Vig, Norman J., and Michael E. Kraft. 2006. *Environmental Policy: New Directions for the Twenty-First Century.* 6th ed. Washington, DC: CQ Press.

CHAPTER 10

Navigation, Dredging, and Protection: The Checkered History of Channel Maintenance

Jerry Fruin and Richard D. Stewart

A lthough Minnesota is part of the Upper Midwest, located in the center of the North American continent, waterborne commerce is very important to the region's economy. Combined, the Minnesota ports on the Mississippi River system would rank in the top 50 of all U.S. ports by tonnage of cargo handled. In 2007, Duluth–Superior ranked 16th among U.S. ports in tonnage handled, Two Harbors ranked 46th, and Silver Bay ranked 73rd.

Minnesota is served by two great waterways: the Mississippi River system and the Great Lakes–St. Lawrence Seaway system. The Mississippi River has a federal 9-foot navigation channel maintained by the U.S. Army Corps of Engineers (USACE). The channel starts at the head of navigation (river mile 857) above the St. Anthony Falls in Minneapolis and continues until it reaches the 45-foot ocean vessel channel at Baton Rouge, Louisiana. The Mississippi River is the backbone of a navigation system that connects the 9-foot channels of the Illinois, Ohio, Missouri, Tennessee, Arkansas, and other rivers. The system allows navigation companies to provide reliable, low-cost transportation to inland cities such as Pittsburgh, Chicago, St. Louis, Memphis, Kansas City, and Catoosa, Oklahoma. The 45-foot navigation channel on the lower river allows access for oceangoing vessels to ports as far upstream as Baton Rouge and river barges to the Gulf Intracoastal Canal West Waterway. This allows movement of river barges to Houston and other Gulf ports as far south as Brownsville, Texas. (The Gulf Intracoastal Waterway connects the Mississippi River system with ports as far east as the Florida panhandle, but this area is not generally served by inland shallow-draft equipment.)

The Great Lakes–St. Lawrence Seaway (GLSLS) system consists of the five Great Lakes and connecting rivers and channels. It provides a water route to the Gulf of St. Lawrence and the Atlantic Ocean for ships with drafts of up to 26 feet, 6 inches, although in times of high water, ships with deeper drafts sometimes sail the GLSLS.

From Minnesota, ships pass from Lake Superior to Lake Michigan and the lower lakes via the Soo Locks at the St. Mary's River in Sault Ste. Marie, Michigan, which can take ships carrying 72,000 tons of cargo (GLSLS Study 2007).

EARLY USE OF WATERWAYS FOR TRANSPORTATION

The earliest transport mode developed in the United States was the maritime mode. Waterborne transportation has played such an important role in the nation's economic development that 75% of the nation's population lives in communities near the coasts or navigable rivers, canals, or lakes (Becht 1970). Although passenger use is almost exclusively for tourist or recreational purposes today, waterborne transportation originally was the dominant form of long-distance passenger travel and freight movements. Predating European settlement of Minnesota, Native Americans made extensive use of the rivers and lakes. Their routes were followed by Europeans and in some cases became borders as well as corridors of trade. During the early settlement period, major Minnesota towns were ports with scheduled steamboat travel extending as far upriver on the Mississippi as Grand Rapids. Boats navigated on the Minnesota, St. Croix, Red, Rainy, and many other rivers in Minnesota. In addition to river navigation, the major inland lakes such as Rainy, Lake of the Woods, Minnetonka, and Bemidji had extensive commercial vessel operations.

The Great Lakes and the rivers of the interior, including the Mississippi River system, provided the earliest routes for Europeans to tap the natural resources of the Midwest and Canada in the 17th century. Native Americans had used these waterways for transportation and trade for centuries prior. At that time, numerous well-trafficked overland portages in Minnesota and the Upper Midwest connected the rivers with lakes and rivers in adjacent watersheds. Examples include the voyageurs' Grand Portage route and the portage from the Brule to the Snake River, where the Minnesota Historical Society maintains a reconstructed North West Company fur-trading post.

Rivers have always been critical to the transportation system of Minnesota. Fort Snelling was constructed to provide navigation control of the upper ranges of both the Minnesota and Mississippi Rivers. The first steamboat to reach Fort Snelling from downriver was the *Virginia* in 1823. The steamboat *Governor Ramsey* started service upstream between Fort Snelling and Sauk Rapids as early as 1849.

Steamboat navigation began on the Minnesota River about 1850. By 1856, five boats were constantly operating on the river, and in the 10-year period from 1855 to 1865, almost 3,000 arrivals were recorded in St. Paul. Although steamboats went as far upriver as Traverse des Sioux, the practical head of navigation in the 1800s was Mankato, about 100 miles upstream from its mouth. This provided water access to much of southern and southwestern Minnesota (MRBDC 2003). The Minnesota River had a steamboat service to New Ulm until 1871, when the railroads reached the river town and provided faster service (MNHS 2009).

River transportation was also important in the third Minnesota watershed, which flows north to the Arctic via the Red River. Both passengers and freight were moved by steamboat on the Red River from 1859 to 1912. In 1858, after a prize was offered to the first person who could put a steamboat on the Red River, the steamer *North Star* went upstream as far as Sandy Lake, Minnesota, where it was dismantled and moved overland to provide service along the Red River of the North as far downstream as Fort Garry in Winnipeg, Manitoba. Another contestant used floodwaters to sail up the Minnesota River in 1859, but the vessel went aground and was abandoned, though it was later refurbished. In the 1870s, steamboats on the Red traveled regularly between Moorhead and Fort Garry between April and October. Typical one-way trips took about 10 days, with the record time being 5 days and 18 hours. The largest Red River boats could carry nearly 300 passengers and 365 tons of freight (NDSU Institute for Regional Studies n.d.). Although most inland commercial navigation channels for freight today provide 9-foot drafts, this is a 20th-century development. During the early history of navigation on the Upper Mississippi, much of the travel was very difficult during periods of floods and low water. Between 1884 and 1895, the USACE constructed five dams and pools at the headwaters of the Mississippi to provide navigable water during low-water periods. A sixth dam and reservoir were added in 1911.

THE MODERN MISSISSIPPI RIVER SYSTEM

It was not until 1930 that Congress funded the building of 29 locks and dams[1] from St. Louis to the Falls of St. Anthony, an action that created today's trade corridor. Locks and dams are necessary because the head of navigation in Minneapolis is 799 feet above sea level, and the river elevation drops 440 feet by the time it gets to St. Louis. The locks and dams are very important to navigation because the pools behind the dams provide deep water and reduce or eliminate the need for dredging channels for most of the distance. Of the 29 locks and dams between Minneapolis and St. Louis, 10 are in Minnesota. These include the Upper and Lower St. Anthony Falls Dams in Minneapolis and Lock and Dam 1 (the Ford Dam) in St. Paul.[2] Other dams in Minnesota waters are near Hastings, Red Wing, and Winona, Minnesota; and Alma and Lacrosse, Wisconsin.[3]

The Minnesota River has a 9-foot channel from its confluence with the Mississippi to mile 14.7 in Savage and is maintained by the USACE. In addition, a 9-foot channel with restricted widths is authorized to mile 21.8 near Shakopee, Minnesota, and this portion has been maintained by private interests. The Minnesota River terminals are major shippers of grains and soybeans. The Minnesota River has no locks and dams to limit tow size, so tow configuration depends on river conditions.

The St. Croix River has an authorized but little-used 9-foot channel from its mouth to Stillwater, Minnesota. The channel was important for the movement of coal until the shift to unit trains and western coal began in the 1980s.

Cargo shipped on the Upper Mississippi River system is moved in several different types of barges. Dry-cargo covered hopper barges holding 1,500 to 1,600 tons are used for hauling weather-sensitive bulk commodities such as grains and salt. Open hopper barges of similar sizes carry sand, gravel, ores, and coal.

Tank barges hold liquids such as chemicals and petroleum products, and there are also deck barges. Towboats push—not tow—a configuration that is three barges wide by five barges long and transports over 22,500 tons of cargo. More than 12 million tons of cargo was moved by barge on Minnesota rivers in 2007 (Lambert 2009).

NAVIGATION ON LAKE SUPERIOR AND THE GLSLS

The second of the great waterways servicing Minnesota is Lake Superior and the Great Lakes–St. Lawrence Seaway system. The value of the water connection provided by Lake Superior to a large portion of the continent has been recognized since the earliest days of marine navigation. Native American settlements and later fur-trading posts were situated where the inland waterway systems connected to Lake Superior. Trade could move from the St. Croix, Mississippi, or Rainy River directly to the lake by marine craft. Those craft were small, however; it was not until the opening of the locks at Sault Ste. Marie and the development of rail service, linking Duluth and other North Shore harbors to key inland markets, that channelization and harbor improvements began in earnest in Lake Superior ports. Duluth was settled around 1854, and the Soo Locks linking Lake Superior to the lower lakes opened in 1855.[4] The majority of waterborne trade moved through the larger and older city of Superior, Wisconsin, because the mouth of the St. Louis River—the natural entrance to the harbor—was on the Wisconsin side. Much of what is waterfront in present-day Duluth was then marshland.

In 1857, the federal government funded a lighthouse on the end of Minnesota Point, which was the western side of the entrance to Duluth Harbor. In 1873, the first contract for dredging the Duluth–Superior Harbor was awarded by the USACE. Prior to this award, a private firm called the Minnesota Canal and Harbor Improvement Company, financed by the legendary J.J. Cooke, had built a controversial canal crossing Minnesota Point (Beck and Labadie 2004). This privately built second entrance to the Duluth–Superior Harbor later became the gateway for thousands of ships. Public and private docks and channels were developed in Duluth, Two Harbors, Silver Bay, and Taconite Harbor from the 1870s to present. Between 1958 and 1964, the USACE deepened all the federal channels of the major Minnesota ports on Lake Superior from 25 to 27 feet. Smaller lake ports such as Knife River, Grand Portage, and Grand Marais also had harbor improvements and dredging.

In 1959, the St. Lawrence Seaway was completed and allowed ships with cargoes over 60,000 tons to operate on the four upper lakes and ocean ships with cargoes over 25,000 tons to sail as far inland as the Duluth–Superior Harbor at an elevation of 601 feet. Ships can sail some 2,300 miles to the Gulf of St. Lawrence in 8.5 days and to northern Europe in 15 days.

The St. Lawrence Seaway is a binational system operated jointly by the United States and Canada that allows access to the Atlantic Ocean. It extends from Montreal to Lake Erie and is made up of two segments. The Montreal–Lake Ontario segment is located partially in Canadian and partially in international waters; it includes a series of seven locks (five Canadian, two U.S.). It enables ships to navigate 190 miles between the lower St. Lawrence River (elevation 20 feet) and Lake Ontario (elevation 243 feet). The Welland Canal segment is 36 miles long, located in Canada, and includes eight locks. It connects Lake Ontario and Lake Erie (elevation 569 feet). Unlike U.S. waterways, which historically have not charged tolls, Canadian Seaway locks do have tolls, which are a major source of revenue for their operation.

There are three general categories of deep-draft cargo vessels on the lakes: ocean vessels, or "salties," and Canadian and U.S. "lakers." Salties enter the lakes with cargo, generally manufactured goods but sometimes bulk commodities such iron ore. Recently, salties have used the route to move oversize items such as wind turbine components to Minnesota and oilfield equipment to Canada. Because the locks are 766 feet long by 80 feet wide, with a depth of 30 feet, salties are limited to maximum dimensions of 740 by 78 by 26.6 feet in the lakes, and their cargo capacity is limited to 18,000 to 25,000 tons.

Canadian lakers have hulls that are specially designed to maximize the amount of cargo (about 34,000 tons) that they can transport through the lower locks. Because of the cabotage provisions of the U.S. Jones Act, these Canadian-owned vessels can move cargo only among Canadian ports or between a U.S. and a Canadian port. They are not allowed to take goods between two U.S. ports but are frequently used to take grain from Minnesota for transshipment from Canadian ports beyond Montreal. (Laker hulls are designed for the wave conditions on the Great Lakes rather than the ocean, so they generally do not go beyond the Gulf of St. Lawrence.)

The U.S. lakers are larger than either salties or Canadian lakers. The largest were built to maximize cargo capacity and can carry more than 60,000 tons when loaded to their maximum draft. Vessel dimensions preclude the largest vessels from transiting the Welland Canal, but they are able to transit the Soo Locks, thus limiting their use to the four upper lakes: Superior, Michigan, Huron, and Ontario. These massive vessels are employed primarily to transport western coal that is transshipped by rail at the Midwest Energy Resources terminal in Superior, Wisconsin, to lakeside power plants in the Midwest, as well as taconite from Lake Superior ports to steel mills on the lower lakes.

ENVIRONMENTAL AND ECONOMIC ADVANTAGES OF WATER TRANSPORT

Waterborne transportation has many advantages over other transportation modes. It is the lowest-cost mode compared with truck, rail, or air transport for many types of cargo. The large capacity of a vessel or barge (a single barge can carry as many as 15 jumbo railcars or 60 semi trucks) means fewer trips, less traffic congestion, and lower risk of accidents and collisions. Water typically is the most energy-efficient mode of transport and has the lowest fuel consumption per ton-mile. As shown in Table 10.1, a national study by the Texas Transportation

Table 10.1 *Transportation mode environmental characteristics*

Parameter	Transport mode		
	Highway truck	Rail bulk car	River barge dry bulk
Cargo capacity (tons)	25	110	1,750
Ton miles per gallon of fuel	155	413	576
Emissions of HC (grams per ton-mile)	0.02	0.024	0.017
Emissions of CO (grams per ton-mile)	0.136	0.064	0.046
Emissions of NO_2 (grams per ton-mile)	0.732	0.654	0.469
Emissions of PM (grams per ton-mile)	0.018	0.016	0.012
Fatalities per billion ton-miles	4.351	0.649	0.028
Injuries per billion ton-miles	99.044	5.814	0.0045
Major oil spills (gallons per million ton-miles)	6.06	3.66	3.6

Notes: HC = hydrocarbons, CO = carbon monoxide, NO_2 = nitrogen dioxide, PM = particulate matter.
Source: TTI (2007).

Institute determined that inland towing moves 576 ton-miles on a gallon of fuel, compared with 413 ton-miles for rail and 155 ton-miles for large trucks. This lower energy consumption results in less air pollution. The table also shows that river barge transportation has lower fatality and injury rates and fewer significant oil spills per ton-mile than trucks or rail transport. And because water routes are generally removed from population centers, waterborne movements reduce highway and rail congestion and noise (TTI 2007).

Fruin and Fortowsky (2004) demonstrated the economic and environmental impacts of water transportation by a simulation of the closure of the uppermost reach of the Mississippi River, an area that included the Minneapolis Upper Harbor area. They estimated the monetary costs and public externalities that would be imposed by the resulting "modal shift" from barge to truck if the area above the confluence with the Minnesota River were closed to barge traffic. Impacts that were estimated included additional transportation costs, increases in fuel consumption, changes in air emissions, highway congestion impacts, highway accident impacts, and changes in highway maintenance requirements. Results from the "most likely" scenario indicated that the closure would require an additional 66,000 truck trips totaling 1.2 million miles in the Minneapolis–St. Paul metro area each year, while increases in transport costs to shippers and customers would exceed $4 million annually. Additional costs to the public were computed using estimates of pavement maintenance, congestion, crash, air pollution, and noise cost coefficients from the Federal Highway Administration Highway Cost Allocation Studies (FHWA 2000). These costs totaled more than a $1 million per year, of which more than $100,000 were environmental.

NAVIGATION AND DREDGING: HISTORY AND LEGISLATION

Since the earliest days of the nation, the USACE has been charged with developing and maintaining navigable waterways. In 1824, it was placed in charge of inland navigation, and the decades that followed saw a surge in canal building and channelizing rivers and harbors to improve navigation and allow safe access by commercial and recreational vessels. Prior to development of the rail system, an extensive canal system was designed and built to connect the navigable rivers and lakes of the United States (Shank 1982). Benefits of barge transportation over horse and wagon were such that the U.S. Congress and state governments funded canals through mountain passes. Until rail provided a more economic mode, many mountain canals were profitable operations (LeRoy 1980). Many lower-elevation canals continue to operate.

A gravity canal system requires a steady supply of water to operate the locks; this can necessitate building dams with pools for a regular water supply. The dams often served the secondary purposes of flood control in wet seasons and providing water resources during periods of drought. Developing navigation systems was considered an essential infrastructure in nation building, and Minnesota successfully lobbied for federal assistance to have its marine infrastructure developed. After completion of the original channel improvements, maintenance dredging was required to remove sediment.

Principal Federal Laws Pertaining to Navigable Waters and Dredging

The principal federal laws impacting dredging include the 1899 River and Harbors Appropriations Act, the 1969 National Environmental Policy Act (NEPA), the 1972 Clean Water Act (CWA), the 1977 Marine Protection Research and Sanctuaries Act, the Endangered Species Act (ESA), the 1958 Fish and Wildlife Coordination Act, the 1966 National Historic Preservation Act, the Ocean Dumping Acts of 1973 and 1977, and the 1990 Oil Pollution Act. The policy significance of this collection of laws is that any dredging operation is subject to scrutiny by a number of federal agencies and a public review process. Because dredging has impacts on so many parties, and the federal and state regulations have evolved over the past 200 years, the institutional dredging process has been described as "complex, cumbersome, unpredictable and fragmented" (NRC Marine Board 1985, 77).

The term "navigable waters" has been defined differently by admiralty courts, state courts, and nonadmiralty federal courts (Mangone 1997). The legal definition of navigable waters is important because the 1899 U.S. River and Harbors Appropriations Act gave authority to the USACE to grant permits for any building or change of harbors or deposits "in any navigable waters." Although originally intended to ensure free movement of commerce, the concept of navigational servitude has been applied to dams, flood control, water pollution, and any other activity that might affect navigable waters. Navigation servitude for the USACE applies to past, present, and future interstate or international trade.

In 2002, the U.S. Environmental Protection Agency (EPA) attempted to broaden the definition of navigable waters as it applies to dredging for use in the CWA. The 2002 definition of navigable waters was vacated by the U.S. District Court in 2008, however, and the original 1973 regulatory definition of navigable waters was restored.[5] The currently approved regulatory definition of navigable waters for EPA is as follows:

(1) all navigable waters of the United States, as defined in judicial decisions prior to the passage of the 1972 Amendments of the Federal Water Pollution Control Act, (FWPCA) (Pub. L. 92-500) also known as the Clean Water Act, and tributaries of such waters as;
(2) interstate waters;
(3) intrastate lakes, rivers, and streams which are utilized by interstate travelers for recreational or other purposes; and
(4) intrastate lakes, rivers, and streams from which fish or shellfish are taken and sold in interstate commerce.

Additional federal permits are required when proposing dredging or flood control activities that may impact navigable waters, including Coast Guard Bridge permits, as part of Section 10 of the Rivers and Harbors Act, and Federal Energy Regulatory Commission permits. A National Pollutant Discharge Elimination System (NPDES) issued by EPA may also be needed.

States that have attempted to apply public trust rights to navigable waters have not always been successful, because some courts have held that waters made navigable through private dredging may not provide navigational servitude.[6] The state of Minnesota has ownership of the bed of a navigable waterway "below the natural ordinary low water level." In Minnesota, the courts have applied the federal test of navigability when determining the state's rights to riverbeds. A minority of the state's nearly 12,000 lakes and more than 5,500 streams are considered navigable waters by the USACE (n.d.).

The Dredging Permitting Process

In order to comply with Section 10 of the Rivers and Harbors Act of 1899, the USACE created a waterway permit program. Under that regulation, a USACE-issued permit is required to do any work in, over, or under a navigable water of the United States. Activities such as dredging and construction of docks, bulkheads, and utility lines require review under Section 10 of the act. Dredging is termed by the USACE as either capital or maintenance dredging. Capital dredging, also known as new-work dredging, is used to create new channels, whereas maintenance dredging is for maintaining existing navigation channels at the authorized depth or to assist in keeping waterways open for flood control. The state of Minnesota works with the USACE Detroit District for Lake Superior dredging and with the St. Paul District for non-Great Lakes dredging, including the Mississippi River.

The USACE began regulating more than traditional navigable waters of the nation when Congress passed the Federal Water Pollution Control Act (FWPCA) Amendments of 1972. The FWPCA was amended in 1977 and became the Clean Water Act. Under Section 404 of the CWA, the USACE regulates all "waters of the US," which include jurisdictional wetlands, ponds, streams, and lakes, including those on private land. Congress also gave EPA oversight of the Section 404 regulatory program and required the agency to develop what has become the Section 404(b)(1) guidelines. In general, the guidelines require that an activity be the least environmentally damaging alternative that is feasible, and that adverse impacts be avoided, minimized, then compensated (mitigated) for, such as by creating or restoring wetlands to replace those that would be filled. The USACE determines whether proposed activities will be contrary to the public interest. Certain discharges for farms, forests, maintenance activities, and other purposes are exempt from Section 404 regulation. Exempt discharges must be for defined purposes and satisfy certain conditions outlined in the guidelines.

The USACE has a nonreporting, general permit authorization for some minor activities. Other general permits can apply if the proposed work also requires authorization from the state Department of Natural Resources (DNR) or the Minnesota Wetlands Conservation Act (WCA). In 2000, the USACE St. Paul District implemented a combined statewide regional permit and letter-of-permission evaluation to replace the nationwide Section 404 permits. The district still uses Section 10 but not Section 404 permits in Minnesota, whereas the Detroit District still uses Section 404 permits. This means that a Minnesota entity that needs to apply for a permit must first determine which district has jurisdiction over the Minnesota body of water for which a permit is needed.

Minnesota Laws and Regulations Related to Dredging and Flood Control

According to the CWA, anyone who wishes to obtain a federal permit for any activity that may result in a discharge into U.S. navigable waters must first obtain a state Section 401 water quality certification to certify that the project will comply with relevant state water quality standards (MPCA 2010). In Minnesota, a state permit issued under authority of Section 401 of the CWA is required in addition to the 404 permit. For example, if someone proposes to discharge dredge or fill material into U.S. navigable waters in the state of Minnesota, including wetlands, he or she must obtain a Section 404 permit from the USACE and a Section 401 water quality certification from the Minnesota Pollution Control Agency (MPCA).

An applicant applying for a dredge permit in the state of Minnesota must contact either the relevant local unit of government for a combined project application form or the appropriate USACE district office. The applicant submits the proposed project location and information to satisfy the federal and state water quality requirements to the USACE. The district evaluates the project and, if an individual permit is warranted, incorporates the proposed project into a public notice. The public notice announces official receipt of the application, describes the project, and serves as the official notice for Section 401 water quality certification. "If the project proposal qualifies for a Corps General Permit or Letter of Permission (GP/LOP) which the MPCA has pre-certified, no further certification action by the MPCA is required" (Stollenwerk et al. 2009).

Both the 401 and 404 permits are subject to public review and comment. The review and comment process can be extensive and time-consuming, depending on the proposed project. Any conditions required to meet water quality standards included in Section 401 certification become conditions of the 404 permit. That means that if MPCA denies the 401 water quality certification, the USACE must deny the 404 permit. After MPCA grants a 401 certification, the USACE completes a public interest review before granting or denying the 404 permit. The interactive process means that there must be close communication and cooperation between agencies, and that permit processors must be aware of the need to include authorization from both state and federal agencies. The depth of state agency involvement in the permit process may change in the future, depending on state budgets, but the need for close interaction will remain.

Projects that may impact Minnesota's water resources may also be regulated by the Minnesota Department of Natural Resources (DNR). DNR water use permits are required for all users withdrawing more than 10,000 gallons of surface water or groundwater. Any projects below the ordinary high-water level (OHWL) that alter the course, current, or cross section of public waters or public wetlands require a permit. These projects may also be required to obtain an aquatic plant management or fisheries-related permit, or both. Minnesota does not use the OHWL established by USACE, but uses its own definition, as described in Minnesota Statutes Section 103G.005, subdivision 14. The statute states that the OHWL is the highest water level maintained for a sufficient period of time to leave evidence on the landscape. This is commonly the point where the vegetation changes from aquatic to terrestrial.

In addition to Minnesota state permits, local permits or an environmental review, or both, may be required under the appropriate soil and water conservation district, tribal, county, or local unit of government. An example of how the complex guidelines for dredging of navigable waters can be cooperatively developed by multiple government agencies is detailed in procedures related to dredging in the Minnehaha Creek watershed, which includes the navigable waters of Lake Minnetonka (DNR/MCWD/LMCD 1993). The board of managers has the following multiple guidelines to consider:

- natural resource and open space plans;
- capital improvement projects and load reduction to impaired waters;
- education communication goals;
- community redevelopment; and
- transportation planning.

Obtaining a dredging permit from the board requires successfully completing six steps, and if hydraulic dredging is undertaken, five additional criteria need to be met (MCWD Board of Managers 2005).

Dredging projects and related permits may be denied because of financial reasons, not just environmental issues. In 2003, the city of Tower in northeastern Minnesota applied for a permit to dredge and restore the East Two Rivers navigable channel. The channel was first created in 1896 and improved in 1913 and 1939, but it had filled with silt, making it useful for only shallow-draft boats. The city wanted to restore the river to historic depths, relocate and modify highways, and provide a marina for boaters. A feasibility study was initiated to determine the viability of restoring navigation to the East Two Rivers. The study indicated a positive cost–benefit ratio, but most benefits were derived from non-navigation sources. The project was denied in 2004.

DREDGING: ENVIRONMENTAL ISSUES AND MANAGEMENT

Dredging not only affects the immediate area, but also disturbs fines (small particulate matter), changes water flow, and negatively affects aquatic life. For example, disturbed fines may remain in suspension for some time, reducing plant photosynthesis or settling on fish spawning grounds and smothering eggs. Numerous studies have documented impacts of dredging, and the issues are generally well understood. However, each dredging operation is unique in terms of hydrology, sedimentation, and topography; the composition of dredged material varies depending on the site from which it was dredged and the type of dredging equipment in use. This means that each proposed operation must be evaluated for potential adverse impact.

Over a five-year period in the 1970s, the USACE spent $30 million on a Dredge Material Research Program (DMRP). The DMRP was a collection of more than 250 studies, conducted by numerous scientists from a wide variety of disciplines. The program was designed for national application and included research on all major types of dredging activities, all regions of the country, and many environmental settings. The research resulted in cost-effective methods and guidelines for assessing and minimizing impacts of conventional disposal alternatives. The importance of ongoing research in dredging issues was recognized by the creation of the Dredging Operations and Environmental Research (DOER) Program, which supports the USACE Operation and Maintenance Navigation Program.

The DMRP found environmental problems, especially in areas with contaminated materials, and proposed a number of options to address them. These options were termed "mitigation sequencing" and are now required under Section 404 of the CWA. Mitigation sequencing includes avoiding dredging and disturbing wetlands, minimizing impacts of projects on wetlands, and mitigating unavoidable damage with establishment of new wetlands (compensation) in a nearby area.

Past dredging activities that have affected the watershed and immediate area of the dredged site may continue to cause problems. Proposed dredge areas in or adjacent to historically polluted sites may have dredged material that is contaminated (material that has been demonstrated to cause an unacceptable adverse effect on human health or the environment). However, a 1999 study found that approximately 90% of the dredged material in the United States was uncontaminated (Winfield and Lee 1999).

Contamination may include chemicals, heavy metals, exotic species, or any other substance that can harm humans or the ecosystem. Disposal of dredged material presents challenges, but the inclusion of contaminated material creates significant difficulties. For example, dredge materials in the Port of New York were found to be contaminated with dioxin and could not be capped or contained within the harbor limit. This required shipment to an approved inland site in Utah thousands of miles away, at a cost 12 times that typical for dredge material disposal (Stewart 1997). Containment of dredged material does not automatically mean the material is contaminated. Material may be contained for a variety of reasons, including the potential of future beneficial reuse. The decision to confine dredge material is a matter of the relevant laws of the government or governments in the dredge area.

On average, approximately 4 million cubic yards of sediment are dredged annually from the Great Lakes. About half of the material removed each year is considered polluted or otherwise unsuitable for open-water disposal and therefore placed in confined disposal facilities (CDFs). An estimated 32% of the material removed each year is disposed in the open lake, 12% is used in beach nourishment projects (hydraulically pumping clean sandy sediment onto eroded beaches for the purpose of restoration), and the balance is used in other beneficial use projects, such as habitat creation, enhancement, and mine land reclamation. Because of environmental concerns, Minnesota restricts or prohibits open-water disposal under state law (GLDT 2005).

Containment of contaminated dredge material is governed by several federal laws: the Marine Protection Research and Sanctuaries Act, the Endangered Species Act of 1973, the Clean Water Act, and the Comprehensive Environmental Response, Compensation and Liability Act (CERCLA) of 1980, as amended by the Superfund Assessment and Reauthorization Act (SARA) of 1988. Depending on the level of contamination, there may be a wide variety of acceptable and effective dredge management techniques, including the option of no action (Herbich 2000).

In the Great Lakes region, dredge material is either placed in open-water disposal sites or confined disposal facilities (CDFs), site-specific locations designed to be on the water or in upland locations. CDFs are usually diked areas used to contain dredged material. Any open-water disposal projects in the state of Minnesota must comply with the Minnesota DNR Lake Superior Coastal Program, which provides guidelines for in-water projects that excavate or fill below the OHWL. In 1998, Duluth–Superior Harbor completed its Dredging Materials Management Plan (DMMP) and identified dredged materials management options for the harbor through 2018. The DMMP addressed three management methods:

- beach nourishment;
- continued use of the Erie Pier confined disposal facility, including removal of clean material from the CDF for beneficial uses; and
- placing dredged material in the five deep holes within the harbor.

The most cost-effective option is disposal of the material in the deep holes; however, both the states of Wisconsin and Minnesota have generally prohibited in-water disposal. Yet in some cases, Minnesota is receptive to allowing in-water disposal, provided that habitat enhancement or creation would result.

The Erie Pier CDF contains dredge material from channels in both Minnesota and Wisconsin. The CDF is on the Minnesota side of the St. Louis River, but most decisions regarding the facility are made through a bistate process that involves governmental agencies from both states, the USACE, the Duluth–Superior Metropolitan Interstate Council, and the Harbor Technical Advisory Committee (HTAC). The mission of HTAC is to do the following (DSMIC 2009):

- provide a forum for the discussion of issues and concerns pertaining to the Duluth–Superior Harbor;
- promote the harbor's economic and environmental importance to the community; and
- provide planning and management recommendations to the Metropolitan Interstate Council regarding the harbor.

The life of a CDF can be extended by removing dredged material for other beneficial uses. This is commonly termed "borrow material," which refers to soil or sediment taken from a site for use in structure construction, such as sandy sediment dredged and pumped to restore an eroded beach or clay taken to build a levee or dike. At the Erie Pier CDF in Duluth, the USACE uses "soil washing" to separate sediment particles by size (GLDT 1999). Approximately 20% to 25% of the original dredge material is available after washing for beneficial use. To reclaim borrow material from dredged material in a CDF so that it can be applied to beneficial use, the debris and trash must be separated from the dredged material, which is technically and economically challenging (Myers and Adrian 2000). In the Erie Pier CDF, the majority of material typically is composed of fines that are not contaminated but do not have current market value. Recent studies supported by the Great Lakes Maritime Research Institute have examined alternatives including reducing transportation costs and seeking new markets (Chen 2009).

RECREATIONAL AND COMMERCIAL VESSELS: ISSUES IN USING A SHARED RESOURCE

Recreational watercraft is a major beneficial use of Minnesota navigable waters. In many cases, such as the prior example of Tower, Minnesota, waterways that previously had commercial uses now seek access to dredging funds for recreational purposes. Minnesota has nearly 900,000 registered boats, the largest number per capita in the United States. In fact, there is about one boat for every six people in the state (DNR 2009). Boating is the most popular and highly valued recreational activity on the Upper Mississippi River System (UMRS). Effects of recreational boating were assessed by the USACE as part of the Upper Mississippi River–Illinois Waterway Navigation Study. The USACE estimated in 1999 that the Upper Mississippi had 6.9 million boater-days per year, 2.6 million boat trips per year, 600 developed boat access sites, 18,000 marina slips, and 217,364 recreational boat lockages. The USACE forecasts that by 2050, more than 600,000 boaters per year will use pool 3, the area that extends from Lock and Dam 3 near Hager City, Wisconsin, upstream to Lock and Dam 2, located near Hastings, Minnesota (Wilcox 2001).

The USACE found that recreational boats had significant effects on the environment from wake waves, propeller jets, noise, and exhaust. These in turn have secondary impacts, including sediment suspension and bank erosion, as well as biological effects on aquatic plants and disturbance of fish and wildlife.

The impacts of recreational boat wakes are a major issue in the Upper Mississippi River. The Minnesota DNR reported in 2004 that "because of multi-jurisdictional authorities along the considerable extent of the UMRS, and in many cases a lack of water surface use regulatory authority, a system-wide solution will likely not occur within a timeframe that will protect against further impairment of the values and uses of the Mississippi River" (DNR 2004). The DNR report sees a lack of coherent multistate policies as the principal impediment to stemming damage from recreational boating. Recreational boat owners who live in Illinois can launch their Illinois-registered craft on the Wisconsin side of the Mississippi River, travel to the Minnesota shore and cause high wake damage there, and not be subject to any regulation of their activities in regard to this damage. The policy issues are significant. The DNR proposed a multijurisdictional solution in its 2004 report but expressed little optimism that the suggestions would be adopted.

The cold water, large waves, and limited safe harbors on Lake Superior deter some recreational boat users. However, a 1989 study found that even before the advent of the global positioning system (GPS) and improved navigation systems, about one-third (34%) of Minnesota boat owners (100,000 pleasure boat owners) had operated or been a passenger on a boat on Lake Superior at least once between 1983 and 1987 (Lime et al. 1989). The Great Lakes Commission completed a study in 2000, which estimated that recreational boating spending in Minnesota was nearly $515 million (Thorp and Stone 2000). Since those studies were completed, new marinas and small-boat safe harbors have been built at Silver Bay and McQuade Road, allowing access along Minnesota's Lake Superior shoreline. These marinas were built to further encourage the safe use of Lake Superior by recreational boaters. Use of recreational boats on the Great Lakes has the same impacts and policy issues as does their use on the UMRS.

Although commercial and recreational vessels use the same waterways, traffic congestion is not a problem, and navigation laws exist to address vessel operations so as to mitigate conflicts. One of the policy issues facing both recreational and commercial users of navigable waters is the vital ability to connect to the shore. In the case of recreational craft, this could mean the ability to launch and retrieve vessels or park autos and trailers. Normal operations of commercial vessels require adequate depth of water, rail and road access, and the ability to operate heavy equipment during loading and unloading.

The accelerating development of urban waterfronts for non-navigation purposes is raising questions about the appropriate balance among competing uses. Urban governments seeking tax income from vacant waterfront land often encourage development that is not compatible with commercial and recreational navigation. Gentrification of the waterfront can reduce or eliminate commercial uses such as cruise line terminals, marinas, and docks. Construction of residential or retail buildings near commercial operations can also block infrastructure access to the waterfront as rail rights-of-way disappear and movement of heavy and oversize cargoes becomes difficult if not impossible. Planning efforts at the city, regional, and state levels often do not take into consideration future development of marine transportation as part of the comprehensive planning process.

The Duluth–Superior Metropolitan Interstate Council has an excellent example of attempting to involve as many stakeholders as possible in the planning process through the Harbor Technical Advisory Committee (HTAC) (DSMIC 2009). HTAC is a bistate process that helps address many cross-border and cross-jurisdictional policy problems that arise when working on navigable water issues. One of the issues relating to navigation and gentrification is the doctrine of public trust. States are charged with upholding the public trust in regard to navigable waterways. Wisconsin's public trust provisions in its state constitution have historically been supported by the Wisconsin Supreme Court. Minnesota does not have public trust provisions as well defined as Wisconsin's, but it does have a number of statutes and rules that address navigable waterways and public trust issues (DSMIC 2005). Minnesota law requires the state to intervene to protect public rights in commercial or recreational use of navigable waters. In Minnesota, the DNR is the state agency charged with enforcing the public trust doctrine. Application of this doctrine can preserve lands for the public at large and provide additional opportunities for public use of navigable waterways, as well as offer more efficient use of federally designated shipping channels, which all federal taxpayers pay to maintain. The public trust doctrine can be and has been applied to stop gentrification of the waterfront to protect the people's right of access for navigation.

CONFLICTS AND CONTROVERSIES

Virtually all navigable waterways have multiple uses leading to water policy issues that affect navigation besides dredging. Some have been sources of conflict and controversy. For example, conflicts may develop between commercial fishing and recreational boating or increased shoreline erosion or noise levels. These conflicts are generally local and resolved locally. In recent years, noteworthy national controversies have arisen over whether navigation improvements such as locks and dams, channel maintenance, or harbor deepening should be funded from general revenues or by user charges.

Because of U.S. dependence on water transport, policy here since colonial days has been that rivers should be "forever free;" tolls were anathema to policymakers. This policy has origins in classical Roman law and was reaffirmed by the Magna Carta in 1215. After the American Revolution, the founding fathers moved quickly to ensure the public right to navigate on all navigable rivers and streams. The very first law passed by the U.S. Congress said navigable waters shall be common highways and be forever free to the public, without any tax, impost, or duty therefore (NORS 2010). Consequently, virtually all navigation improvements—including harbor development and maintenance at ocean ports and on the Great Lakes, the locks and dams and channels on navigable rivers, and the U.S. portion of the St. Lawrence Seaway—have been financed with general revenues.

That remained true until late in the 20th century. In the 1970s, a coalition of competing transportation firms, environmental groups, and budget hawks challenged—and halted—construction of a replacement for Lock and Dam 26 on the Mississippi River above St. Louis. The eventual compromise from that action, which allowed construction of a new Lock and Dam 26, resulted in the Inland Waterways Trust Fund Act of 1978 and partially changed the "forever free" policy. This act included a specific tax on fuel used for shallow-draft commercial navigation on most navigable rivers. Proceeds of this $0.20/gallon tax are retained in a trust fund dedicated to paying half the costs of future river improvements, such as constructing replacement locks and dams. The fuel tax is a user fee on barges and is not applicable to fishing boats, passenger vessels, deep-draft oceangoing vessels, or government vessels on official business. River channel maintenance and dredging continue to be financed from general revenues.

Harbor channel maintenance historically was accepted as a federal responsibility and funded by general U.S. Treasury revenues. This changed after much debate when Congress passed the 1986 Water Resources Development Act and created the Harbor Maintenance Trust Fund to pay for a portion of deep-draft channel maintenance dredging at ocean and lake harbors. Revenue for the Harbor Maintenance Trust Fund was to be generated by assessing a fee, the harbor maintenance tax (HMT), on the value of export, import, and domestic cargo moving through the nation's deep-draft ports. However, the U.S. Supreme Court ruled in 1998 that the tax was not a user fee and was unconstitutional when applied to exports. This ad valorem tax, currently at 0.125%, continues to be charged on imports and domestic cargo but remains controversial; proposals to eliminate or replace it are being considered (AAPA 2009).

CONCLUSIONS

The waters of Minnesota are valuable resources that provide its citizens with multiple benefits, including recreation, fisheries, municipal and industrial water supplies, sanitation, hydropower, and the transportation of people and goods. Although dredging is often opposed on environmental grounds, maintenance dredging is frequently necessary to maintain or increase these societal benefits. In addition, new projects that require dredging are sometimes justified for economic, safety, or environmental reasons. Historically, the navigable waters of the United States were a federal responsibility, and the U.S. Army Corps of Engineers was charged with their development and maintenance, including dredging to remove sediments, and the Coast Guard with regulating safety. In the modern era, however, a number of additional agencies have a role in managing Minnesota's water resources and are empowered to grant permits, enforce laws, and set standards. These include the U.S. Environmental Protection Agency, the

Minnesota Pollution Control Agency and Department of Natural Resources, other state agencies, local and tribal planning authorities and governments, and interstate agencies.

The recent history of dredging and flood control clearly indicates that the USACE and the state of Minnesota are committed to environmentally sound dredging and placement or management of dredged material as defined by applicable laws and policies. The sometimes conflicting interests of economic development and environmental sustainability will best be served when dredged material placement proceeds according to a well-conceived management plan. The keys to a sound dredging management plan are communication and active involvement of all stakeholders during the entire process.

Minnesota's open-water disposal policy creates tension between federal and state entities. Disposal of dredged material into deep holes in Lake Superior would be the most cost-effective method under existing federal guidelines. Because habitat enhancement or creation with dredge spoils is not the least costly alternative, which is required under the federal standard, the USACE requires additional costs to be paid by the project's nonfederal sponsor. This means that the state must provide partial funding for the dredging process if it is to adhere to its policies. A process is needed that requires all parties involved to develop and implement alternative management options when the least costly alternative is not feasible from environmental or engineering standpoints. The life of a confined disposal facility can be extended by removing dredged material for its application to beneficial uses. The cost–benefit ratio for removing uncontaminated dredge material should be weighed against the long-term costs of building and maintaining new CDFs.

NOTES

1. There are 27 numbered dams, with no dam 23 but both 5 and 5a dams.
2. The lock chambers at the Ford and Upper and Lower Harbor Dams are 400 by 56 feet. Tows through these locks generally are done with two barges.
3. These locks have chambers 600 by 110 feet, large enough to lock 9 barges in a 3-by-3 configuration or a towboat and 6 barges (2 long and 3 wide) at one time. Most tows on the Upper River consist of 15 barges, requiring double lockages.
4. Wikipedia, s.v. "U.S. Soo Locks," http://en.wikipedia.org/wiki/Soo_Locks#U.S._Soo_-locks (accessed October 3, 2009).
5. *United States District Court for the District of Columbia (D.D.C.) in American Petroleum Institute v. Johnson*, 571 F.Supp.2d 165.
6. *Dardar v. LaFourche Realty Co*, 55 F.3d 1082 (5th Circuit, 1995).

REFERENCES

AAPA (American Association of Port Authorities). 2009. Harbor Maintenance Tax. www.aapa-ports.org/Issues/USGovRelDetail.cfm?ItemNumber=891 (accessed January 19, 2010).

Becht, J.E. 1970. *A Geography of Transportation and Business Logistics*. Dubuque, IA: W.C. Brown Company.

Beck, B., and P. Labadie. 2004. *Pride of the Inland Seas: An Illustrated History of the Port of Duluth–Superior*. Afton, MN: Afton Historical Society Press.

Chen, H. 2009. Erie Pier Re-Use Facility Phase II: An Optimized Cost-Effective Strategy for Increased Transport and Handling of Dredged Materials. Great Lakes Maritime Research Institute. www.glmri.org/downloads/2009Reports/affiliatesMtg/Chen.pdf (accessed December 5, 2010).

DNR (Minnesota Department of Natural Resources). 2004. Shoreline and Water Quality Impacts from Recreational Boating on the Mississippi River. Mississippi River Landscape Team. http://files.dnr.state.mn.us/aboutdnr/reports/boating/impacts_mississippi_2004/impacts_mississippi_2004.pdf (accessed April 5, 2010).

———. 2009. Boating. www.dnr.state.mn.us/boating/index.html (accessed December 5, 2010).

DNR/MCWD/LMCD (Minnesota Department of Natural Resources/Minnehaha Creek Watershed District/Lake Minnetonka Conservation District). 1993. Lake Minnetonka Dredging Joint Policy Statement. www.minnehahacreek.org/documents/LakeMinnetonka DredgingJointPolicyStatement.pdf (accessed December 5, 2010).

DSMIC (Duluth–Superior Metropolitan Interstate Council). 2005. Duluth Port Land Use Plan. www.dsmic.org/Default.asp?PageID=435#220 (accessed December 5, 2010).

———. 2009. Harbor Technical Advisory Committee (HTAC). www.dsmic.org/Default.asp?PageID=190 (accessed December 5, 2010).

FHWA (Federal Highway Administration). 2000. Addendum to the 1997 Federal Highway Cost Allocation Study, May 2000. www.fhwa.dot.gov/policy/hcas/addendum.htm (accessed December 5, 2010).

Fruin, Jerry, and J. Keith Fortowsky. 2004. Modal Shifts from the Mississippi River and Duluth/Superior to Land Transportation. University of Minnesota, Department of Applied Economics. Staff Paper P04-10. http://purl.umn.edu/14057 (accessed December 5, 2010).

GLDT (Great Lakes Dredging Team). 1999. *Dredging and the Great Lakes*. Ann Arbor, MI: Great Lakes Commission.

———. 2005. Open Water Disposal of Dredged Materials in the Great Lakes–St. Lawrence River Basin. Ann Arbor, MI: Great Lakes Commission.

GLSLS (Great Lakes Saint Lawrence Seaway) Study. 2007. Final Report. www.glsls-study.com/Supporting%20documents/GLSLS%20finalreport%20Fall%202007.pdf (accessed December 5, 2010).

Herbich, John B., ed. 2000. *Handbook of Dredging Engineering*. 2nd ed. New York: McGraw-Hill.

Lambert, Richard. 2009. Personal communication between Richard Lambert, director of Ports and Waterways Section, Minnesota Department of Transportation, and the authors. October.

LeRoy, E.D. 1980. *The Delaware and Hudson Canal and Its Gravity Railroads*. Honesdale, PA: Wayne County Historical Society.

Lime, David W., Leo H. McAvoy, Curtis Schatz, and David G. Pitt. 1989. Recreational Boating on Lake Superior. Research Summaries No. 5. Minnesota Extension Service, University of Minnesota. http://web1.msue.msu.edu/imp/modtd/33840107.html (accessed January 15, 2011).

Mangone, Gerard J. 1997. *United States Admiralty Law*. Boston: Kluwer Law International.

MCWD (Minnehaha Creek Watershed District) Board of Managers. 2005. Revisions Pursuant to Minnesota Statutes § 103D.341, Adopted January 13, 2005. Rule E: Dredging, 2005. http://www.minnehahacreek.org/pdf/AdoptedRuleE-011305.pdf (accessed March 1, 2010).

MNHS (Minnesota Historical Society). 2009. History of Inland Water Transportation in Minnesota. www.mnhs.org/places/nationalregister/shipwrecks/mpdf/inship.html (accessed December 5, 2010).

MPCA (Minnesota Pollution Control Agency). 2010. Clean Water Act Section 401 Water Quality Certifications. www.pca.state.mn.us/water/401.html (accessed December 5, 2010).

MRBDC (Minnesota River Basin Data Center). 2003. Steamboating. http://mrbdc.mnsu.edu/mnbasin/fact_sheets/steamboating.html (accessed October 7, 2009).

Myers, T.E., and D.D. Adrian. 2000. Equipment and Processes for Removing Debris and Trash from Dredged Material. Dredging Operations and Environmental Research Technical Notes DOER-C17. http://el.erdc.usace.army.mil/elpubs/pdf/doerc17.pdf (accessed December 5, 2010).

NDSU (North Dakota State University) Institute for Regional Studies. No date. Fargo, North Dakota: The Final Days of Red River Steamboats. www.fargo-history.com/transportation/steamboats2.htm (accessed October 1, 2009).

NORS (National Organization for Rivers). 2010. Rivers and the Public Trust Doctrine: Public Ownership of Rivers in the United States. www.nationalrivers.org/us-law-public.htm (accessed January 19, 2010).

NRC Marine Board. 1985. *Dredging Coastal Ports: An Assessment of the Issues.* Washington, DC: National Academies Press.

Shank, William. 1982. *Towpaths to Tugboats: A History of American Canal Engineering.* York, PA: American Canal and Transportation Center.

Stewart, Richard D. 1997. A Tale of Two Ports: A Study in Dredging Dispute Resolution. *Corporate Environmental Strategy: The Journal of Environmental Leadership.*

Stollenwerk, Jeff, Jeff Smith, Brett Ballavance, Julianne Rantala, Dale Thompson, and Sandy McDonald. 2009. Managing Dredged Materials in the State of Minnesota. Minnesota Pollution Control Agency. www.pca.state.mn.us/publications/wq-gen2-01.pdf (accessed December 5, 2010).

Thorp, Steve, and John Stone. 2000. Recreational Boating and the Great Lakes–St. Lawrence Region, Great Lakes Commission. www.glc.org/advisor/00/recboating.pdf (accessed December 5, 2010).

TTI (Texas Transportation Institute). 2007. A Modal Comparison of Domestic Freight Effects on the General Public. www.marad.dot.gov/documents/Phase_II_Report_Final_121907.pdf (accessed December 5, 2010).

USACE (U.S. Army Corps of Engineers, St. Paul District). No date. Navigable Waters of the United States in Minnesota. www.mvp.usace.army.mil/docs/regulatory/mn_nav_waters.pdf (accessed March 15, 2010).

Wilcox, Dan. 2001. Effects of Recreational Boating on the Upper Mississippi River System. U.S. Army Corps of Engineers, St. Paul District. www.mvp.usace.army.mil/docs/nav/effects.pdf (accessed March 03, 2010).

Winfield, L.E., and C.R. Lee. 1999. Dredged Material Characterization Tests for Beneficial Use Suitability. Dredging Operations and Environmental Research Technical Notes DOER-C2. http://el.erdc.usace.army.mil/elpubs/pdf/doerc2.pdf (accessed December 5, 2010).

CHAPTER 11

Water Quality: Trading and the Effects of Agricultural and Energy Policy

Jay S. Coggins and Steven J. Taff

*W*ater quality in Minnesota is not particularly high, and agriculture is said to contribute significantly to this problem. This chapter examines some of the many interconnections among water quality, agricultural practices and policies, energy, and the various policies that currently or might be put in place to protect water quality in the state. It pays special attention to claims that market mechanisms such as water quality trading schemes might improve the situation.

Many of the assertions about the links among agriculture and energy policies, crop production and management, and ambient water quality are at best untested or unproven in a specific context. Although most people believe, for example, that U.S. ethanol policy has led to more corn production and hence degraded water systems, few or no data actually support such a causal claim. By considering empirical evidence relating the constituent parts, we can gain insights into the role of agriculture (including biofuels) in determining water quality in the state.

For water quality to be influenced by policy, several pathways could be at play. The chapter begins by examining the assertions and the (sparse) evidence that can be brought to bear in testing them. The extent to which policies, crop production, and water quality are related is not altogether self-evident. We often lack sufficiently detailed data disaggregated to a meaningful geographic level that are consistent across a reasonable span of years. Consequently, we can at best gain glimpses of the relationships of interest. This chapter examines the interplay of these three components. The key questions, along with our provisional answers, are these:

- Are Minnesota lakes and streams dirty? Yes.
- Are they getting worse over time? Some are and some are not.
- Is water quality affected by crop selection and management? Yes, somewhat.

- Are these cropping decisions influenced by agricultural policies? A little.
- Are there better policy regimes that might improve water quality? Probably.

PRICES AND PLANTING

The main assertion is that agricultural policy influences farm decisions, which in turn influence pollution and, ultimately, observed water quality. In this high-level examination of farm policy and production data, we cannot tell for sure whether policies always influence crop selection. Certainly, they do in specific instances. A major piece of the puzzle is the fact that corn prices doubled between 2005 and 2007, then doubled again in the next year, then halved between 2008 and today. The story is complicated by the parallel fact that prices for soybeans showed a similar dynamic, although this pairing is not surprising, given historic correlations among farm commodity prices. If producers had expected these price increases to persist, they logically would have increased corn and soybean planting. They did not, as the planted acres clearly show. Over the past four decades, the number of acres planted to soybeans increased and those planted to wheat decreased, due in large part to a new system of wheat-soybean rotations in the northwestern part of the state. But these changes occurred largely prior to the year 2000, long before the runup in crop prices (Figure 11.1). The more recent price fluctuations were associated with a onetime increase in corn production and a parallel decrease in

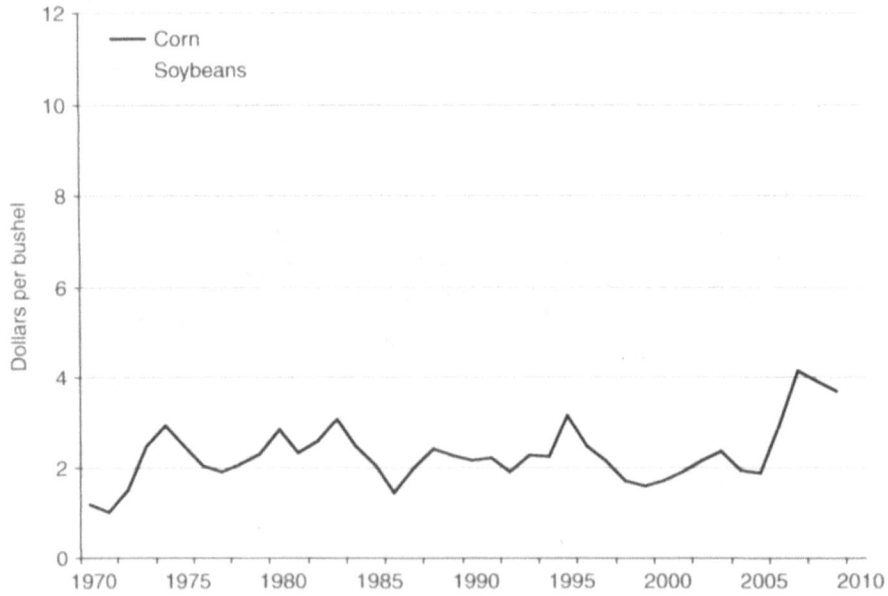

Figure 11.1 *Minnesota annual average grain price*

Source: USDA NASS (2010).

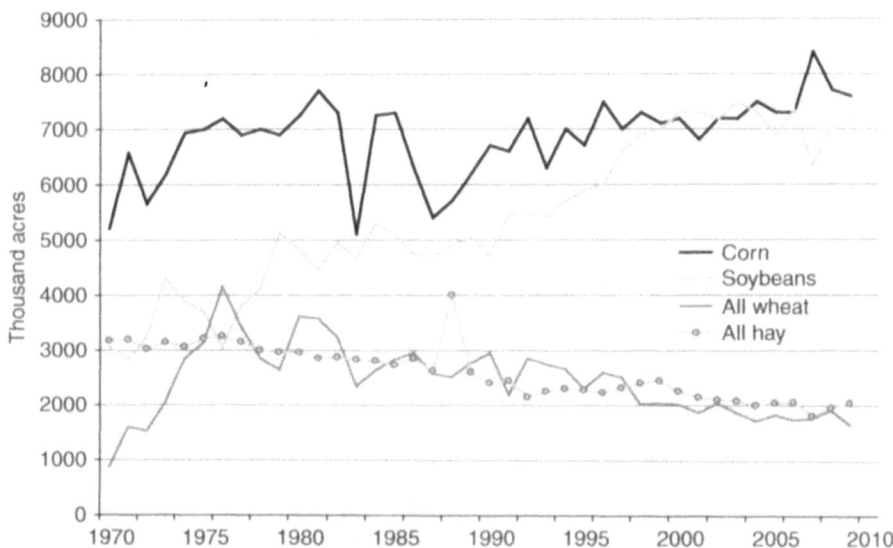

Figure 11.2 *Minnesota planted acres*

Source: USDA NASS (2010).

soybean production. But the acres planted to traditional corn-soybean rotation do not seem to have changed (Figure 11.2).

What about noncropland decisions? A different temporal glimpse at policy–planting linkages is provided by a land cover change matrix (Table 11.1) that shows statewide changes between 1992 and 2002, the most recent year for which such data are available. This period obviously does not cover the more recent commodity price swings, but there were large (but not enormous) price movements in the 1990s as well. No major changes are evident, despite the sizable changes in crop prices and agriculture policy over the period. One might argue, however, that policies in place over the period prevented changes that might otherwise have occurred, that the presumably "pro-corn" policies kept some land in production that might otherwise have switched to urban uses (or, to a lesser extent, to the vaguely specified "grass" category).

AGRICULTURAL POLICIES AND WATER QUALITY

Do changes in agricultural policies, especially long-running crop subsidies and more recent biofuel incentives, result in those modest changes in crop production shown in Table 11.1? Certainly, the direction of that kind of effect is accepted by most economists: the more subsidies are increased for any given crop, the more of the crop will be grown, all else being equal. But how big are the subsidies, really, and how much do they actually influence cropping decisions?

American agricultural policy has shifted focus in the past decade, moving away from specific crop support toward overall farm revenue support. And the impact of

Table 11.1 Changes in land cover, 1992–2001

		Agriculture	Barren	Forest	Grassland/shrub	Open water	Urban	Wetlands	1992 totals
					Acres				
From	Agriculture	24,120,804	2,021	97,638	32,365	86,360	43,362	110,464	24,493,015
	Barren	120	63,541	1,458	182	3,807	31	897	70,036
	Forest	89,799	6,689	14,393,111	19,130	14,044	22,424	112,967	14,658,164
	Grassland/shrub	36,275	149	88,885	2,093,448	382	9,278	21,756	2,250,173
	Open water	12,487	3,127	21,627	3,287	3,032,070	742	25,059	3,098,400
	Urban	6,523	13	2,189	661	3,134	2,656,976	4,360	2,673,857
	Wetlands	71,394	557	167,702	23,349	17,577	6,966	6,453,303	6,740,849
	2001 totals	24,337,402	76,098	14,772,610	2,172,424	3,157,374	2,739,779	6,728,806	

		Agriculture	Barren	Forest	Grassland/shrub	Open water	Urban	Wetlands
				Percent of 1992 Value				
From	Agriculture	0.985	0.000	0.004	0.001	0.004	0.002	0.005
	Barren	0.002	0.907	0.021	0.003	0.054	0.000	0.013
	Forest	0.006	0.000	0.982	0.001	0.001	0.002	0.008
	Grassland/shrub	0.016	0.000	0.040	0.930	0.000	0.004	0.010
	Open water	0.004	0.001	0.007	0.001	0.979	0.000	0.008
	Urban	0.002	0.000	0.001	0.000	0.001	0.994	0.002
	Wetlands	0.011	0.000	0.025	0.003	0.003	0.001	0.957
	2002/1992	0.99	1.09	1.01	0.97	1.02	1.025	0.998

Source: Fry et al. (2009).

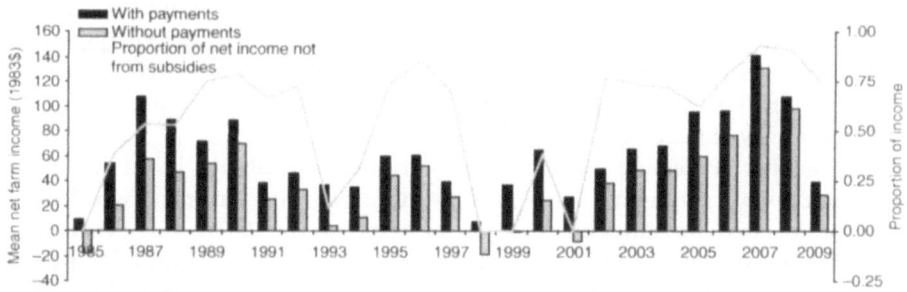

Figure 11.3 *Effect of government payments on mean net farm income per acre in southwestern Minnesota*

Source: Chart prepared by authors from data in Nordquist et al. (2010).

this financial support has diminished substantially in recent years, as Figure 11.3 illustrates. It is simply incorrect to state that the U.S. government subsidizes corn production. The bulk of the current U.S. Department of Agriculture (USDA) subsidies take the form of direct payments, the level of which is not related to present planting decisions, and countercyclical payments, which are inversely related to crop prices but by design are known only at harvest, so they too cannot influence planting decisions. One might claim that the existence of these payments, even though they are not directly related to planting decisions, provides a certain level of income assurance to farmers (indeed, that is what they are intended to do) and thereby result in farmers expanding production—and bearing risk—more than might otherwise be the case.

Examination of farm program payments over recent years does not lend much credence to the subsidy–overproduction link: Minnesota corn and soybean planted acreages have remained essentially the same, while the amount of countercyclical subsidy has varied widely over the years. It has essentially vanished (because of high market prices) since 2006, as examination of Southwestern Minnesota Farm Business Management Association records (Figure 11.3) clearly shows. The portion of net farm income from subsidy is, at least for now, at an all-time low.

Why do farmers decide not to switch to some other crop when corn prices drop? An obvious explanation is that these producers have an enormous amount of physical and intellectual capital invested in the growing of corn and soybeans. They have been growing these crops for a long time, and they are good at it. Too, publicly funded agronomic research has resulted in production systems that are increasingly efficient, able to withstand even long-run declines in crop prices.

An ancillary reason for slow or even nonexistent switches to crops that seemingly offer higher profits might be that farmers decide that staying with corn is the only way to retain eligibility for whatever changes in risk-reduction programs like crop or income insurance might develop over the coming years. The argument would be that even if another crop seemingly promises more money in a good year, if that crop is not associated with some risk-reduction program, then the risk-weighted income of the new crop is still lower than the corresponding income of corn. Current federal farm programs are largely designed to be crop-neutral, but farmers may prefer to stay with the known, worrying that federal programs may change again—as they have in the past.

So neither increasing crop prices nor decreasing subsidy levels are clearly associated with Minnesota farm planting decisions over the past two decades, except for that onetime bump in corn plantings in 2007. But what about energy policy? Has the onset of biofuel subsidies and the diversion of Minnesota corn from livestock feed to ethanol feedstock had an observable effect? Certainly, Minnesota's mandated 10% ethanol blend and 1% to 2% biodiesel blend rates, coupled with longtime state (early, but relatively minor) and federal (more recent, but substantial) subsidies for biofuel production, have led to more corn being used for ethanol. (We will follow the corn-ethanol line here, but the soybean-biodiesel line is similar.) Has this production shift led to increased prices, which have led to even more corn being produced, which in turn has led to decreased water quality? (For a detailing of this storyline, see GAO 2009.)

We do know that at least 25% of the state's corn is now being used for ethanol production, but this has resulted in less corn being exported, not in a large increase in Minnesota corn production. Minnesota's soils and climate do not support much continuous corn cropping. Farmers generally use rotation with soybeans and, to a much lesser extent, small grains to break certain disease cycles as well as provide a different palette of plant nutrients. Even if farmers wanted to plant a great deal more corn, responding to what turned out to be fleeting high prices, there just is not much more suitable land for corn in the state, given the available genetic stock constrained by climate.

In a dynamic now being explored by researchers throughout the country, more corn going into ethanol has resulted in increased corn production in some places—but not on a one-to-one basis. The increase in prices has led to some livestock producers scaling back production because corn has become more expensive. Yet the ways in which prices are transmitted throughout the world's agricultural economy are not well agreed upon. As Iowa State University researcher Bruce Babcock put it:

> Farmers base their decisions about what and how much to plant on numerous factors, including rotation considerations, production costs, expected market prices, availability of crop insurance, and expected benefits from farm programs. ... The role that these programs play in farmers' planting decisions varies across crops, regions, and crop years. Simple "rules of thumb" that use total payment levels as a guide or the belief that the programs work ... are inadequate measures of the impacts of farm payments on U.S. supply and international commodity prices. (2006, 3)

While biofuel policy remained essentially unchanged between 2005 and 2010, corn prices fluctuated widely over the period. For present purposes, for Minnesota crop production, we can find no strong evidence that high crop prices, driven by state and federal crop and biofuel production incentives (which may have many and sometimes perverse effects in nonagricultural settings), have resulted in more crop production than would otherwise be the case.

Effects of Farming Decisions on Water Quality

We have shown that the link between policies and planting decisions is tenuous, at least on a broad scale. What about the link between farming decisions and ambient water quality?

There is no dispute that Minnesota water quality is not as good as it should be. The state has a long list of officially "impaired" waters. That these impairments are due in part to human activity is attested to by research examining presettlement and current water quality in 55 Minnesota lakes. As Ramstack et al. reported, "30% of urban and 30% of agricultural region lakes record a statistically significant increase in total phosphorus between 1800 and the present. These changes, which are attributed to road salt and nutrient runoff, are strongly correlated with the percentage of watershed area that is developed (residential or urban) in the case of chloride increases and the percentage of developed (metropolitan areas) or agricultural (agricultural areas) land in the case of nutrient increases" (2004, 561). More recent quality data show 455 lakes with improving trends, but still 231 with decreasing trends. Another 515 monitored lakes show neither improvement nor decline over the years (MPCA 2008).

Are these impairments changing over time? Two studies look at water quality in a large river that drains a significant portion of Minnesota's agricultural region. Figure 11.4 draws from a study that looks at variability over time of total

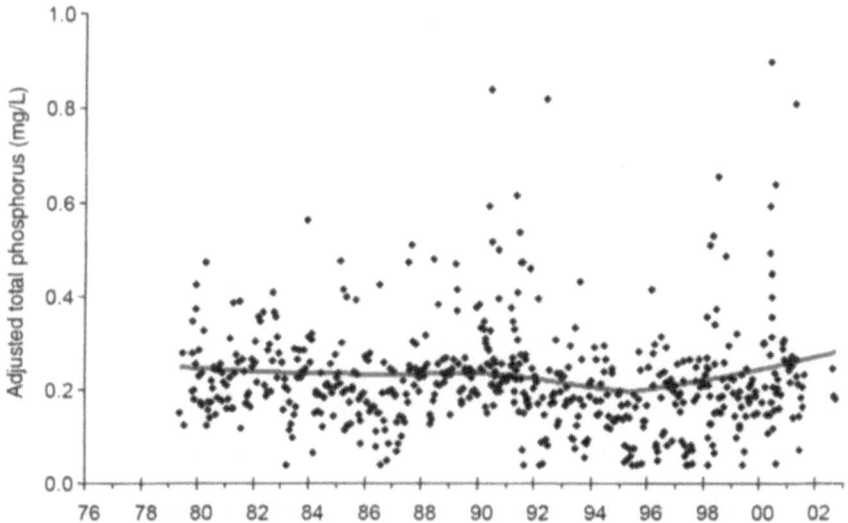

Figure 11.4 *Total phosphorus trends in Minnesota river mainstem*

Source: Metropolitan Council (2004).

phosphorus measured near the mouth of the Minnesota River at Jordan (Metropolitan Council 2004; these data are available only through 2002). Although the authors make a point of the initial decline followed by a more recent increase in the adjusted average readings, it is clear that the average concentrations have changed little since 1980, whatever the statewide changes in crop coverage (especially the increase in soybean acres) over that period.

Using many of the same observations, with data available only through 2004, Johnson et al. looked at annual averages and found "decreasing trends in flow-adjusted concentrations of total suspended solids (TSS), total phosphorus (TP), and orthophosphorus (OP) and a generally increasing trend in flow-adjusted nitrate plus nitrate-nitrogen (NO_3-N) concentration." The authors caution that even these relatively modest improvements might be due to "dilution effects from wet weather or additional tile drainage" (2009, 1018).

Such single-site measurements can hide wide variation in water quality. Minnesota State University–Mankato research (MSU/MPCA/MDA 2009) shows the large range in orthophosphorus (OP) and nitrate-nitrogen (N) concentrations across several tributaries (Figures 11.5 and 11.6). Certainly, different policy–planting–water quality stories can be told for different portions of the agricultural landscape.

Clearly, not all the sediments or nutrients observed in these studies come from agricultural production. For example, as research has shown, a large portion of the Minnesota River's observed sediment load comes from streambank erosion, not

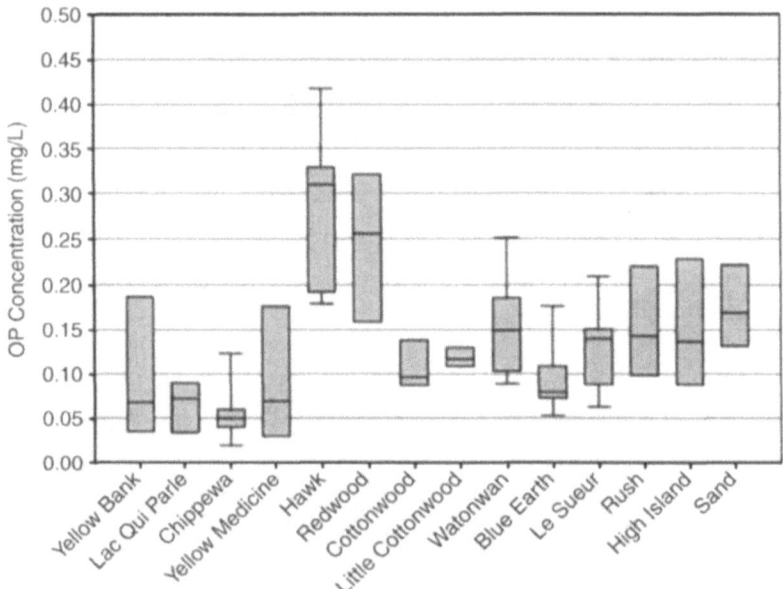

Figure 11.5 *Flow-weighted mean orthophosphorus concentrations at major tributaries of Minnesota river*

Source: MSU/MPCA/MDA (2009).

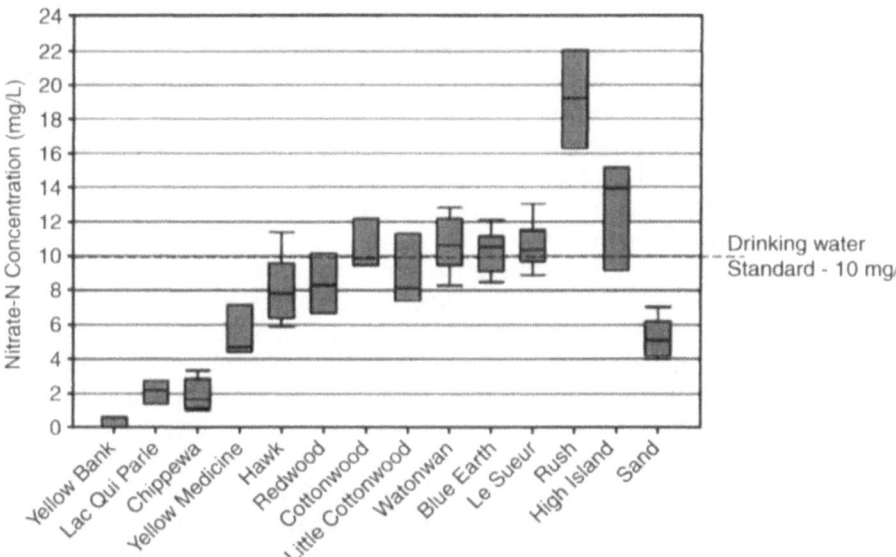

Figure 11.6 *Flow-weighted mean nitrate-nitrogen concentrations at major tributaries of Minnesota river*

Source: MSU/MPCA/MDA (2009).

from field erosion, although that bank erosion might be exacerbated by riverflows increased by field drainage (UMN Institute on the Environment 2008). (Drainage policy is discussed separately below.) Phosphorus observed in the water might come from crop production, livestock production facilities, or municipal sources.

Two other types of agricultural policies deserve special, if brief, attention. One kind of policy, drainage, influences the way fields are managed; the other, land retirement, results in a direct change in land cover.

Drainage

Several million acres of Minnesota farmland are tiled, provided with subsurface drainage that creates a potential pathway for water to carry pollutants. Tile lines consist of perforated pipes laid a few feet below the surface of the field. These pipes collect infiltrating water and channel it to larger tiles, ditches, and ultimately streams and lakes. The bulk of these systems were installed prior to the mid-1980s (see Chapter 7 for further discussion of drainage issues), but there has been a recent resurgence of tiling in the state, brought about in part by higher crop prices and in part by reductions in tile installation costs.

Tile drainage in Minnesota is regulated by state law but is under the immediate control of local drainage authorities, usually county government. The drainage authorities have jurisdiction over the placement and management of ditches and the larger collector tile systems, but they do not have much say over the installation or management of the field tile systems. The laws, which predate statehood, are principally designed to facilitate drainage, providing an outlet for field drainage. Consequently, state drainage policy cannot be said to influence farm-level decisions to any great extent: landowners have a largely unencumbered right to drain wetter lands (not wetlands, making the distinction here as discussed in Chapter 7), and local drainage authorities have a largely mandatory requirement to arrange for the disposal of water from drained fields. Provisions exist for ditch abandonment, but these are hardly ever employed, because they require the agreement of all landowners served by the ditch.

In the past 20 years, more attention has been paid to the design of new ditch systems and the "improvement" of existing systems, with respect to ditch wall sloping, in-channel barriers, ditch-end wetlands, and even reconfiguration of the ditch channel. But these projects are expensive and not usually undertaken. The major environmental regulation affecting field drainage is the requirement that most ditches be buffered by a grass strip a few dozen feet along each side. These strips are intended to intercept overland water flows, trapping sediment and nutrients prior to their entry into the ditch system. Whether they actually serve this function effectively is open to debate, as even casual observation of Minnesota ditch systems shows that not all ditches are buffered, and that even those that are buffered are frequently notched to relieve field backup during high rain events. And ditch buffers do nothing to influence the quality or quantity of water delivered to ditches via tile lines.

Most ditches in Minnesota, regardless of size, are not regulated as public waters and consequently are not subject to the water quality regulations that cover "natural" watercourses. Recent discussions surrounding total maximum daily loads (TMDLs) are calling this distinction between artificial and natural watercourses into question, but no major regulatory changes have been proposed in recent years.

We conclude that current Minnesota drainage policy neither affects the selection or management of crop systems in the state nor intentionally influences the quality of water in the state significantly. It could, but it does not.

Land Retirement

If any public agriculture policy influences water quality, it ought to be land retirement. The federal Conservation Reserve Program and the state Reinvest in Minnesota Reserve Program have removed from production more than 1.5 million acres of cropland since 1985. The federal program is larger by an order of magnitude, but its effects are by design only short-term, usually 10 years for a given contract. The state program consists largely of perpetual easements. The aggregate size of these retirement programs is impressive, but they must be compared to the 20 million to 25 million acres of cropland in the state: their effect on total farm production has been modest, except in areas where enrollment has been proportionately large.

It is to be expected that changing cropland into permanent vegetation should have significant water quality effects. And multiple studies have shown that field-edge erosion and nutrient infiltration are both reduced substantially by a field shifting from corn, say, into grass. However, the effects of these retirements are much more difficult to find in changes in stream or lake water quality. It could be that there is an effect but we just cannot find it because of the mixing within large river systems. Or perhaps we have not done enough land retirement in the upstream areas for a discernible-effects threshold to be crossed. Or perhaps river systems take longer to "clean out" than we might wish; patience might be the prescribed policy action.

Usually, land retirement makes up only a relatively small proportion of a given watershed, so even large per-field reductions in pollution are difficult to observe in the stream itself. Some recent U.S. Geological Survey research hints, however, that the effects are real and measurable if sufficient land is retired in a watershed. According to Christensen and Lee (2008), "Preliminary results show that nitrite plus nitrate concentrations were highest ... in the subbasin with little to no land retirement, and lower in ... subbasins with more riparian or upland land retirement."

Water Quality Models and Monitoring

We are left with a noisy picture of water quality. Large annual fluctuations are evident, even when the data are adjusted for seasonality and flow levels. No easy link can be made between what we see in crop production—rising, then level, corn and bean production—and ambient water quality.

These observed levels and trends in water quality are more ambiguous than the farming–water quality stories told in research models. We rarely have direct evidence of specific practices and water pollution levels; consequently, we resort to models. The models use a common story: the more crop selection and management result in soil being open to spring snowmelt and precipitation events throughout the year, the more sediment is moved from that field to the nearest water body. Similarly, the more a crop is treated with fertilizers and pesticides, the more these will move to nearby water bodies, through either water runoff or tile drainage. There is no disputing that well-calibrated water quality models suggest a strong link between land management and stream water quality. Although many crop-water models exist, these are only laboriously run and infrequently calculated in small watersheds.

Despite the ups and downs of crop prices and commodity program rules over the past several years, observed planting decisions and water quality have remained fairly constant. Why do Minnesota farmers stick with corn, beans, and wheat—whether prices are high or low, or policies are lucrative or not? It is because these three crops, year in and year out, are the best for most Minnesota farmers on most soils. If a policy can be shown to substantially change the crops being grown or how they are grown, that policy might be linked to water pollution. It is not enough to show that growing corn on certain soils can result in nutrient runoff: we must show that the policy has caused a change in the amount of corn being grown or the way it is being grown. And this we have not demonstrated so far with any degree of certainty, even though evidence from both models and monitoring suggests that agricultural practices do indeed have direct, albeit local, effects.

MARKET-BASED SOLUTIONS TO WATER QUALITY PROBLEMS

Do agricultural policies influence water quality in Minnesota? Yes, probably, but not to the extent that we can clearly label some policies good for water quality and others bad. Agricultural policies are designed to meet other perceived needs. As a result, any desired changes in water quality will have to come from programs much more targeted at environmental services than are traditional agriculture programs. If water quality is bad because of agricultural practices, then more directed policies, loosely labeled "conservation policies," need to be applied in far broader (hence more expensive) and far more effective (hence politically difficult) manners. Simple removal of existing commodity or ethanol subsidies will not result in either massive changes in Minnesota cropping patterns or significant changes in management practices that are used on those lands. We have decades of proof that the problems of crop agriculture and water quality are far more complex than this.

Under the current policy regime, it is not easy to see how marginal changes in agricultural policy could deliver significant improvements to Minnesota's water quality. How about water policy itself? Because of the way agriculture is largely excused from reducing its contribution to water pollution, marginal changes in water policy, at least under the current Clean Water Act (CWA), are also unlikely to lead to major reductions in water pollution from agriculture. Solutions that achieve large improvements from agricultural lands will likely require changes to both agricultural and water policies, at both the federal and the state levels.

An essential fact about U.S. water quality policy is that the main statutory foundation of the nation's policy regarding water, the CWA, does not reach agriculture. With the exception of concentrated animal feeding operations (CAFOs), the contribution of agriculture to water quality problems is uncontrolled. Another essential fact about U.S. water quality policy is that most regulation of water quality in the United States is executed through command-and-control policies, such as technology mandates or source-specific quantitative emissions restrictions. Market-based approaches such as pollution taxes or water quality trading (WQT) play a minor role.

The National Pollutant Discharge Elimination System (NPDES) program requires facilities that emit pollutants to surface waters to hold a permit under the CWA. Applied only to point sources such as industrial facilities or wastewater treatment plants, the NPDES system specifies numerical limits based on allowable discharges that would result from the use of state-of-the-art or "best conventional" control technology. Permit holders are free to choose a different technology, so long as the resulting emissions meet the specified limit (Boyd 2003).

A discussion of the potential effects of market-based solutions on the agricultural sector, then, requires two leaps. One is to imagine how environmental policy might one day be revised to bring agricultural sources under control of the regulatory apparatus. The other is to imagine how market-based approaches might be implemented for agriculture and other sources of water pollution.

Water pollution is difficult to control effectively, in part because it is difficult to understand. Surface waters are complex biological and physical systems in which the assimilative capacity of water and the harm done by a given concentration of a pollutant can vary dramatically across space. Also, it is usually difficult to entirely understand the flow of pollutants over the land, into and through groundwater aquifers, and through the surface water body itself. Especially significant is the difficulty associated with controlling water pollution from nonpoint sources such as farm fields and residential lawns. The science involved in estimating the flow of nutrients off a field in a rainfall event is relatively well understood, but understanding the way this flow continues into and down a waterway is much more difficult, and understanding the profile of concentrations and changes in concentrations throughout an entire watershed is even more arduous. In short, a unit of pollutant discovered in a stream or lake cannot be traced back to its source with any certainty. This is what makes it a nonpoint-source problem. Yet a policy aimed at controlling pollution from nonpoint sources would necessarily require a considerable amount of certainty about each source's level of responsibility for the pollution levels detected in a shared waterway. An alternative would be to place restrictions on pollution-causing inputs, such as fertilizer, rather than the emissions of pollutants themselves (see Griffin and Bromley 1982).

Advantages of Market-Based Policies

From an economic perspective, in a textbook setting the use of emissions taxes or trading to control pollution is attractive. An agency charged with administering environmental policy cannot know enough about the cost structure of polluting firms to achieve an optimal distribution of abatement. With either an emissions tax or a permit-trading system, so long as it is set up correctly, aggregate abatement costs can typically be lowered relative to the administered alternative, sometimes significantly. The regulator does not need to know each firm's abatement costs; firms themselves make decisions about how much to emit.

To illustrate the claim that market forces, suitably applied, can lead to lower aggregate abatement costs, consider the following example of two point sources of a given pollutant on a single waterway. Before regulation, each plant emits 10 units of the pollutant into the stream. Plant 1 has abatement cost function $C(a_1) = 2.5a_1^2$ and thus marginal abatement costs of $MC(a_1) = 5a_1$. Plant 2 has abatement cost function $C(a_2) = 5a_2^2$ and thus marginal abatement costs of $MC_2(a_2) = 10a_2$, where a_i is plant i's abatement relative to the uncontrolled initial emissions level.

Suppose an environmental regulator has decided that a total of 12 units of abatement must be achieved by the two plants. A policy to require uniform reductions would mean each plant must reduce by $a_i = 6$, which would cost the plants 90 and 180, respectively, for a total cost of 270. A tax on emissions of 40 per unit would induce Plant 1 to abate at $a_1 = 8$, where its marginal cost equals the tax. Plant 2 would abate at $a_2 = 4$. These levels would lead to abatement costs of 160 and 80, respectively, for a total cost of 240. The firm with lower costs achieves more of the abatement, and this leads to a savings of over 10% in aggregate costs. With the tax, each plant must also pay 40 for each unit of remaining emissions. This comes to 80 for Plant 1 and 240 for Plant 2, a total of 320 in tax payments that could be used to reduce other tax burdens in the economy or for many other purposes.

With a water quality trading policy, the regulator would distribute 8 permits to the two firms in some manner. Suppose each plant receives 4. The plants can buy and sell permits from one another, the only requirement being that at the end of the year, each must have a number of permits at least equal to its emissions. The outcome would be that Plant 1 would again reduce by $a_1 = 8$ and sell two permits to Plant 2, which would reduce by $a_2 = 4$. If the market were competitive, the price would be 40, equal to both firms' marginal abatement costs. Thus Plant 2 would pay 80 to Plant 1. If the permits were given out free, plants would be much better off financially than with the tax. If the regulator were to sell them to the highest bidder, the price of permits would reach 40, and the cost of the permits would equal the tax described above. Auctioned permits and emissions tax therefore are equivalent in this simple example. Free permits achieve the same environmental outcome at the same total abatement cost but are better for the polluting firms. Uniform reductions achieve the same environmental outcome but at a higher total abatement cost.

Water Pollution Taxes and Agriculture

Water policy in the United States is not based on emissions taxes to any significant degree.[1] In the year in which the 1965 Water Quality Act was passed, a panel of the president's Science Advisory Committee recommended that a system of emissions taxes be studied. This idea was not implemented, however; the 1972 Federal Water Pollution Control Act established the NPDES system of permits, and emissions taxes were no longer considered seriously.

The example above illustrates, for a simple setting, how a system of emissions taxes would work. That example is too simple, though. Because the location of emissions in a watershed is important to its impact on water quality, to be most effective a tax system must be sensitive to where emissions occur and how a pollutant load at one place affects concentration at all points downstream. Incorporating this idea into any policy to protect environmental quality, whether market-based or not, requires science-based information about transfer coefficients that describe, for a given unit of pollution emitted at one point, how much will be experienced at downstream points. A river, for example, might be divided into zones, and the transfer coefficients t_{ij} between each pair of zones would describe how much of a unit of a pollutant emitted in zone i remains when the water reaches zone j.

If a fully optimal system of emissions taxes is to be implemented for point sources only, the tax assigned to each source must equal the marginal damages cause by emissions from that source in its own zone and in each downstream zone. Getting this right requires a great deal of information and computation on the part of the regulator. A uniform emissions tax, which would ignore transfer coefficients and differential damages along a river, is not guaranteed to achieve a given environmental standard at least cost. Boyd (2003) provides an example in which a uniform standard can achieve a given level of environmental quality at lower cost than can a uniform emissions tax.

A system of emissions taxes that also incorporated nonpoint sources would be much more complicated to implement and would add the uncertainty of where a given farmer's emissions end up. Taxes applied to inputs such as fertilizer, as suggested by Griffin and Bromley (1982) and others, typically require tax rates far exceeding the price of the fertilizer input. This approach is unlikely to be feasible politically, and no policy discussions appear to be under way that would lead in the direction of emissions taxes for agriculture.

Water Quality Trading and Agriculture

The U.S. Environmental Protection Agency (EPA) is serious about promoting and encouraging the use of emissions trading to achieve improvements in the nation's waters. In 2003, EPA released a document regarding water-trading policy, which provides states with guidance regarding the development of trading programs. It also encourages them to do so, stating as its purpose: "to encourage states, interstate agencies and tribes to develop and implement water quality trading programs for nutrients, sediments, and other pollutants" (2003, 2). Among the objectives stated by EPA, the agency supports trading where it does the following:

- reduces the cost of implementing TMDLs through greater efficiency and flexible approaches;
- establishes economic incentives for voluntary pollutant reductions from point and nonpoint sources within a watershed;
- offsets new or increased discharges resulting from growth in order to maintain levels of water quality;
- achieves greater environmental benefits than those under existing regulatory programs;
- secures long-term improvements in water quality through the purchase and retirement of credits by any entity; and
- combines ecological services to achieve multiple environmental and economic benefits.

EPA's policy statement on water quality trading describes several requirements, some more firmly stated than others, that any trading program should satisfy. Trading programs must be consistent with the CWA, and trading should occur within a watershed. EPA supports trade in nutrients and sediment loads but is more guarded in its support of trade in other pollutants. Such trades, it says, "may pose a higher level of risk and should receive a higher level of scrutiny to ensure that they are consistent with water quality standards." Trading in persistent bioaccumulative toxics is not supported at all. Any party wishing to sell nutrient credits must reduce its nutrient loading below a baseline established by an approved TMDL or, if before a TMDL is established, "the applicable water quality based effluent limitation" (2003, 4). The credits generated for sale will be measured relative to the seller's baseline, computed in this way. Trading may be used to maintain high water quality in nonimpaired waters, prior to a TMDL in impaired waters, or under a TMDL.

EPA lists several elements that should be part of a "credible" trading program. The first of these is that the CWA provides authority for EPA and the states and tribes to establish trading programs. No new statutory authority is needed. Together with the fundamental notion that authority to establish trading already exists, the following list in summary form is the heart of the EPA policy:

- Trading programs should establish clearly defined "units of trade." In the case of thermal trading, it is not entirely clear what these units should be. The most likely candidate is joules or some other measure of energy appropriate for establishing a thermal cap.
- Credits should be generated either before or during the period in which they are used to comply with the buyer's limitation requirement. The policy appears to be silent on whether a credit generated in one year could be sold and used the following year.
- States and tribes should establish standardized protocols for quantifying load reductions and credits. These should be included in NPDES permits so that trades and compliance can be tracked.
- Where trade involves nonpoint sources, states must account for increased uncertainty in estimates of nonpoint-source loads. EPA requires using trading ratios greater than 1:1 for trades involving nonpoint sources, as well as other methods that reflect increased uncertainty.
- States should establish mechanisms for ensuring compliance and for enforcement of a trading program. If a permittee purchases credits and the seller defaults on its commitment, the permittee remains responsible for the effluent limitation that was to be satisfied by the seller's reduction.
- The public should be invited to participate in the development of a trading program and, once it is in place, should be notified of all trades.
- Trading programs should be evaluated periodically, with a focus on environmental and economic effectiveness. Revisions should be made as necessary.

Bringing agricultural sources into any new regulatory initiative aimed at improving water quality in Minnesota will be challenging. Agricultural interests are powerful politically and could be expected to resist new controls on farming practices. The science of nonpoint-source pollution is imperfectly understood, which would also make enforcement difficult.

Because EPA and other regulatory agencies have stated a clear desire to implement water quality trading in the United States, how might this work in Minnesota, and how would agriculture likely be affected? Considerable evidence exists against successful trading schemes for water quality, and thus trading is unlikely to be effective, for agricultural and other sources alike. Though EPA appears to be eager to see trading programs emerge, the extensive literature related to water quality trading is, for the most part, discouraging. In most of the trading markets in the United States, few trades have taken place. Morgan and Wolverton (2005), for example, list 36 nutrient-trading and offset projects in the United States.[2] Of these, 4 have seen 20 or more trades, whereas the remaining projects have had 3 or fewer trades. There is a reason for the low trading activity seen in many cases: the challenges facing water quality trading programs are real and significant. Uncertainty is great, the number of traders tends to be small, and contributions from nonpoint sources are difficult to measure. Still, the potential costs savings and the possibility of environmental improvement beyond that achievable by traditional regulations mean that further attempts to make trading work are justified.[3]

What does success mean in the context of water quality trading? A useful definition of success would be that the following three conditions are met:

1. *Market activity.* Trades occur in numbers that indicate an active, vibrant market.
2. *Reliability.* Participants know what to expect from the program and find that they can trust its functioning; it has a minimum of surprises.
3. *Transparency.* Parties to a trade, potential buyers and sellers, are able to understand the way the market works and find one another with relative ease.

These three conditions are all difficult to achieve, for a variety of reasons. The first requirement, that the potential market involve significant numbers of willing buyers and willing sellers, is often quite difficult to achieve. Indeed, King and Kuch (2003) believed that the primary barrier to successful trading is the small number of parties who stand ready to create credits for sale, to buy the credits, or both. In many cases, an important regulator goal is to control water quality locally. Because water quality can vary spatially, and because "hot spots" of high pollutant concentrations in small areas can often create significant harm to the local biology, regulations typically require precise spatial control of pollution levels. This means that many trading program must be circumscribed geographically. The result is that relatively few buyers and sellers qualify as potential participants.

A crucial component of the requirement that several potential buyers and sellers exist is that various sources of the pollutant in question must have different costs of abatement. When point sources have already achieved a great deal of abatement, the cost to them of further reductions can be quite high. For certain pollutants, such as phosphorus in some cases, the cost of achieving reductions on farmland can be much lower. This is a case where buyers (the point sources) and sellers (owners of farmland) might both be plentiful. Inducing farmers and other owners of farmland to participate in a trading program has proved to be difficult (Morgan and Wolverton 2005).

Further restrictions on who may trade with whom are often built into trading programs. The draft rules for water quality trading developed by the Minnesota Pollution Control Agency (MPCA) in 2008, for example, prohibit sales from a downstream source to an upstream source, except within a minor subwatershed. This means that sources on the upstream reaches of the river system who wish to "buy" credits will face very few "sellers".

The second requirement, that participants can rely on the trading program to function the same way over time, is also often a barrier. This is due at least in part to the inherent difficulty of understanding a system at the requisite level of spatial resolution. Even the most advanced scientific study of a river system, for example, will leave regulators and participants with considerable uncertainty about water quality after a trading program is installed. In order to justify investment in abatement practices that generate pollution credits for sale on the open market, polluters need to feel some certainty that the sale will be durable. Any risk that this is not the case, that future changes in regulations might cause the number of credits to be reduced, will naturally hinder a potential seller from making the investment at all.

The third requirement, that the market function in a transparent fashion and traders can find each other easily, can also be a substantial barrier to trading. In the case of the U.S. SO_2 allowance market, all participants understand that an allowance confers the right to emit a ton of SO_2, from any point in the country. Water quality trading is much more demanding informationally, as has been noted above. In order to obtain approval from the regulating agency, a given proposed trade might require a relatively complex scientific study of the effects of the trade on water quality throughout the watershed. This is especially the case if one of the traders is a nonpoint source, such as a farmer. If the agency chooses to operate as a centralized clearinghouse for all trades, perhaps maintaining a public list of potential buyers and sellers of credits and conducting the scientific study of each potential trade, then regulatory costs can be considerable. If the program relies on buyers and sellers to find each other, perhaps with the help of brokers, and also to provide the scientific study themselves, the transaction costs can be considerable.

In short, if it is to work effectively, a trading program must offer flexibility for participants, characterized by the ability to make off-site trades with as many potential trading partners as possible. Any trading program will be pulled in two opposite directions by the desire to offer broad market opportunities, on one hand (which will increase the number of potential participants and help lead to a successful program), and to control water quality precisely along the length of the river, on the other. Resolving these competing goals has so far proved to be a challenge for environmental regulators.

In 2008, the MPCA developed a set of draft guidelines for water quality trading in the state. The guidelines explicitly include the possibility of trade between point and nonpoint sources, so agriculture could potentially be affected. As of this writing, though, and for the foreseeable future, participation by farmers will be voluntary. The NPDES permit system does not affect agriculture apart from CAFOs, and the MPCA has no authority to limit agricultural practices.

The draft WQT rules emphasize the importance of protecting Minnesota's water resources in the face of population growth and increasing development pressure, while at the same time creating incentives for sources of pollution to reduce their emissions in a cost-effective manner. However, prospects for nonpoint sources including agriculture to participate voluntarily in trades with permitted point sources may not be bright. Participation by nonpoint sources is voluntary, as noted, and the likely financial gain to selling permits does not appear to be great relative to the cost and administrative burden of meeting the requirements of the rules.

Water Quality Trading in Practice and in Theory

Water quality trading programs that have achieved at least some success are not easy to find. Two that provide instructive lessons for Minnesota agriculture involve the Rahr Malting Company and the Southern Minnesota Beet Sugar Cooperative.

In 1997, the MPCA issued a permit to Rahr Malting that allowed the company to expand the wastewater treatment plant at its Shakopee, Minnesota, facility. The Minnesota River is impaired by dissolved oxygen at the Shakopee reach, so acquiring an industrial user fee would have been very expensive for Rahr. The company proposed and was granted a point–nonpoint NPDES permit to obtain reductions in dissolved oxygen in agricultural areas upstream of the plant. Rahr installed four erosion control sites upstream of the plant: two at the confluence of the Minnesota and Cottonwood Rivers near New Ulm, Minnesota, approximately 70 miles from Shakopee, and one each, closer, at 8-Mile Creek and the Rush River. At the New Ulm sites, riparian vegetation was restored, resulting in reductions in sedimentation, which in turn led to reductions in phosphorus loading. At the remaining two sites, eroded banks were stabilized and livestock were excluded from the stream. These practices, which involved payments from Rahr to agricultural landowners in the upstream areas, in aggregate achieved a daily reduction of 212 pounds of chemical biological oxygen demand (CBOD) at an estimated cost of around $5 per kilogram (Fang and Easter 2003), much lower than alternative mitigation approaches. In return, Rahr was allowed to increase emissions of CBOD at the Shakopee plant by 150 pounds per day.

In 2000, the MPCA issued a permit to the Southern Minnesota Beet Sugar Cooperative to expand its Renville, Minnesota, facility. Expansion meant increased emissions of total phosphorus to the Minnesota River. Facing the prohibitively expensive alternative of expanding its spray irrigation treatment capacity, the cooperative agreed instead to engage in point–nonpoint trades with upstream landowners. The cooperative paid for more than 58,000 acres of shareholders' land that was previously under row crop cultivation to be planted to spring cover crops. It also funded stream stabilization and livestock exclusion projects. The resulting reduction in runoff and attendant phosphorus loading, according to estimates by scientists, amounted to 2.6 times the increase in phosphorus loading that the plant's expansion created.

These two trades are among the few WQT successes nationally, and each involved only one buyer of credits. Several other studies of potential trading programs in Minnesota have not yielded promising results. Two summarized here are a study of groundwater nitrates in Minnesota's Olmsted County and a study of temperature trading on the Vermillion River in Scott and Dakota Counties.

Nitrate pollution exceeds EPA standards at more than half of municipal and rural domestic wells in the country. Leaching of fertilizers from farm fields is the primary source of groundwater nitrates in rural areas, and treating contaminated water supplies can cost households hundreds of dollars per year. In a 2000 study, Cynthia Morgan developed a detailed hydrologic-biological-economic model of an agricultural region in Olmsted County near Rochester, Minnesota. This region has compromised groundwater quality. Morgan's model attempted to assign responsibility for each unit of nitrate pollution found at a representative local municipal well to the farmers in the surrounding area. This ambitious project included a model of the movement of nitrates through the root zone and into the aquifer below. It also included a hydrologic model of the movement of nitrates through the aquifer and in the direction of the well. Finally, it included a model of the effects of different farming practices on nitrate leaching and, ultimately, nitrate concentrations at the well.

Morgan compared two different regulatory approaches to controlling nitrate leaching: direct regulatory restrictions on farming practices and a WQT program that would allow farmers who reduced their nitrate contributions more than necessary to sell the excess nitrate credits to neighbors who reduced less than necessary. The comparison between the two alternative approaches was not favorable to the trading scheme. Although the trading scheme realized modest cost savings relative to a uniform reduction, it would have involved a great deal of uncertainty and intensive modeling efforts by a regulatory agency that wished to implement it. The policy was never introduced, but as an empirical study of WQT in a groundwater setting, the findings were not encouraging. Very few trades occurred, and the transaction costs of executing those few trades in practice would likely be prohibitive.

The Vermillion River watershed in Minnesota drains 388 square miles in the southern part of metropolitan Minneapolis–St. Paul. The river contains more than 45 miles of designated trout stream, where a vibrant population of brown trout make their home. Scott and Dakota Counties, in which the watershed lies, are among the fastest-growing in the state. The population of the watershed increased by 40% during the 1990s and is expected to double between 2010 and 2030. There is evidence that the river's temperature has been rising in recent years, at least in part due to increasing development pressure, raising the possibility that it could be listed as impaired for temperature by the MPCA.

Changes in land use due to development influence stream temperatures in several ways. Primary among them is the increase in the temperature of runoff that flows into the river and its tributaries from warm, impervious surfaces during some summer rain events. Water warmed by commercial rooftops, streets, and paved parking lots can also find its way into the groundwater, leading to warming of the recharge to the stream. Various best management practices (BMPs) that mitigate the warming effect of developed landscapes are available, but they are expensive.

In 2007, the Vermillion River Watershed Joint Powers Organization, along with several partners, secured a targeted watershed grant from EPA to study the potential for a temperature-trading program to achieve reductions in the warming trend in the Vermillion. A team of scientists and economists examined the problem for more than two years, studying the biology of the trout population, the hydrology of the aquifers in the region, the stream itself, surface runoff, and the economics of trading. A great deal was learned about the watershed and the Vermillion. We now know more about how different land uses influence temperature in a trout stream than was known before. Various mitigation practices, such as rain gardens, permeable pavement, infiltration beds, and green roofs, were studied for their installation costs and their capacity to reduce thermal loading in the river.

Wang (2010) synthesized this information and conducted a study of the potential for a temperature trading program in the Vermillion watershed to achieve thermal mitigation at a moderate cost. The model posited a development project in each of several trading zones in the watershed, each of which included a BMP that would mitigate the contribution of the developed area to thermal loading in the river. For a variety of parameter values regarding the cost of BMP installation and land acquisition, and for two different sets of trading zones (5 or 15), only one trade occurred in Wang's model.

In summary, although Minnesota has witnessed some important successes in WQT trading that have a direct bearing on agriculture, several studies of possible trading programs have yielded discouraging results. For reasons related to the underlying science, the economics, and various political constraints, prospects do not appear to be good for increases in water quality trading involving agriculture. This is too bad, because point sources have already achieved a considerable amount of reductions for many key pollutants, and further reductions by those sources will incur relatively high costs. Agriculture and other nonpoint sources appear now to be the low-hanging fruit, but bringing them into the regulatory apparatus will not be easy.

CONCLUSIONS

This chapter has explored two primary sets of questions. One regards the complex causal connection that links federal and state agricultural policy and water quality. The other concerns the potential for market-based environmental policy that includes agricultural sources to improve water quality in the state. In the first case, we do not find compelling evidence that agricultural policy has, in and of itself, led to significant degradation of Minnesota's valuable water resources. Although agricultural sources do pose an important threat to water quality, especially with respect to nutrients such as phosphorus and nitrogen, it is not at all clear that agricultural policies are important drivers of farmers' cropping decisions. Corn- and soybean-planted acreages in the state do not appear to be responsive to the level of countercyclical subsidy payments. Dramatic changes in corn, soybean, and wheat prices have not led to significant shifts in acreages planted to these crops.

Monitoring and modeling studies alike do demonstrate that agricultural practices have an important effect on water quality in the state. But changes in agricultural policies do not appear likely to lead directly to improvements in water quality.

If changes in the current suite of agricultural policies are not likely to lead to major improvements in Minnesota's water quality, might other types of policy changes improve matters? Little evidence suggests that water quality trading, a policy alternative that appears to be a favorite of federal and state regulators alike, has the potential to bring about significant improvements in water quality. Within the current regulatory framework, in which agricultural nonpoint sources are given a free pass with respect to their effect on environmental quality, agriculture is unlikely to be a part of the solution. A variety of institutional, informational, and economic forces work against the success of any trading scheme that seeks to bring farmers into the picture.

If changes to agricultural policies, on their own, and if market-based programs, on their own, are similarly unlikely to provide significant improvements to

Minnesota water quality, what sort of alternative *might* work? Applying ever more stringent limits to point sources such as wastewater treatment plants cannot yield significant additional pollution mitigation at a reasonable cost. Most point sources are already dramatically less polluting than they were a generation ago. Big improvements will require contributions from agriculture. But those contributions will not be forthcoming unless and until the policy-induced incentives are changed.

Direct and aggressive new limits on management practices could potentially induce farmers to reduce their levels of fertilizer application and change tillage practices. New measures to restrict crop production in riparian zones might also produce improvements. Perhaps the most straightforward means of bringing about pollution reductions from agricultural lands would be a change in the Clean Water Act that required nonpoint sources to mitigate their loading into waterways. If states had the authority to regulate pollution from nonpoint sources, including agriculture, as they do point sources, then a rationalization of mitigation across the two realms would be possible. With state environmental authorities hobbled by their inability to reach nonpoint agricultural sources, potentially inexpensive reductions are not achieved.

Were it possible under current regulations to place serious requirements on agriculture, several potential approaches might yield good results. Taxes levied on pollution-creating inputs such as fertilizer might do double duty in bringing about improvements in water quality and in providing revenues for the government that could be used to reduce distorting taxes elsewhere in the economy. In the end, though, because the two are inextricably linked, solutions to agriculture's contribution to pollution in Minnesota waters will likely require changes to both agricultural policy and water policy.

NOTES

1. European countries have shown a greater willingness to impose emissions taxes. See, for example, Sterner and Köhlin (2003). Water quality trading appears to be a more attractive alternative to many in the United States.
2. See also the summaries in Environomics (1999); King and Kuch (2003); and Breetz et al. (2004).
3. The cost savings that trading can deliver are potentially quite large. EPA has estimated that the annual cost of meeting TMDLs nationally would be $900 million lower with trading than without (as cited in King and Kuch 2003, 10359). Doering et al. (1999) estimated that the cost of nutrient reductions required to eliminate the hypoxia problem in the Gulf of Mexico would be $14 billion lower with trading than without.

REFERENCES

Babcock, Bruce A. 2006. Cheap Food and Farm Subsidies: Policy Impacts of a Mythical Connection. *Iowa Ag Review* 12:1–3.

Boyd, James. 2003. Water Pollution Taxes: A Good Idea Doomed to Failure? *Public Finance and Management* 3 (1):34–66.

Breetz, Hanna L., Karen Fisher-Vanden, Laura Garzon, Hanna Jacobs, Kailin Kroetz, and Rebecca Terry. 2004. *Water Quality Trading and Offset Initiatives in the U.S.: A Comprehensive Survey.* Hanover, NH: Dartmouth College.

Christensen, V.G., and K.E. Lee. 2008. Effects of Agricultural Land Retirement in the Minnesota River Basin. Paper presented at American Water Resources Association (AWRA) Summer Specialty Conference, June–July 2008, Virginia Beach, VA.

Doering, Otto, Francisco Diaz-Hermelo, Crystal Howard, Ralph Heimlich, Fred Hitzhusen, Richard Kazmierczak, John Lee, Larry Libby, Walter Milon, Tony Prato, and Marc Ribaudo. 1999. Evaluation of the Economic Costs and Benefits of Methods for Reducing Nutrient Loads to the Gulf of Mexico. www.epa.gov/owow_keep/msbasin/pdf/hypox_t6final.pdf (accessed December 11, 2010).

Environomics. 1999. A Summary of U.S. Effluent Trading and Offset Projects. www.epa.gov/owow/watershed/trading/traenvrn.pdf (accessed December 11, 2010).

EPA (U.S. Environmental Protection Agency). 2003. Final Water Quality Trading Policy. www.epa.gov/owow/watershed/trading/finalpolicy2003.html (accessed December 11, 2010).

Fang, F., and K.W. Easter. 2003. Pollution Trading of Offset New Pollutant Loadings: A Case Study in the Minnesota River Basin. Proceedings of the American Agricultural Economics Association Annual Meeting, Montreal, Canada.

Fry, J.A., M.J. Coan, C.G. Homer, D.K. Meyer, and J.D. Wickham. 2009. *Completion of the National Land Cover Database (NLCD) 1992–2001 Land Cover Change Retrofit Product.* Washington, DC: U.S. Geological Survey.

GAO (U.S. Government Accountability Office). 2009. Energy-Water Nexus: Many Uncertainties Remain about National and Regional Effects of Increased Biofuel Production on Water Resources. Report to the Chairman, Committee on Science and Technology, House of Representatives. GAO-10-116. Washington, DC: GAO.

Griffin, Ronald, and Daniel Bromley. 1982. Agricultural Runoff as a Nonpoint Externality: A Theoretical Development. *American Journal of Agricultural Economics* 64:547–552.

Johnson, Heather O., Satish C. Gupta, Aldo V. Vecchia, and Francis Zvomuya. 2009. Assessment of Water Quality Trends in the Minnesota River using Non-Parametric and Parametric Methods. *Journal of Environmental Quality* 38:1018–1030.

King, Dennis M., and Peter J. Kuch. 2003. Will Nutrient Credit Trading Ever Work? An Assessment of Supply and Demand Problems and Institutional Obstacles. *Environmental Law Reporter* 33:10352–10368.

Metropolitan Council. 2004. Regional Progress in Water Quality: Analysis of Water Quality Data from 1976 to 2002 for the Major Rivers in the Twin Cities. http://es.metc.state.mn.us/eims/related_documents/repository/630414225.pdf (accessed December 11, 2010).

Morgan, Cynthia. 2000. Tradable Permits for Controlling Nitrates in Groundwater. PhD diss., University of Minnesota.

Morgan, Cynthia, and Ann Wolverton. 2005. Water Quality Trading in the United States. Working Paper #05-07. Washington, DC: U.S. Environmental Protection Agency.

MPCA (Minnesota Pollution Control Agency). 2008. Secchi Transparency Trend Summary. www.pca.state.mn.us/index.php/view-document.html?gid=6238 (accessed December 11, 2010).

MSU/MPCA/MDA (Minnesota State University, Mankato Water Resources Center/Minnesota Pollution Control Agency/Minnesota Department of Agriculture). 2009. State of the Minnesota River: Summary of Surface Water Quality Monitoring, 2000–2008. http://mrbdc.mnsu.edu/reports/basin/state_08/2008_fullreport1109.pdf (accessed December 11, 2010).

Nordquist, Dale W., James N. Kurtz, Donald L. Nitchie, Garen J. Paulson, Janet M. Froslan, and James L. Christensen. 2010. *Southwestern Minnesota Farm Business Management Association 2009 Annual Report.* Staff paper P10-2. St. Paul, MN: University of Minnesota, Department of Applied Economics.

Ramstack, J.M., S.C. Fritz, and D.R. Engstrom. 2004. Twentieth Century Water Quality Trends in Minnesota Lakes Compared with Presettlement Variability. *Canadian Journal of Fisheries and Aquatic Science* 61:561–576.

Sterner, Thomas, and Gunnar Köhlin. 2003. Environmental Taxes in Europe. *Public Finance and Management* 3 (1):117–142.

UMN (University of Minnesota) Institute on the Environment. 2008. Minnesota Statewide Conservation and Preservation Plan. www.lccmr.leg.mn/statewideconservationplan/Final_plan/SCPPFinalReport.pdf (accessed December 11, 2010).

USDA NASS (U.S. Department of Agriculture, National Agricultural Statistics Service). 2010. Statistics by Subject: Crops and Plants. www.nass.usda.gov/Statistics_by_Subject/index.php (accessed December 11, 2010).

Wang, Bin. 2010. Using the Hung-Shaw Model for Vermillion River Watershed Pollution Control. MS thesis, University of Minnesota.

PART V

EMERGING ISSUES

CHAPTER 12

The Impacts of Climate Change on Distribution and Use of Water

Chuck Dayton and Don Pereira

T here is strong consensus in the scientific community today that the climate system of the earth is undergoing significant change, as key sources of uncertainty have continued to decline over time. Furthermore, anthropogenic forcings (i.e., increases in atmospheric concentrations of greenhouse gases that originate from human activities) from greenhouse gas emissions and large-scale landscape change explain far more of the trend observed in the last two decades than do natural forcings (Solomon et al. 2007; Stott et al. 2000). The rate and degree of future change will therefore depend in large part on emerging mitigation and adaptation policies. However, even with aggressive mitigation at a global scale, we will likely experience change by the middle of this century that will result in significant impacts to both natural and human systems (Solomon et al. 2007). This chapter examines potential impacts to water resources in Minnesota from ongoing and expected climate change and suggests policy implications that will require adaptation of management systems and infrastructure.

CLIMATE CHANGE IN MINNESOTA: THE LIKELY SCENARIO

In recent decades, Minnesota has experienced a significant increasing trend in air temperature (Figure 12.1); this trend is expected to continue for at least several decades, perhaps most of this century, depending on future mitigation policy and emissions scenarios (Solomon et al. 2007). Figure 12.1 shows that rates of increase have been greater in the northern part of the state, and that in both north and south, they have been much greater during the past two and a half decades in comparison with the entire time series.

Although the rate of future increase will depend on mitigation policies and resultant greenhouse gas emissions, we will likely experience an increase of

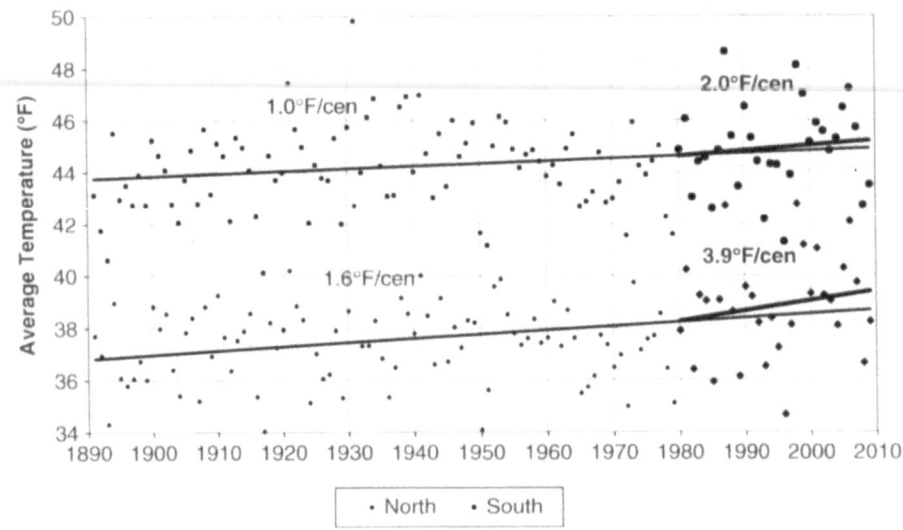

Figure 12.1 *Annual average temperatures for southern and northern Minnesota*
Source: Based on state climate date from Zandlo (2010).

2 to 4 degrees C by 2050. The rate of this increase will vary both seasonally and spatially, as follows:

- The rate of change will be greater at higher latitudes and during the winter (IPCC 2007; Kling et al. 2003).
- Summer rates of change will be greater for daily minimums than maximums, thus nights will warm at a greater rate than days (Kling et al. 2003).
- Growing seasons will lengthen, with earlier spring thaws, runoff, and ice-out dates, and later frost and fall freeze-up dates (Johnson and Stefan 2006; Kling et al. 2003; UCS 2009).

The quantity and spatial and seasonal distribution patterns of precipitation are also expected to change as a result of anthropogenic climate forcings. Although the general circulation models are in less agreement about changes in precipitation at continental scales, they are in reasonable concordance for the polar and subpolar regions of North America, and also for the northeastern region of the United States (Solomon et al. 2007). The general forecasts for the north-central United States range from little to no change to moderate increases (probably ≤ 20%) in total annual precipitation. However, the seasonal patterns and intensity of precipitation events are forecast to change as follows:

- Precipitation will be concentrated in stronger events, with increased thunderstorm activity, most likely in late winter, spring, and fall (note that emerging evidence in the current climate record indicates a recent trend in heavy precipitation events starting around 1975 for 2-inch and larger, and

4-inch and larger events; see Figure 12.2). This forecast is generally more robust than other precipitation changes as a result of increased evapotranspiration rates and higher overall energy in the atmosphere (Solomon et al. 2007; CCSP 2008; UCS 2009).

- Concomitantly, summers will be drier, with increased drought frequency (Kling et al. 2003).
- There will probably be less snowpack overall, with earlier spring runoff (see Johnson and Stefan 2006, for current trends), and thus changes in seasonal hydroperiods, the seasonal timing and height of water-level changes in water bodies. However, a higher probability of rain-on-snow events in late winter could increase the probability of local flooding.

The incidence of heavy rainfall events has increased and is now double the rate of a century ago (Kunkel et al. 1999). The top graph in Figure 12.2 shows the percentage of all recorded rain events of 2 inches or more, and the bottom graph shows those of 4 inches or more. Both graphs include a lowess smooth (locally weighted regression) to show changes in general trends over time. The figure shows that the climate system produced relatively large rain events at the turn of

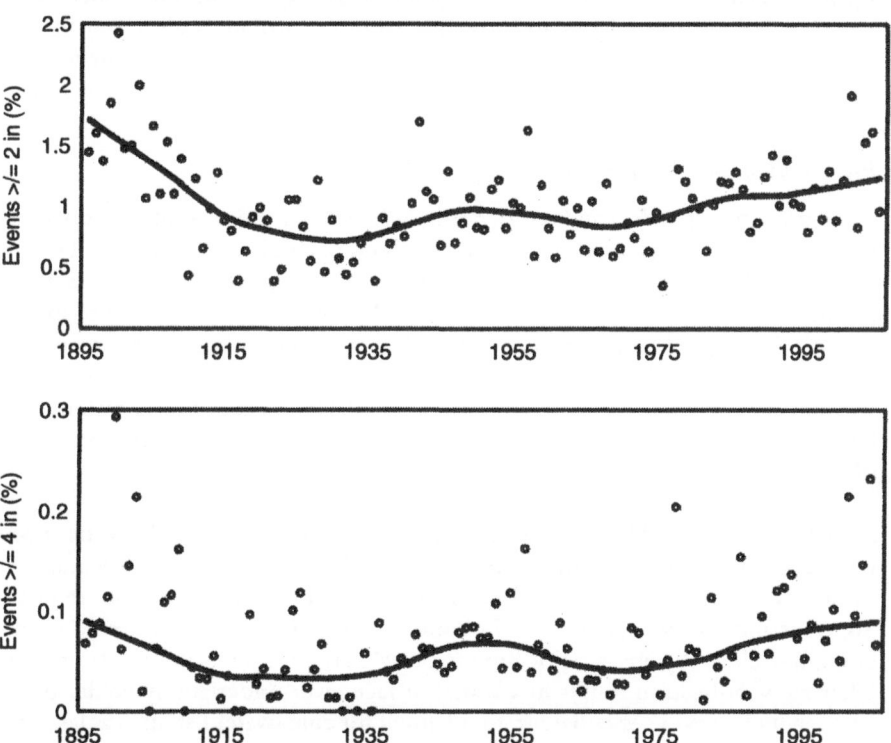

Figure 12.2 *Trends in heavy rainfall events in Minnesota*

Source: Based on state climate date from Zandlo (2010).

the last century, was relatively quiescent during the middle of the 20th century, and now the frequency of heavy rain events appears to be increasing again. Climate change forecasts indicate that the current increasing trends will continue into the future and have the potential to exceed frequencies recorded around 1900. We also note that the data used by Hershfield (1963) included the time period from the 1930s and through most of the 1950s, a window of time heavily influenced by the extreme drought during the Dust Bowl era. Given that Hershfield (1963) has been used to determine the frequency of storm "design" events for important infrastructure such as urban stormwater conveyance, a significant change in magnitude and frequency of storm events may result in such systems being underdesigned for the future.

EFFECTS OF CLIMATE CHANGE ON MINNESOTA WATER RESOURCES

More intense winter and spring storms and hotter, drier, longer summers will have a number of adverse effects on Minnesota's valuable water resources and their beneficiaries: floods, droughts, aquifer volume, pollution, wetlands, peatlands, aquatic organisms, terrestrial wildlife, agricultural production, forest species mix, and the quality of our lakes and streams. Attention to adaptation and planning should not be in lieu of efforts to minimize climate change, but climate-driven changes are already taking place; foresightful actions can help reduce their impacts.

Precipitation and Temperature Impacts

Changes in precipitation and temperature will both have direct impacts on aquatic resources in Minnesota. If heavy storm events do occur, they probably will result in increased surface water runoff into receiving water bodies and potentially less infiltration and slower recharge of aquifers. An increase in frequency of extreme events is likely, including a higher probability of floods in late winter, spring, and fall and drought in summer. Temperature increases will have impacts on surface water bodies, including lakes, rivers, and streams. Cold-water stream resources will be affected, particularly if cold base flows are reduced as a result of decreased rates of aquifer recharge.

Aquatic Habitats. Habitat availability throughout the year for cold-water species, including native trout, could be greatly reduced in streams that rely primarily on surface water, including most of the stream resources on the North Shore of Lake Superior and the trout streams of the southeast. Many lakes in Minnesota are deep enough to allow thermal stratification to occur in the spring as surface waters warm. Thermal stratification results in a warm surface layer underlain by colder bottom strata, which provide essential summer habitat for cold-water fish such as lake trout. However, because stratification often seals off the deeper, cold water, microbial decomposition in the sediments uses up dissolved oxygen in more productive lakes, forcing fish and other organisms that are not adapted to low–oxygen conditions to move to warmer, upper layers where plant life maintains higher levels of dissolved oxygen. In deep, relatively infertile lakes, deeper, colder waters will maintain relatively high levels of oxygen and may support cold–water species throughout the warmer summer months.

Currently in Minnesota, thermal stratification occurs by late spring in the south and early summer in the north. The period of thermal stratification is expected to lengthen considerably as the climate warms (Fang and Stefan 2009). A longer period of stratification will allow more time for dissolved oxygen to be consumed in deep water. This effect will be exacerbated if increased surface water runoff results in higher nutrient loading to lakes, increasing the biological productivity of the lake and the rate of oxygen depletion in deep water. This "temperature–oxygen" squeeze (Coutant 1990; Fang and Stefan 2009) can cause severe habitat loss for cold–water fish species, including lake trout, lake whitefish, burbot, and cisco, and limit habitat for optimal growth of primary cool–water predators, such as walleye and northern pike.

As the climate warms, the state's lakes and rivers will experience significant decline in habitat for cold–water fish, especially in southern Minnesota and in smaller lakes. In parallel, there will be significant habitat increases for warm–water fish, primarily in northern Minnesota (Stefan et al. 1995). These forecasts include increases in cool–water habitat for such popular species as walleye, but while walleye may gain habitat in northern Minnesota, there could be significant loss of habitat in the southern part of the state. For instance, forecasts for Lake Pepin in southern Minnesota suggest that elevated summer water temperatures will cause larger walleye to experience metabolic deficits, resulting in declining yields in this important sport species. Some changes in fish communities throughout the state may be perceived as undesirable by the angling public. For example, the non–native black bass in northern Minnesota is currently suppressed by short growing seasons and therefore does not have significant impacts on native species. However, as the climate warms and growing seasons lengthen, it is likely that expanding black bass populations will start to compete with or prey on native species. Indeed, emerging evidence indicates that this ecological fish community shift may already be under way (Fronhauer 2010; Fayram, 2010). Finally aquatic resource managers can anticipate potential losses of other sensitive, native aquatic taxa, notably amphibians.

Wetlands and Shallow Lakes. The ecology of Minnesota's shallow lakes and wetlands will also be affected by changes in temperature and precipitation. Seasonally flooded wetlands, an important resource for waterfowl production, may be reduced with decreases in snow depth and spring runoff. The probability that this will occur will be determined in part by patterns in spring precipitation, which are actually expected to increase in the Prairie Pothole Region (see UCS 2009, 6). Fish in shallow basins, which have the potential to indirectly decrease water quality and directly compete with waterfowl for food resources (Hanson et al. 2005), have historically been limited by winterkill, which causes extensive extirpation following depletion of dissolved oxygen in these highly productive basins. The frequency of winterkill has been declining in Minnesota because of shorter periods of ice cover as well as greater light penetration that occurs with decreased snow cover and supports more photosynthesis and oxygen production under the ice. This trend of reduced winterkill is expected to increase with climate change (Fang and Stefan 2000).

Furthermore, the depth of shallow basins has increased over time as a result of consolidation of surface waters and increased inputs from drainage. Given the uncertainty surrounding future precipitation forecasts, such as more spring storms or more summer droughts, it is difficult to precisely conjecture future hydrologic budgets for these systems. If drier conditions prevail in the western and northern areas of the Prairie Pothole Region, Minnesota's shallow basins will be important for the future of waterfowl (Johnson et al. 2005). Additionally, plant communities in upland areas exert considerable influence on the hydrologic budgets of wetland systems (Voldseth et al. 2009) because of significant differences in infiltration and transpiration rates. Ecological resilience of shallow basin ecosystems in the future will be even more closely tied to plant management strategies and agricultural practices in their surrounding uplands.

Increased Runoff and Pollution

As a result of increased storm frequency, higher rates of surface water runoff can be expected. In the absence of best management practices in any type of landscape, such as urban, agricultural, or forest, this will result in greater loading rates of sediment, nutrients, and other pollutants to receiving water bodies, with resultant negative impacts on water quality.

Groundwater Aquifers. Regardless of climate uncertainty, the interaction of groundwater and surface waters is a key research need. One scenario is that increased surface runoff and decreased infiltration will decrease groundwater recharge rates, exacerbating challenges to sustainable groundwater use. That scenario is spatially complex because in some areas of Minnesota, such as the southeastern karst region, surface water bodies may interact with and directly discharge to groundwater.

Water managers also need to consider climate change impacts to local water resources from groundwater withdrawals as well as downstream impacts. For example, Valley Creek in Washington County, once referred to as a blue ribbon trout stream, is a tributary to the St. Croix River, a national scenic riverway. Currently, water managers are trying to determine sustainable rates of extractions from the aquifers that feed Valley Creek, which are needed for municipal development. It is distinctly possible that recharge rates to these aquifers will decrease in the future as a result of climate change, causing seasonal dewatering of Valley Creek. In this case, not only will Valley Creek proper become impoverished, but the St. Croix River ecosystem will lose a relatively clean tributary. Given that the St. Croix River has already been adversely affected by urbanization (Triplett et al. 2009), additional impacts from losses of high-quality tributaries could contribute significantly to declining water quality in this important resource. Incorporating climate change uncertainties into current and future water policy will be essential for achieving sustainable water quality and quantity goals.

Changes in Water Levels

Predicted changes in precipitation patterns, higher summer temperatures, and reduced periods of ice cover will result in lower water levels in lakes, rivers, and streams (Karl et al. 2009), with ecological and economic impacts to many lakes. For Lake Superior, Austin and Colman (2008) show that lake water temperatures have increased more rapidly than air temperatures, which they attribute to decreased periods of ice cover. Also, extreme changes in water levels have negative impacts on the shipping industry, as recently experienced on the Mississippi River during the 1993 floods and 1988 drought. Thus, increased frequency of extreme hydrologic events brought on by climate change will have significant implications for many economic sectors.

Increases in Invasive Species

The predicted summer droughts, longer growing seasons, and milder winters due to climate change are likely to lead to increased outbreaks of pest or invasive species. An example of this is the recent increase in infestations of curlyleaf pondweed, an invasive rooted aquatic plant that is winter-hardy but also associated with algal blooms following senescence in the spring. More vigorous growth of this plant occurs with milder winters, probably from reduced snow cover and more light penetration through the ice.

AGRICULTURAL IMPLICATIONS

Management of Minnesota farmland will be a primary factor in determining the impacts of climate change on the quality and availability of the state's waters. However, as future climates evolve, agricultural impacts of climate change on water resources in Minnesota can be described in three terms: uncertainty, too much water, and not enough water. Large storm events will be more frequent and less predictable in terms of magnitude and timing, summers will be drier, and variance among years will be greater.

The impacts of climate changes on water resources for Minnesota agriculture are likely to be profoundly adverse. The number and intensity of large rainfall events is expected to increase. Summers will experience more periods of drought and extreme heat. More intense storms, particularly in early spring, will flood rivers and fields, interfering with spring planting and damaging young crops. Increased runoff from larger storms will carry more sediment, nutrients, fertilizer, and pesticides into streams, lakes, and groundwater. Higher runoff velocities will decrease groundwater infiltration. Beyond but related to water resource impacts, the warmer winters and longer growing seasons will enable invasions of additional pest species, such as the rootworm and European corn borer. The resulting inevitable increase in pesticide use will affect streams and groundwater. Additionally, heat stress is predicted to adversely affect crop yields and livestock (Rosenzweig et al. 2000; UCS 2009).

Spring Downpours and Floods

A 1-degree increase in temperature increases the moisture-holding capacity of the air by 7% (the Clausius–Clapeyron relation). It is estimated that atmospheric water vapor over the oceans has increased over the last century by about 5%. This has resulted in more intense storms, both as rain and as snow (IPCC 2007). As discussed earlier, intense rainfall events have increased in Minnesota over the past decade. Projections are that late winter and spring will see a significant increase in precipitation, with autumn rainfall also increasing, but summers becoming drier (UCS 2009).

Spring flooding of fields results in planting delays. Production of high-value crops, such melons, sweet corn, and tomatoes, will be in trouble if that occurs. If fields flood during the growing season, crop losses will occur from anoxia and root diseases. Crop damage from heavy storms and concurrent strong winds will increase as well (CCSP 2008). If the response of agricultural landowners is simply to increase tile draining, heavy erosion of streambanks and increased flooding will result.

Recent empirical evidence in Minnesota has shown that large storms cause the bulk of soil loss, and that vegetative cover is very effective in reducing losses. For example, at two agricultural research sites south of the Twin Cities, 4 inches of rain fell in a few hours on June 26, 1998. One site had been conventionally plowed with moldboard plowing and planted to corn, which was then about a foot high. A neighboring cornfield had been chisel-plowed, which disturbs less of the surface and leaves more residue. Soil losses during this storm were huge: the moldboard-plowed field lost 8.3 tons of soil per acre in that one day; the chisel-plowed field lost about half that. Another research plot nearby, which was planted to alfalfa and pasture, lost only 53 pounds (less than 0.03 tons) per acre during the storm event (DeVoire 2001).

Hot, Dry Summers

Extreme summer heat will become more common. Three-day periods of temperatures greater than 95 degrees F have historically occurred only about once a decade. Under the higher-emissions scenario of the Intergovernmental Panel on Climate Change (IPCC), however, the projection is that by 2050, this will occur every other year. By the end of the century, these heat waves will be common, occurring in 3 of every 4 years. Seven consecutive days of greater than 95 degrees F have occurred historically only once in 30 years. In the higher-emissions scenario, projections are that conditions comparable to the European heat wave of 2003 will occur about once in 5 years by 2050, and every other year by the end of the century (UCS 2009).

Summer rainfall is projected to decrease by about 15%, although this is less certain than is increased spring rainfall. Combined with the stress of increased heat, the impact on crops could be severe. Corn and wheat may fail at 95 degrees F, the risk increasing with duration (UCS 2009). The summer drought of 1998 resulted in economic losses of $56 billion, a large portion of which was from agriculture. With increasing exposure to extreme temperatures and precipitation changes, farmland in affected areas could decrease in value substantially (Rosenzweig et al. 2000).

EFFECTS ON FORESTS AND ASSOCIATED WATERS

Clearly, forests in Minnesota will change as a result of global warming, and these changes could have major impacts on the state's waters. Healthy forests are important to a clean water supply in forested watersheds (USFS 2010). When trees are removed or cut back, reduced forest cover changes the hydrology of the area, with the increase in runoff proportional to the amount of cover removed. Increased forest fires and more devastating insect infestations will result from climate change. These events will remove trees, affecting water quantity and quality. Water volume and pollution in years immediately after a severe fire may increase by orders of magnitude. Insect epidemics can increase flows as well as the likelihood of forest fires (CCSP 2008). Water temperatures also may increase, which will affect the capacity of waters to sustain cold-water fish species, as discussed above. Also, changes in the species mix of terrestrial vegetation will change temperature and water chemistry, particularly dissolved carbon.

Boreal Forest

Minnesota's boreal forest extends through the northeastern third of the state, north of a northwest–southeast diagonal line across the top third of the state. It is likely to retreat several hundred kilometers to the north with increasing temperatures (Frelich and Reich 2010), with spruce and pine giving way to deciduous species, including oaks (Axelson 2010). Such forest conversions, or conversion of forestlands to other uses as the boreal forest edge retreats, would bring large changes. Although wide-scale conversion to agriculture is unlikely, even relatively small changes would result in significant impacts.

When millions of acres of forestland and native prairies in upper midwestern states were converted to cropland by European settlement, this resulted in significant increases in runoff and corresponding decreases in evapotranspiration. Forests have higher evapotranspiration rates than prairies or cropland, resulting in lower moisture levels in forest soils and greater infiltration. During winter, the forest canopy intercepts snow and slows melting by providing shade. In a simulation of water-related impacts of conversion of the midwestern forest to cropland, runoff increased in deforested areas by 20% to 40% as a result of conversion to agriculture by settlers (Mao and Cherkauer 2009).

It is true that climates have changed in the past. However, under presettlement conditions, as the forest retreated, native prairies replaced them over thousands of years. Now native prairies are only 1% of their original extent, and the exotic species found in roadside ditches, crop borders, and cleared unplanted land could become dominant (Galatowitsch et al. 2009a).

Chemistry, Metabolism, and Temperature of Forest Lakes

Lakes are expressions of the watersheds in which they lie. Changes in forest composition, runoff, amounts of precipitation, and forest fires will all affect the watersheds, resulting in impacts to the chemistry and biota of forest lakes in ways that are not currently well understood. Some impacts are predictable. More forest fires will increase runoff of nitrogen and phosphorus to lakes, but those effects will be short-lived; fires will also increase mercury loading in lakes (Schindler 2009). Because of the reduced forest cover and changes from forests to grasslands or savannahs, greater runoff will likely occur in previously forested areas from any given precipitation amount.

Forested regions in Minnesota have lower summer maximum daily temperatures than do those of more open landscapes at the same latitude, as a result of the absorption of incoming solar radiation by the tree canopy and the cooling effects of evapotranspiration. The heat budgets of lakes may be directly affected by climate change, because of increased solar energy received, and also indirectly affected as forests lose canopy and resulting higher evening temperatures and higher dew points decrease lake-surface cooling.

Other Water-Related Forest Landscapes

Boreal peatlands, which occupy more than 2.4 million hectares in northern Minnesota, are one of the greatest storehouses of carbon in the state. They include peatland vegetation, bogs, black spruce and tamarack swamps, and fens that have accumulated over thousands of years. Abundant water and low temperatures are essential to their maintenance, because these attributes reduce decomposition rates. With lower water levels, shrub growth could increase at the expense of bog vegetation. Peatlands are sensitive to disturbance. If a drier climate causes peatlands to dry out, decomposition will be accelerated, fires will become more likely, and there could be changes in vegetative composition, releasing large amounts of carbon into the atmosphere. Of particular concern are Minnesota's approximately 100 calcareous fens, unusual ecosystems that are rich in minerals and contain many rare plant species. Loss of groundwater recharge will reduce incoming water and change species composition (Galatowitsch et al. 2009a, 2009b).

PRINCIPLES AND POLICIES TO ADDRESS IMPACTS ON WATER RESOURCES

The 2007 Next Generation Energy Act,[1] enacted by the Minnesota legislature, established statewide goals of reducing greenhouse gases to 30% below the 2005 levels by 2025 and 80% by 2050. The act also bans new sources of coal-fired electrical generation until a cap-and-trade system or its equivalent is established. A broad-based stakeholder group recommended a plan to meet the greenhouse gas reduction goals (MCCAG 2010). However, although Minnesota is among

28 states to have formulated plans to reduce greenhouse gas emissions, it is not among the 8 that have created proactive plans to minimize adverse impacts of changing climate (Galatowitsch et al. 2009a). The legislature should consider directing the preparation of a climate adaptation plan.

Although policy changes to combat climate changes have been under discussion for decades, global greenhouse gas emissions continue to increase. Global temperatures will continue to rise even if emissions reductions are soon achieved. If we wish to minimize adverse impacts and increase the probability of a relatively smooth transition to a future climate and its resulting ecosystems, we must begin planning and acting to minimize future damage to water and other natural resources from an uncertain, more variable, warmer climate.

A great deal of water resource policy planning activity is currently under way in Minnesota, timed to culminate with the 2011 legislative session. The Clean Water Council, created through the Clean Water, Land, and Legacy Amendment of 2006, is designed to prioritize state activities to restore and protect Minnesota waters. Interagency work groups are preparing reports on groundwater and surface water strategies, and the 2009 legislature appropriated funds for the Minnesota Water Sustainability Framework, a participatory process administered by the University of Minnesota's Water Resources Center, to develop a framework for sustainable water management. Its goal is to preserve the state's waters for this century and beyond, while protecting current needs, water quality, and natural ecosystems. Despite the legislature's recognition of the dangers of climate change, however, demonstrated by ambitious goals for carbon emissions reduction, the water-planning legislation does not specifically mention climate change, and it is not clear that ongoing planning activity aimed at restoring and protecting water resources will consciously and pointedly take into account the likely and potentially dramatic impacts of climate change and the need for carbon reduction and sequestration. Failure to do so is likely to cause the state's extensive efforts to be of diminished benefit in the long term.

Chapter 16 of this book sets forth five relevant principles for decisionmaking. Water resource policymaking in the era of global climate change requires close attention to those basic tenets and adds further technical complexity, as discussed in more detail later in this chapter. The following points must be seriously considered in future planning, given the complications arising from the uncertainties of future climate:

- *Past is not prelude.* Although it is very likely that hydrologic conditions will change in the future as a result of climate change, many factors are unknown and unpredictable, including the extent to which humankind will develop ways to reduce and sequester carbon. The possibility of unforeseen feedback mechanisms that could intensify the impacts is under study (CFS 2010). Clearly, policy formation cannot rely exclusively on past experience.
- *Plan for various scenarios.* Water management will be an exercise in risk management, involving socioeconomics, crop and timber production, fish and wildlife, species preservation, and biological diversity (IPCC 2007). Water

planners should evaluate a range of precipitation levels and varied drought intensities in order to develop a suite of potential policy responses for each possible scenario.

- *Flexibility is critical.* It is imperative to be able and willing to change with changing conditions. Changing climate will drive other changes as well as conflicts over water. Planning on the governmental, corporate, and individual levels must have the capacity and willingness to change rapidly.
- *Think cash for carbon.* The economics of adapting to climate change are likely to include incentives to avoid greenhouse emissions and sequester carbon from the atmosphere. Economic incentives are necessary to accomplish enough adaptation to protect water quality, prevent flooding, and retain groundwater infiltration. Economic incentives will be necessary to transition some of the land now devoted to row-crop agriculture to other, more diverse cropping systems.
- *Invest in the future.* Research funding is needed to develop new techniques and products that will facilitate food and fiber production in ways that will help reduce atmospheric carbon while minimizing adverse water impacts (McIntyre 2009). Finding and employing new techniques to minimize water use, particularly for irrigation, is critical (Gleick 2002).
- *Get to know the watershed neighborhood.* One of climate change's primary detrimental impacts to water quality will be more runoff and potentially less infiltration if the frequency of heavy storms increases and long, steady soaking rains decrease. A key conservation response to this threat will be to apply concepts of integrated watershed management (discussed in Chapter 14; see also DNR 2010; CWP 2010). Management at this level should focus on four primary activities. First, conduct detailed spatial analysis at the watershed scale to determine current and future sources of surface runoff and infiltration areas. Often only a minor fraction of the pollutant load to a surface water body comes from the immediate riparian zone, with significant nonpoint-source loads emanating upstream in the watershed. Such detailed assessments, often conducted with sophisticated hydrologic-based models, should incorporate precipitation scenarios likely to occur with climate change. Second, identify areas on the landscape that may be strong candidates for promoting infiltration. This activity may be done at various scales, depending on the type of human development in the watershed. For example, municipal and residential rain gardens could be developed in urban watersheds, while larger parcels in agricultural areas could be targeted for conservation programs or acquisition and subsequent restoration of the plant community. Third, identify surface water resources that are prone to runoff events and identify key areas in riparian corridors for restoration of buffer zones. Seavy et al. (2009) provide a good case for the role of riparian restoration in climate change adaptation. Finally, engage local citizens and units of government in developing watershed plans. Experience with watershed management at the local scale indicates greater success when local involvement and ownership are established (see Chapter 6).

MITIGATION AND ADAPTATION STRATEGIES FOR WATER MANAGEMENT

Water resource management agencies may deploy a number of potential policy options to address both mitigation of and adaptation to climate change. In the climate change literature, "mitigation" refers to practices that reduce greenhouse gas emissions, such as the development of clean, renewable energy sources and carbon sequestration through reforestation, and "adaptation" refers to changes to natural resource and human infrastructure management programs because pending climate change will render today's management systems ineffective or obsolete.

Galatowitsch and colleagues (2009a, 2009b) described three potential strategies for adaptation to climate change that could be applied in various ecosystems. The first is *resistance*, finding natural systems likely to survive and maintain most of their native biota, and preserving them to the extent possible to allow time for research to determine whether permanent preservation is possible.

The second is *resilience*, the preferred strategy, designed to help existing natural ecosystems adjust to changed climates so they can maintain biological diversity. We know that ecosystems will change; resilience seeks to facilitate the transition to a different and healthy ecosystem that includes a diversity of native species. Diverse ecosystems are protected through buffers on adjacent lands. This requires identifying high-value natural areas and acquiring title or conservation easements or creating incentives for resilience management on private lands. Managing lands to promote resilience involves planting a variety of species after disturbances create opportunities, so that the resulting diversity allows natural selection to determine the species best able to withstand climate change, while protecting important natural areas from stresses that prevent recovery, such as deer overbrowsing and insect damage. An example of a new research and monitoring program focusing on identifying key components of ecological resilience in lakes is the Minnesota Department of Natural Resources' sentinel lakes program (discussed in Chapter 3).

The third is *facilitation*, which involves importing entirely new species and is the most controversial. Some scientists argue that we cannot be knowledgeable enough to know the results of trying to engineer new ecosystems, and that importing new species may interfere with natural migration and seed dispersal of native species. Others say that importing native species is better than colonization by nuisance species like buckthorn (Axelson 2010). As discussed later in this chapter, facilitated migration may also be a reasonable adaptation strategy for conserving aquatic biodiversity.

In the face of climate change, management will have to deal with mitigation and adaptation challenges in a number of areas, including groundwater resources, water conservation, carbon sequestration, and fish and other aquatic fauna. They will also have to consider the potential implications for the Endangered Species Act as species' ranges shift.

Managing Groundwater

Protecting groundwater resources and ensuring sustainable withdrawal rates present challenges in Minnesota today. The challenge will only increase in the future if changes in precipitation patterns decrease infiltration and recharge rates. Therefore, the need is fairly urgent to advance our understanding of surface water and groundwater interactions. Such advancement will help better identify critical recharge and infiltration areas for protection or restoration and will improve our ability to inventory and forecast future changes to groundwater resources. Better integration across the numerous agencies, units of government, and nongovernmental organizations (NGOs) involved in water management would support this goal.

Water Conservation

Regulatory agencies working with various partners should consider reasonably aggressive reform of policies that permit wasteful water use. Such policy reform should apply to all sectors of society, including municipal extractions, irrigation for agriculture, and private use, where lawns can account for as much as 50% of a municipality's annual water usage. It is also important to acknowledge the sobering possibility of impacts on water resources from synergistic stressors resulting from climate change. Greater frequency of storms could decrease infiltration and recharge rates of groundwater, while greater frequency of summer droughts may increase withdrawal rates in both agricultural and municipal sectors.

Carbon Sequestration in Lakes, Wetlands, and Peatlands

Summed together globally, lakes, impoundments, and peatlands are estimated to sequester carbon at a rate that may be threefold the rate that occurs in the oceans (Dean and Gorham 1998). Wetlands may also play significant roles in carbon sequestration through burial in bottom sediments. Carbon will be sequestered following the restoration of wetland systems that have been drained and currently hold little standing water at any time during the year. Based on estimates by Euliss et al. (2006), restoration of wetlands in the Prairie Pothole Region of North America has the potential to sequester 378 teragrams of carbon in 10 years. This is over 400 million tons of carbon, nearly 2.5 times Minnesota's emission of about 150 million tons of greenhouse gas emissions per year. Thus, restoration of drained basins should be considered a viable and productive carbon sequestration policy. Moreover, policy changes should consider the enormous benefit that restoring Minnesota wetlands would return to the migratory bird resource, which could be greatly diminished if the prairie pothole wetlands of the "duck factory" in the Dakotas and Manitoba are lost to drier conditions. The value of this important ecosystem service should be incorporated into any cost–benefit analysis applied to wetland restoration programs. The picture for carbon sequestration in existing wetlands is considerably more complex and is currently being studied. For example, emerging research indicates that the rate of carbon burial may be greater for wetlands in clear water than during turbid times (Domine and Zimmer 2009). Further research is needed to quantify these fundamental dynamics of wetlands.

Adaptation for Fish and Other Aquatic Fauna

With climate change, Minnesota's aquatic species will experience local extirpations and potentially significant range shifts. The most sensitive species will be obligate cold-water species, such as cisco, lake trout, burbot, and lake whitefish, as well as a variety of other fish, reptiles, and amphibians that are intolerant of ecosystem degradation. Expanding populations of warm-water species will increase competition and predation on species less adapted to warming conditions. To prepare for such changes, aquatic resource managers may consider assisted migration, translocating cold-adapted species from southern Minnesota to lakes in the north where they do not exist now but where conditions may remain cold enough for them to persist. This would be analogous to the facilitation techniques of assisted migration of plant species for terrestrial ecosystems.

As discussed above, maintenance and restoration of ecological resistance, the capacity of an ecosystem to absorb change without undergoing significant shifts in structure and function, will also be an important adaptation practice. The Minnesota Department of Natural Resources (DNR) and the University of Minnesota are currently conducting research to identify lakes that are strong candidates to provide refuge for cold-water species in a warmer future. After identifying these potential climate change refuges, resource managers can direct conservation efforts to these systems with the goal of preserving ecological resilience, such as by minimizing external nutrient loading with the aim of maintaining dissolved oxygen during the period of thermal stratification. Conversely, the same methods can be used to identify cold-water systems today that likely will not be able to maintain cold-water species in the future, taking a triage approach to management.

Changes in precipitation patterns along with timing and magnitude of spring runoff may cause significant changes in river hydrographs. For some key riverine species, the height of today's spring hydrograph is critical to allow access to backwater habitats for spawning (Ickes et al. 1999). Numerous other sensitive aquatic species have key aspects of their life history that depend on the natural flood–drought cycle, a cycle that will be different in the future. Aquatic resource managers should devote energy to identifying such key relationships and perhaps consider appropriate river and stream restoration techniques that would provide continued access to critical riverine habitats.

Implications for the Endangered Species Act

Climate change also has potential implications for the Endangered Species Act (ESA). Reform of this important act will be a significant policy challenge as species range shifts become inevitable. The recent decision by the U.S. Fish and Wildlife Service not to list coaster brook trout under the ESA is an interesting case to consider. Coaster brook trout are native to Lake Superior, other areas of the Great

Lakes, the Appalachian region, and the Atlantic coast. This species could be imperiled if cold-water stream habitat becomes limited. For example, because the streams along Minnesota's Lake Superior North Shore are primarily surface water fed, cold-water habitat exists only during the colder months of the year. If the climate warms enough and warm weather persists long enough into the fall, coaster brook trout could lose critical spawning and nursery habitat. This raises questions about the utility of listing this species as threatened if the options available to a management agency for saving it are extremely limited. It may be wiser to direct conservation efforts toward finding suitable habitat in more northerly locations where climatic conditions may allow the species to persist.

Support for Research and Implementation of Adaptation Strategies

Units of government can provide support for research and the implementation of adaptation strategies in a variety of ways. Lakes are excellent ecological sentinels because they integrate almost everything going on in a watershed, from atmospheric changes and deposition to groundwater and surface water dynamics (Schindler 2009). Development of sentinel monitoring systems in Minnesota clearly requires strong collaboration among key agencies, including the DNR, Minnesota Pollution Control Agency, and U.S. Geological Survey. Government agencies can also provide support for downscaling the general circulation models (GCMs) used to forecast climate change. Outputs from GCMs are arranged in a grid system that spans the entire globe, but a grid cell occupies a very large area. For example, the state of Minnesota is almost entirely covered by only two grid cells. Unfortunately, much finer spatial resolution is required for natural resource managers to be effective in their pursuits; well-developed downscaling technologies should receive strong support right away.

Also, given that the climate system is currently in a highly nonstationary state, development of new policies and specific technical guidance must be done in a way that allows efficient revision as new climate data and forecasts become available. A good example of this is Technical Paper 40 (Hershfield 1963), which defines probability-based storm events used by water resource managers and engineers for designing water management infrastructure. This important source of technical information is currently being revised by a joint-agency stormwater committee in Minnesota. If this revision process is not capable of producing an easily updated, living document, the committee risks having its work becoming prematurely outdated. These technical challenges clearly indicate that design and implementation of effective climate change adaptation strategies require institutional barriers to come down, across all levels of government.

AGRICULTURAL POLICY RESPONSES

There is an important relationship between greenhouse gas reduction and adapting U.S. agriculture to the impacts of climate change, but a policy gap exists between what government supports and the need to reduce greenhouse gases in the atmosphere (see Chapter 11). An integrated approach is needed that takes into account water, vegetation, wildlife, and economics of agriculture. Such an approach would compensate farmers for "ecosystem services" that would benefit society through reduction of greenhouse gases and carbon sequestration and provide protection to the land and water from adverse effects of increased storm intensity, summer drought, and flooding.

Cap-and-trade legislation now pending in Congress could produce funding for agricultural practices that sequester carbon and reduce greenhouse gases. In crafting legislation and administrative rules to evaluate the efficacy of sequestration, policymakers should take account of the other environmental benefits of agricultural practices that will reduce runoff, pollution, and flooding, as well as restore and save wetlands. Moreover, agricultural practices that produce fewer greenhouse gases from fuel use, nitrogen fertilizer, and methane should be favored. Research on the efficacy of various farming practices in reducing runoff, reducing carbon emissions, and sequestering carbon should be intensified in anticipation of the need for such data in constructing new economic relationships (McIntyre 2009). However, finding ways to increase appropriate vegetative cover on farmland in a way that makes economic sense is a big challenge.

Some government programs in place for years have helped convert marginal cropland or pastureland for the purposes of prairie or wetland restoration (MDA 2010). The 2008 Farm Bill enhanced the Conservation Reserve Program (CRP) (USDA FSA 2010), and many hope this will overcome the kind of trend observed in South Dakota between 1983 and 1987, when more land was converted from grassland to cropland than was added to the CRP. Much more effort is needed to encourage the types of farming that will help in adapting to climate change. Government support programs that encourage only production of row crops, primarily corn and soybeans, are antithetical to the goal of adaptation to climate change. Government policy needs to be aligned with the goals of protecting public goods, supporting agricultural diversification, and rewarding farmers for environmental benefits (Boody et al. 2009). Although row-crop farming will continue to dominate Minnesota agriculture, there are methods and practices that can be economically productive while helping reduce runoff, pollution, and floods, as well as enhancing groundwater infiltration to restore aquifers. New techniques that will enable profit from production of ecological services as well as agricultural products are being explored and should be encouraged (Jordan and Warner 2010).

A well-known farming practice that can dramatically reduce runoff and produce other benefits is to either interplant cover crops with row crops or plant them before the row crops. The cover crops are not intended for harvest. They improve soil quality by increasing biotic activity, soil percolation, and organic content, all of which increase infiltration and reduce runoff (Boody et al. 2009). They also reduce soil loss.

Another practice is grass farming for animal and energy production. Fertilized annual row crops, particularly corn, produce the greatest volume of runoff and pollutants. Converting some cropland that was planted in row crops to grassland agriculture has multiple benefits, including reduction of runoff, cleaner water, greater wildlife habitat, carbon sequestration, groundwater infiltration, scenic values, and profitability for farmers. Riparian lands are prime candidates for such treatment.

A Minnesota study by Boody et al. (2009) demonstrated that converting sloped fields from row-crop farming and best management practices to management-intensive rotational grazing or perennial energy crops can produce reductions in sediment and nitrogen, with additional benefits in the form of greater habitat, reduced greenhouse gases, and profitability for farmers. A breakthrough in converting prairie grasses to fuel could result in large-scale conversion of corn to native prairie or other grasses, with concomitant benefits in water quality, energy saving, carbon sequestration, reduced runoff, increased groundwater infiltration, and reduced fertilizer and pesticide use.

In developing agricultural policy, efforts to preserve and restore riparian lands should be paramount, particularly in choosing lands to convert from row crops to prevent runoff. Riparian ecosystems are naturally resilient and able to recover from disturbance. They provide the connection between terrestrial and water habitats and enhance the productivity of both. Riverine habitats serve as ecological corridors for many species and will allow movement in response to climate change. Maintenance of these ecosystems should help reduce the impact of flooding, improve water quality, and preserve microclimates for a range of species (Seavy et al. 2009).

WATER POLICY IMPLICATIONS OF CLIMATE-CHANGED FORESTS

Impact on water quality and quantity will be only one of many factors that will drive policy decisions on the future management of Minnesota forests in response to climate change, which will also include concerns for preservation of biological diversity and wilderness, commercial timber harvesting, wildlife species, legacy ecosystems, native plants, recreation, and scenic beauty. One thing is clear: scientific research is needed concerning the effects of climate change on water quality and quantity in the forested areas of the state in order to help guide future management decisions. It is not known how different replacement species following disturbances will affect water quality and quantity. For example, how will dissolved carbon in lakes be affected by particular types of tree species in their watersheds? The species of trees and other vegetation is only one of many factors affecting these waters; other factors include ice cover duration, temperature, nutrient content resulting from runoff, precipitation levels (wet years or drought), and pH levels. Some of these may not be controllable, but awareness of the likely effects is needed to aid decisionmaking.

In their recent article about climate impacts on Minnesota forests, Frelich and Reich ask:

> Important decisions regarding forest management will need to be made as the climate warms: do we prioritize the maintenance of as much of our

recent forest heritage as is possible? Do we let nature take its (human-aided) course? Or do we proactively use adaptive management strategies to accelerate the shift toward new vegetation, better suited to the conditions of the late 21st century? The answers to these questions are not simple and effective management will probably include a combination of all three approaches, guided by input from a wide array of stakeholders, including citizens, the timber industry, and public and private land management agencies. (2010, 376–377)

The three types of management approaches described earlier can be applied to management of water in forested areas that are adapting under changing climate conditions.

Resistance

Resistance in this context means identifying natural, native forest ecosystems that appear to have a chance of survival. Even if they do not survive long-term (although there is some evidence that balsam and black spruce could persist), preservation for decades buys time to develop other choices and complete research on how best to manage the forest and protect waters and other valuable aspects of the ecosystem.

In the boreal forest, this would mean prescribed burning to maintain fire-dependent species like paper birch, black spruce, and jack pine, enabling them to spread, as well as seeding burned areas to promote natural selection in changed conditions and allowing the best-suited individuals to survive. It would mean not logging the last of these areas. In other forested areas, remnants of old-growth forests amid agricultural areas should be evaluated as possible seed sources to be preserved for use in northern forests as the climate changes (Axelson 2010).

For peatlands, a huge storehouse of carbon, water levels should be maintained to the extent possible by designating buffer areas, preventing drainage of neighboring areas, and restoring nearby previously drained wetlands. The benefits, preventing release of large stores of carbon through fires and accelerated decomposition from drying of the wet peat, may well justify extensive funding through carbon credits. Additional research is needed to identify the techniques most likely to be successful in continuing carbon storage by peat. Calcareous fens should be protected through buffer areas. For areas of peat, fens, and existing wetlands, no water removal by drainage or groundwater appropriation that will affect water levels in those features should be allowed.

Resilience

Resilience, in the boreal forest, means facilitating changes to a desirable hardwood forest with greater diversity than would occur naturally, retaining jack pines and spruces where possible, and replanting to white pines, maples, and oaks rather than letting nature take its course to convert the forest to brush, grass, maples, and aspens (Axelson 2010). In terms of water quantity and quality, forest cover has a greater evapotranspiration rate than grasslands and will facilitate more water infiltration to groundwater. Some areas now in forest will become savannah, simply because we cannot control such large areas with management techniques. However, best management practices should be used to create buffer areas in riparian zones within changing forest areas to protect water quality in lakes and streams from excessive runoff, particularly from large storm events. Other than prescribed burning, forest management techniques of replanting and reseeding are not allowed within the federal wilderness areas. This policy may need to be reconsidered under new climate conditions.

Facilitation

Facilitation means replacing existing disturbed forests that are regenerating with species that do not occur naturally in a given area, but that seem likely to do well in a warmer climate. This is controversial. Some scientists argue that when it is certain that the species will change, we should pick something that is likely to do well, based on experience in other climate zones. Others fear that we are not smart enough to know what is likely to survive and could trigger unintended consequences. They argue that natural migration can still proceed in Minnesota, in view of facilitation provided by the ridges of the St. Croix and Itasca moraines (Axelson 2010). However, artificial facilitation would seem to make sense where we know that a forest will be lost. For example, in large areas of ash, where the emerald ash borer will cause devastation and experience in Michigan and Ohio suggests that buckthorn will invade rapidly, it would make sense to pick replacement species likely to survive under changing conditions and encourage their growth. Replacing the boreal forests along the North Shore of Lake Superior with eastern hemlocks from a remnant stand near Duluth would be another sensible option (Axelson 2010).

CONCLUSIONS

At the beginning of this second decade in the 21st century, the future global political and biophysical climates are interrelated and uncertain. Whether or not decisive and effective action to reduce greenhouse gas emissions is taken, it is clear that anthropogenic climate changes will continue to increase. The extent of the resulting changes will depend on the timing and extent to which we mitigate them, such as by reducing emissions and sequestering carbon. Significant climate changes are already under way and will become more pervasive and increase in magnitude within the next few decades. Adaptation will be needed in any case to minimize adverse impacts. It makes sense to develop new planning and management approaches now, recognizing an uncertain future. An adaptation plan for Minnesota, one that is flexible and recognizes uncertainties and the need to react to changing scenarios, would be a reasonable first step.

Water is obviously a critical resource in all systems, both natural and human-constructed, and thus adaptation for climate change should cover all sectors of human activity. Increased frequency of large storms, summer droughts, and higher evapotranspiration rates will drive changes that will require adaptation. In human-constructed systems, water managers will need to critically review infrastructure for managing stormwater and assess future supplies and delivery of potable water sources; in agrarian landscapes, waters sources for irrigation will be affected, including recharge rates for relatively shallow aquifers, and potential for soil erosion may increase with greater frequency of storm events and resultant runoff. In the prairie zone, natural shallow basins will be subjected to the ecological impacts of greater persistence of resident fish populations as the frequency of winterkill decreases as the natural disturbance regime declines. Future water levels of shallow basins are uncertain but will likely be linked to upland vegetation management and land use. In the forested zone of northern Minnesota, the boreal forest is expected to decline and transition to another, yet uncertain vegetation type, with potential downstream implications for carbon dynamics and nutrient loading. Concomitantly, fish habitat will change, with large-scale loss of cold-water habitat, large-scale increase in warm-water habitat, and a northerly shift of optimal habitat for cool-water species. Because of the strong linkage between surface water attributes and watershed-scale land use, climate change adaptation of water resources will have to be closely linked to adaptation of terrestrial-based human activities.

NOTE

1. Minn. Stat. § 216H.02, subd. 1; § 216H.03, subd. 3.

REFERENCES

Austin, J., and S. Colman. 2008. A Century of Temperature Variability in Lake Superior. *Limnology and Oceanography* 53 (6):2724–2730.

Axelson, G. 2010. Trees Fit for the Future. *Minnesota Conservation Volunteer* (January–February):12–23.

Boody, G., P. Gowda, J. Westra, P. Welle, B. Vondracek, and D. Johnson. 2009. Multifunctional Grass Farming: Science and Policy Implications. In *Farming with Grass: Achieving Sustainable Mixed Agricultural Landscapes*. pp. 171–191. Edited by A.J. Franzluebbers. Ankeny, IA: Soil and Water Conservation Society.

CCSP (U.S. Climate Change Science Program). 2008. The Effects of Climate Change on Agriculture, Land Resources, Water Resources, and Biodiversity in the United States. Washington, DC: U.S. Department of Agriculture.

CFS (Canadian Forest Service). 2010. Fire and Climate Change. http://cfs.nrcan.gc.ca/projects/284 (accessed December 12, 2010).

Coutant, C.C. 1990. Temperature-Oxygen Habitat for Freshwater and Coastal Striped Bass in a Changing Climate. *Transactions of the American Fisheries Society* 119 (2):240–253.

CWP (Center for Watershed Protection). 2010. Home page. www.cwp.org/ (accessed December 12, 2010).

Dean, W.E., and E. Gorham. 1998. Magnitude and Significance of Carbon Burial in Lakes, Reservoirs, and Peatlands. *Geology* 26 (6):535–538.

DeVoire, B. 2001. Same Storm—Different Outcomes. *Land Stewardship Letter* (April/May/June).

DNR (Minnesota Department of Natural Resources). 2010. Watershed Assessment Tool. www.dnr.state.mn.us/watershed_tool/index.html (accessed December 12, 2010).

Domine, L., and K. Zimmer. 2009. Personal communication between L. Domine and K. Zimmer, St. Thomas University, and the authors.

Euliss, N.H. Jr., R.A. Gleason, A. Olness, R.L. McDougal, H.R. Murkin, R.D. Robarts, R.A. Bourbonniere, and B.G. Warner. 2006. North American Prairie Wetlands Are Important Nonforested Land-Based Carbon Storage Sites. *Science of the Total Environment* 361: 179–188.

Fang, X., and H.G. Stefan. 2000. Projected Climate Change Effects on Winterkill in Shallow Lakes in the Northern United States. *Environmental Management* 25 (3):291–304.

———. 2009. Simulations of Climate Effects on Water Temperature, Dissolved Oxygen, and Ice and Snow Covers in Lakes of the Contiguous U.S. under Past and Future Climate Scenarios. *Limnology and Oceanography* 54 (6):2359–2370.

Fayram, A. 2010. Personal communication between A. Fayram, Wisconsin Department of Natural Resources, and the authors.

Frelich, L.E., and P.B. Reich. 2010. Will Environmental Changes Reinforce the Impact of Global Warming on the Prairie-Forest Border of Central North America? *Frontiers in Ecology and the Environment* 8:371–378.

Frohnauer, Nick. 2010. Personal communication between Nick Frohnauer, Minnesota Department of Natural Resources, and the authors.

Galatowitsch, S., L. Frelich, and L. Phillips-Mao. 2009a. Coping with Climate Change: Conservation Planning in Minnesota. *CURA Reporter* 39:3–10.

———. 2009b. Regional Climate Change Adaptation Strategies for Biodiversity Conservation in a Midcontinental Region of North America. *Biological Conservation* 142:2012–2022.

Gleick, P.H. 2002. Soft Water Paths. *Nature* 418:273.

Hanson, M.A., K.D. Zimmer, M.G. Butler, B.A. Tangen, B.R. Herwig, and N.H. Euliss Jr. 2005. Biotic Interactions as Determinants of Ecosystem Structure in Prairie Wetlands: An Example Using Fish. *Wetlands* 25 (3):764–775.

Hershfield, D.M. 1963. *Rainfall Frequency Atlas of the United States for Durations from 30 Minutes to 24 Hours and Return Periods from 1 to 100 Years.* Washington, DC: U.S. Department of Commerce, Weather Bureau.

Ickes, B.S., D.L. Pereira, and A.G. Stevens. 1999. Seasonal Distribution, Habitat Use, and Spawning Locations of Walleye *Stizostedion vitreum* and Sauger *Stizostedion canadense* in Pool 4 of the Upper Mississippi River, with Special Emphasis on Winter Distribution Related to a Thermally Altered Environment. Investigational Report No. 481. St. Paul: Minnesota Department of Natural Resources, Division of Fish and Wildlife.

IPCC (Intergovernmental Panel on Climate Change). 2007. Climate Change 2007: The Physical Science Basis. Contribution of Working Group I to the Fourth Assessment Report of the Intergovernmental Panel on Climate Change. Cambridge, UK: IPCC.

Johnson, S.L., and H.S. Stefan. 2006. Indicators of Climate Warming in Minnesota: Lake Ice Covers and Snow Melt Runoff. *Climatic Change* 75:421–453.

Johnson, W.C., B.V. Millett, T. Gilmanov, R.A. Voldseth, G.R. Guntenspergen, and D.E. Naugle. 2005. Vulnerability of Northern Prairie Wetlands to Climate Change. *Bioscience* 55: 863–872.

Jordan, N., and K.D. Warner. 2010. Enhancing the Multifunctionality of US Agriculture. *Bioscience* 60 (1):60–66.

Karl, T.R., J.M. Melillo, and T.C. Peterson, eds. 2009. Global Climate Change Impacts in the United States. Cambridge, UK: Cambridge University Press.

Kling, G.W., K. Hayhoe, L.B. Johnson, J.J. Magnuson, S. Polasky, S.K. Robinson, B.J. Shuter, M.M. Wander, D.J. Wuebbles, and D.R. Zak. 2003. Confronting Climate Change in the Great Lakes Region. Cambridge, MA: Union of Concerned Scientists and Ecological Society of America.

Kunkel, K., K. Andsager, and D. Easterling. 1999. Long-Term Trends in Extreme Precipitation Events over the Conterminous United States and Canada. *Journal of Climate* 12:2515–2527.

Mao, D., and K.A. Cherkauer. 2009. Impacts of Land-Use Change on Hydrologic Responses in the Great Lakes Region. *Journal of Hydrology* 374 (1–2):71–82.

MCCAG (Minnesota Climate Change Advisory Group). 2010. Final Report. www.mnclimatechange.us/MCCAG.cfm (accessed December 12, 2010).

McIntyre, B.D. 2009. *International Assessment of Agricultural Knowledge, Science and Technology for Development (IAASTD): North America and Europe (NAE) Report*. Washington, DC: Island Press.

MDA (Minnesota Department of Agriculture). 2010. Home page. www.mda.state.mn.us/Home.aspx (accessed December 12, 2010).

Rosenzweig, C., A. Iglesias, X.B. Yang, P.R. Epstein, and E. Chivian. 2000. *Climate Change and U.S. Agriculture: The Impacts of Warming and Extreme Weather Events on Productivity, Plant Diseases, and Pests*. Cambridge, MA: Center for Health and the Global Environment, Harvard Medical School.

Schindler, D.W. 2009. Lakes as Sentinels and Integrators for the Effects of Climate Change on Watersheds, Airsheds, and Landscapes. *Limnology and Oceanography* 54 (6):2349–2358.

Seavy, N.E., T. Gardali, G.H. Golet, F.T. Griggs, C.A. Howell, R. Kelsey, S.L. Small, J.H. Viers, and J.F. Weigand. 2009. Why Climate Change Makes Riparian Restoration More Important Than Ever: Recommendations for Practice and Research. *Ecological Restoration* 27 (3):330–338.

Solomon, S., D. Qin, M. Manning, R.B. Alley, T. Berntsen, N.L. Bindoff, Z. Chen, et al. 2007. Technical Summary. Climate Change 2007: The Physical Science Basis. Contribution of Working Group I to the Fourth Assessment Report of the Intergovernmental Panel on Climate Change. p. 91. Edited by S. Solomon, D. Qin, M. Manning, Z. Chen, M. Marquis, K.B. Averyt, M. Tignor, and H.L. Miller. Cambridge, UK: Cambridge University Press.

Stefan, H.G., M. Hondzo, J.G. Eaton, and J.H. McCormick. 1995. Predicted Effects of Global Climate Change on Fishes in Minnesota lakes. *Canadian Special Publication of Fisheries and Aquatic Sciences* 121:57–72.

Stott, P.A., S.F.B. Tett, G.S. Jones, M.R. Allen, J.F.B. Mitchell, and G.J. Jenkins. 2000. External Control of 20th Century Temperature by Natural and Anthropogenic Forcings. *Science* 290: 2133–2137.

Triplett, L.D., D.R. Engstrom, and M.B. Edlund. 2009. A Whole-Basin Stratigraphic Record of Sediment and Phosphorus Loading to the St. Croix River, USA. *Journal of Paleolimnology* 41: 659–677.

UCS (Union of Concerned Scientists). 2009. Confronting Climate Change in the U.S. Midwest: Minnesota. www.ucsusa.org/assets/documents/global_warming/climate-change-minnesota.pdf (accessed December 12, 2010).

USDA FSA (U.S. Department of Agriculture, Farm Service Agency). 2010. Conservation Programs. www.fsa.usda.gov/FSA/webapp?area=home&subject=copr&topic=crp (accessed December 12, 2010).

USFS (U.S. Department of Agriculture, Forest Service). 2010. Northeastern Area: Watershed Program. www.na.fs.fed.us/watershed/ (accessed December 12, 2010).

Voldseth, R.A., W.C. Johnson, G.R. Guntenspergen, T. Gilmanov, and B.V. Millett. 2009. Adaptation of Farming Practices Could Buffer Effects of Climate Change on Northern Prairie Wetlands. *Wetlands* 29 (2):635–647.

Zandlo, J. 2010. Personal communication between J. Zandlo, Minnesota state climatologist, and the authors.

CHAPTER 13

Management of Invasive Aquatic Species

Frances R. Homans and Raymond M. Newman

I nvasive species are considered to be one of the top threats to native ecosystems. Although not all introduced species become established, and even fewer become invasive, the ones that do become established invasives may be particularly damaging. This chapter follows the state and federal definitions of invasive species as exotic (non-native) species that cause or have the potential to cause economic, environmental, or human health harm or may threaten natural resources or use of natural resources in the state. Exotic or non-native species are those that have not naturally and prehistorically occurred in the state. Because invasive species lack the natural predators to keep them in check, their populations can explode and crowd out native species and their prey, which may find themselves defenseless against an abundant new predator. The abundant populations of invasives can lead to a variety of deleterious effects on ecosystems and the human activities that rely on those ecosystems.

More than a dozen aquatic invasive species pose problems in Minnesota. One of the most prominent is the zebra mussel (*Dreissena polymorpha*), a species that was introduced from Eurasia into the Great Lakes and now has spread far beyond the point of its original introduction. Besides displacing native species, zebra mussel infestations can reduce the quality of water-based recreation and clog water intake pipes. Nationally, direct economic costs to drinking-water treatment and power plants are estimated to exceed $11 million per year (Connelly et al. 2007). Aquatic plants such as Eurasian watermilfoil (*Myriophyllum spicatum*) and hydrilla (*Hydrilla verticillata*) reduce native plant and animal diversity in waterways where they become established and also interfere with water-based recreation. Wetland plants such as purple loosestrife (*Lythrum salicaria*) damage wetland ecosystems by displacing native vegetation and creating conditions unfavorable for many species of native wildlife, waterfowl, and fish. One study estimates that more than $110 million per year is spent nationally on the control of aquatic weeds (Rockwell 2003).

The sea lamprey (*Petromyzon marinus*), a parasitic fish that entered the upper Great Lakes in the 1930s after expansion of the Welland Canal, devastated lake trout (*Salvelinus namaycush*) populations and commercial fisheries. In 1955, because of catastrophic losses to Great Lakes fisheries, Canada and the United States established the Great Lakes Fisheries Commission (Fetterolf 1980) to conduct research and suppress sea lamprey populations. Although it was successful in bringing sea lamprey under control and allowing lake trout populations to rebound, keeping the population in check has required continuing costs of at least $12 million per year by the two countries (Krantzberg and Boer 2008). In the Great Lakes alone, an estimated 180 non-native species have become established (Ricciardi 2006) and an average of 1.8 new species are discovered each year (Ruiz and Reid 2007), presenting new challenges to overburdened agencies contending with the species that already inhabit U.S. waterways.

The challenges to invasive species management are many but generally fall into three categories: preventing new introductions; detecting and attempting eradication or containment of new populations; and managing well-established populations to reduce economic, aesthetic, and ecologic impacts. It is often said that the most cost-effective way to manage invasive species is to prevent their introduction in the first place. However, it has been difficult for states to take unilateral action on issues such as ballast water control, because this seems to be a federal issue. In addition, governments are limited in their ability to prevent introductions, because laws governing commerce make it difficult to impose restrictions on interstate travel and trade. Second, early detection is difficult because it is costly, particularly for small populations, and is most useful for species that are as yet unknown in the new environment. Even when these obstacles to early detection are overcome, effective control methods may not be available, or citizens may be unwilling to sustain the undesirable side effects of the most effective control method. Third, the burden of managing species that become established often falls on those who may not enjoy the full benefits of that action. Aggressive control strategies often lead to undesirable side effects for affected ecosystems. Even if a thorough cost–benefit analysis concludes that invasive species management is worthwhile, issues of cost allocation and responsibility for action become difficult to sort out. Not only do multiple government agencies have different mandates and possible gaps and overlaps when it comes to invasive species management, but individuals and interest groups also play important roles in prevention and containment of aquatic invasives. Coordination among these many groups is necessary, but it is not always obvious where responsibilities should lie.

Minnesota shares these common challenges with all other states and nations but has additional circumstances that have shaped policy toward aquatic invaders. The state's vast water resources have represented an important industrial advantage over the course of history. Its ports on Lake Superior and the Mississippi River (discussed in Chapter 10) provide access to the products of Minnesota's mines and midwestern farm fields, its rivers have been a power source for grain mills, and its thousands of lakes and thousands of miles of rivers are a source of recreational enjoyment for both residents and visitors. These many uses of water have made

Minnesota vulnerable to invasions of aquatic species, which have harmed the very industries and recreational assets that rely so much on the water resources. Yet awareness of the economic and environmental losses occurring as a result of current invaders and the potential damage from future invaders has provided the impetus for a variety of creative and effective approaches to prevent, contain, and manage these species. This chapter discusses Minnesota's responses to the threats posed by aquatic invaders.

DEVELOPMENT OF STATE RULES, LAWS, AND POLICY

Before 1987, no specific laws existed relating to aquatic invasive species in Minnesota. The Minnesota Department of Agriculture (MDA) was charged with responsibility of managing species that caused economic harm to agricultural crops and livestock. One of the primary means of controlling these species was their inclusion on noxious weed lists. Categorizing a plant as noxious allowed the MDA to control possession and transport of the species and, most important, to require landowners to control infested sites through herbicides or harvesting when a listed plant was discovered on private or public property.

In 1987, the Minnesota legislature passed laws specific to a single aquatic invasive: purple loosestrife (Minnesota Interagency Exotic Species Task Force 1991). Purple loosestrife was introduced to the United States in the 1800s via ship ballast and later for human use (Galatowitsch et al. 1999). It was documented in Minnesota in 1924 (Skinner et al. 1994), and its spread accelerated because of intentional ornamental plantings. Purple loosestrife quickly became established in many wetlands throughout the state. Despite initial opposition from some in the horticulture industry who sold the plant, the 1987 legislation prohibited the sale of purple loosestrife, directed the MDA to list it as a noxious weed, and established a funded program led by the Minnesota Department of Natural Resources (DNR) to conduct research on and control purple loosestrife. This was the first state program in the country to curb its spread (Skinner et al. 1994).

The same year the purple loosestrife program was instituted, Eurasian watermilfoil was discovered in Lake Minnetonka, and a freshwater fish called ruffe (*Gymnocephalus cernuus*) was found in the Duluth–Superior Harbor. Two years later, in 1989, additional laws specific to Eurasian watermilfoil possession and transport were passed, and a program was established for control of this new invasive. The chain of new discoveries continued when, in 1989, the first zebra mussel was discovered in Minnesota. Realization that the state should be proactive rather than reactive and the seemingly quickening pace of new discoveries of problematic invasive species prompted the 1989 legislature to mandate creation of the Minnesota Interagency Exotic Species Task Force, which was charged with reporting to the legislature about the extent of harmful exotic (non-native) species threats to Minnesota. The task force was composed of representatives of the DNR, MDA, Minnesota Department of Health (MDH), Minnesota Department of Transportation (MDOT), and Board of Water and Soil Resources (BWSR), as well as three outside experts and ad hoc representation from the Minnesota

Pollution Control Agency (MPCA) and several federal agencies. Its report identified and ranked exotic species of concern, made a number of policy and legislative recommendations, and identified potential funding sources (Minnesota Interagency Exotic Species Task Force 1991).

Based on the task force report, the 1991 state legislature assigned the DNR the responsibility of managing all aquatic invasive species and coordinating efforts among responsible agencies. The charge to prevent and curb the spread of invasive species expanded beyond aquatic plants to include ecologically harmful animal species, including invertebrates, fish, and terrestrial birds and mammals. That authority, in combination with the MDA's authority to regulate plant pests and terrestrial plants, ensured that all invasive species would be considered.

In addition to establishing a comprehensive exotic species program, the legislature enacted laws and provided regulatory authority to the DNR and MDA to prevent and curb the spread of invasive species and mandated inspections of watercraft (ESP 1993). These were some of the first laws in the country to prohibit invasive species transport by boaters and others and were used as exemplary legislation by the Great Lakes Panel on Aquatic Nuisance Species (Glassner-Shwayder 1999). Minnesota's 1989 legislation prohibited transport of two aquatic plants—Eurasian and northern watermilfoil—on public roads or launching of watercraft with these species attached, and it established a structure for inspections, enforcement, and penalties. In the next few years, these statutes were extended to a general prohibition on movement of any species that the DNR identified through rulemaking as a harmful exotic, including zebra mussels (ESP 1993). New rules were also adopted by the DNR to prohibit transfer of any water from "infested waters" and specifically require draining water from bilges, bait buckets, or boat bottoms, as well as removing drain plugs at waters infested with zebra mussel or spiny water flea (*Bythotrephes longimanus*).

Funding to carry out the program was provided to the DNR from a surcharge on watercraft licenses. The surcharge on a three-year license increased from $1 in 1990 to $3 in 1993 and $5 in 1994, generating approximately $1 million per year (ESP 1993). Finally, this early legislation required the DNR to report on the status of exotics management to the legislature each year. The DNR's annual reports are extensive, starting in 1993 and continuing to the present (e.g., ISP 2010).

The assortment of laws related to invasive species of aquatic plants and wild animals, which was in several chapters of statutes and difficult to understand, was completely revised, expanded, and consolidated into a new chapter of statutes (M.S. 84D) in 1996. The new statutes accommodate a classification system that accounts for known and unknown threats (ESP 2000). Non-native species can be classified as prohibited, regulated, unregulated, or unlisted. Species that are known to be harmful in Minnesota or elsewhere can be designated or classified as *prohibited* to prohibit their possession, propagation, sale, importation, or introduction. *Regulated* species, such as aquarium plants, aquaculture species, and carp, are economically important but pose a lesser threat; they may be sold or possessed but not released into the wild. *Unregulated* species are presumed to be a minimal threat, are considered beneficial (e.g., rainbow trout), or are so widespread and established that they cannot be controlled or regulated (e.g., starlings). *Unlisted* species are those that have not been evaluated and classified; in order for them to be introduced into the wild, they must undergo an assessment and be placed into one of the other categories, ensuring that new introductions receive careful assessment prior to release. The DNR is authorized to assess and classify species under emergency and permanent rulemaking procedures without additional legislation. This classification approach, combining a "clean" and a "dirty" list, leaves harmless species unregulated while restricting known risks, but prohibits introduction of unknown risks until a proper assessment can be made. Miller (2004) considered this one of the strongest state approaches to regulating introductions but also noted that its success depends on funding, as well as agency and citizen support to follow the law and conduct assessments.

STATE POLICY WITHIN A FEDERAL SYSTEM

On the federal level, prior to 1991, the main responsibility for invasive species resided within the U.S. Department of Agriculture's Animal and Plant Health Inspection Service (APHIS), whose primary goal was protection of agronomic interests. Like the MDA, APHIS maintained lists of prohibited species, such as the federal noxious weed list, and exercised authority to control imports of species on these lists as well as outbreaks when discovered. The APHIS lists included a small number of aquatic plants, among them hydrilla and water hyacinth (*Eichhornia crassipes*). APHIS focused primarily on regulating introduction and preventing establishment and spread of new invasive species.

The current situation concerning control of invasive aquatic plants can be thought of as a broadening of the effort to control weeds beyond agriculture. In contrast, control of aquatic animals can be traced to the Lacey Act, instituted in 1900 to protect wildlife from harm. Originally designed to protect game species from illegal harvest by regulating possession, transport, and sale of wildlife, the Lacey Act is now used to provide federal agencies (the U.S. Departments of Interior, Commerce, and Agriculture) with authority to regulate introduction of wildlife that might be injurious to agriculture, horticulture, or native ecosystems, including aquatic ecosystems (FWS 2007).

The Lacey Act can be used by the U.S. Fish and Wildlife Service (FWS) to prevent the introduction of potentially harmful fish, such as snakeheads (*Channa* spp.), silver carp (*Hypophthalmichthys molotrix*), or black carp (*Mylopharyngodon piceus*), whether for food, aquaculture, or the pet trade, if the likelihood exists that they could escape into the wild and survive to injure native animals. Introduction and possession of zebra mussels and mitten crabs (*Eriocheir sinensis*) is also prohibited under the act.

Although the federal government retains authority over introduction of new species into the country, the states are largely responsible for management of plants and animals within their borders and have little control over actions of other states (OTA 1993). Thus Minnesota could object to North Dakota introducing the European zander (*Sander luciopercus*), but without federal prohibition via the Lacey Act, it could not prevent the introduction despite the potential for the fish to cause damage in Minnesota. Furthermore, once species escape to the wild, such as occurred with the bighead carp (*H. nobilis*) and silver carp, both introduced legally despite objections of many states, it is very difficult to prevent the spread of an animal into adjacent states, especially via interconnected river systems such as the Mississippi (ISP 2009). Lack of federal action to prevent new introductions and interstate spread of previous introductions place significant burden on states that have to deal with problems they cannot prevent.

With numerous new introductions of aquatic invasive species in the late 1980s, it became clear that federal laws and institutions were ill prepared to deal with aquatic invasive species. The Nonindigenous Aquatic Nuisance Prevention and Control Act (NANPCA) was passed in 1990. In addition to calling for ballast water management, the NANPCA established the Aquatic Nuisance Species Task Force, an intergovernmental organization chaired by the National Oceanic and Atmospheric Administration (NOAA) and FWS, which develops and coordinates a comprehensive federal program to prevent and control aquatic nuisance species introductions. The task force contains representatives of other federal agencies; representatives of state and regional entities participate as nonvoting members. A set of committees and regional panels furthers the interests of the task force. Minnesota has been active in and chaired both the Great Lakes and Mississippi River Basin Regional Panels. Following congressional establishment of the Aquatic Nuisance Species Task Force, the Clinton administration's Executive Order 13112 of 1999 established the National Invasive Species Council to oversee and coordinate federal activities concerning invasive species. In addition to coordinating federal agencies and implementing the order, the council is charged with developing a National Invasive Species Management Plan. An advisory committee with representatives from state, tribal, regional, and local stakeholders was created to provide information and advice to the council.

BALLAST WATER: A FEDERAL AND STATE CHALLENGE

To control stability and buoyancy, cargo ships take on and release water into and from their ballast tanks when the weight of the ship changes as a result of loading and unloading of cargo, as well as to maintain stability in changing weather conditions. Ballast water released from ships into ports in Lake Superior and other Great Lakes has been responsible for the introduction of a substantial number of invasive species. Some of the most troublesome species, including the zebra mussel, spiny water flea, and round goby (*Neogobius melanostomus*), arrived through the ballast water vector (ISP 2009). Ruiz and Reed (2007) have estimated that the average rate of discovery of aquatic invasive species in the Great Lakes between 1960 and 2003 was one new invader every 30 weeks. These species have not necessarily stayed confined to the Great Lakes: the zebra mussel and spiny water flea, for example, have spread to inland waters by several pathways, such as boats and connected waterways, and continue to spread, causing economic and ecological damage. Furthermore, control methods for these species are virtually nonexistent. The hope for species currently present is the prevention of further spread and eventually, through research, development of viable control technologies.

Ballast water has been the subject of federal legislation, including the NANPCA, which was reauthorized as the National Invasive Species Act (NISA) in 1996. These pieces of legislation granted authority to the U.S. Coast Guard to regulate ballast water for the purpose of preventing introduction of invasive species. The Coast Guard established ballast water exchange rules for ships entering the Great Lakes from outside the U.S. Exclusive Economic Zone. Compliance with the rules were initially voluntary but became mandatory in 1992 (Buck 2007).

Despite programs instituted in the 1990s to regulate ballast water, concern about new species introductions remained, because new introductions kept appearing (e.g., Grigorovich et al. 2003). It was known that regulating ballast water exchange in itself was insufficient to prevent all new invasions. Vessels declared as NOBOB ("no ballast on board") have been exempt from the ballast water exchange requirement, yet they may contain species in sediment left behind from previous ballasting activities. These sediments may be released when ballasting occurs with fresh water. Coast Guard rules apply only to vessels entering from outside the United States, yet short trips among U.S. ports and among the Great Lakes may introduce species to new areas. The Coast Guard rules published in June 2004 specified a set of best management practices, primarily ballast water exchange guidance, rather than objective measurable standards of species concentrations in released ballast water (Buck 2007).

Concerns about ineffectiveness of Coast Guard regulations enacted under the prior legislation prompted a consortium of environmental advocacy groups to petition the federal government to use the Clean Water Act (CWA) to set and enforce specific standards for the quality of ballast water released. The CWA and its associated National Pollutant Discharge Elimination System (NPDES) were not originally applied to ballast water. The U.S. Environmental Protection Agency (EPA) interpreted the act to specifically exclude "discharges incidental to the normal operation of a vessel" (EPA 2008a) and denied the petition to apply the act's provisions to ballast water. The groups brought the issue before the courts; several states' attorneys general, including Minnesota's, joined in the lawsuit once it reached the U.S. Court of Appeals. In 2006, the court ruled in favor of the plaintiffs, so the exclusion of ballast water from regulation under the CWA ended on December 19, 2008. To comply with NPDES procedures, EPA developed a vessel general permit (VGP) that incorporated Coast Guard rules on ballast water exchange, added reporting requirements, and specified procedures for incoming NOBOB vessels (EPA 2008b).

Some states, including Minnesota, are authorized to implement and oversee the NPDES permitting process used to implement the CWA. Minnesota reacted to state and federal court rulings on the issue of ballast water discharges by developing its own ballast management plan. The MPCA's vessel general permit for shipping within Minnesota waters was issued in September 2008, consistent with legislation passed by the state in May of that year.

Minnesota's permit specifies biological performance standards for organisms and requires that approved technology be installed on vessels. The state anticipates rapid development of useful technologies—mechanical, chemical, physical, or combinations of the three—that can be installed on vessels to reduce concentrations of biological contaminants. The permit standards must be met by 2012 for new vessels and 2016 for existing vessels. Importantly, these rules apply to all commercial ships, including "lakers," those that travel exclusively within the Great Lakes. This is a key provision, because 95% of ballast water released into Lake Superior comes from lakers (MPCA 2008).

Minnesota's permit reflects an effort to harmonize regulations with 2004 standards adopted by the International Maritime Organization (IMO 2009). This decision is an example of a common pattern by which standards developed by states and countries affected by international shipping are matched by others in a step-by-step fashion. Canadian regulations require that NOBOB vessels entering the Great Lakes be flushed with salt water and that salinity levels in ballast tanks remain high enough to be inhospitable to freshwater species living in sediment (Reid 2008). This regulation was matched by the St. Lawrence Seaway management agencies, effectively requiring this practice for all transoceanic ships entering the Great Lakes. Nevertheless, a patchwork of different regulations is still applied by different states and countries. For example, California's legislation requires adherence to numerical standards several orders of magnitude more strict than the IMO's. New York requires adherence to its permit for vessels passing through state waters even if no water is discharged there. Michigan has its own permit system for water discharged within its borders (EPA 2008b).

The failure of federal legislation or Coast Guard rules to adequately regulate ballast water led to the lawsuit that forced EPA to apply the Clean Water Act and to subsequent state actions to apply more stringent standards. Additional federal ballast water legislation has been considered since the late 1990s. Legislation proposed in 2003, as part of the reauthorization of the National Invasive Species Act of 1996, mandated ballast water treatment with best available technology (Miller 2004). The NISA still has not been reauthorized as of 2010 (ISP 2010), in part because of disagreement as to who should regulate ballast water: the states, the Coast Guard, or EPA (ISP 2009). Industry would prefer uniform federal regulations, but states are reluctant to have their rights preempted. Provisions for ballast legislation preempting the Clean Water Act are equally controversial. Furthermore, complementary U.S. and Canadian regulations would be most effective and are desired by industry (NRC 2008). Still, pressure by states on EPA and the Coast Guard has resulted in some action to more effectively manage ballast water (ISP 2009). For example, in 2009, the Coast Guard proposed standards for allowable concentrations of living organisms in ballast water released into U.S. waters (DHS 2009).

DIVISION OF RESPONSIBILITY

In addition to the complementary but often conflicting state and federal roles, agencies within the state have different mandates, authority, and responsibility.

For example, the MDA has statutory authority for most terrestrial invasive plants and plant pests, including buckthorn (*Rhamnus cathartica*), gypsy moth (*Lymantria dispar*), and emerald ash borer (*Agrilus planipennis*), while the DNR manages state forests, wildlife management areas, parks, trails, and scientific and natural areas on which terrestrial invasives inflict damage. Under state statutes, the MDA is exclusively responsible for management of terrestrial plant pests before they become established. Once invasive species that are plant pests have become established, responsibility for their management largely shifts to the landowner or managing agency. This can result in disagreement over responsibility and expected actions and less effective management if sufficient resources are not provided to manage a species. The DNR's clear role with both aquatic plants and animals reduces these conflicts for aquatic systems. Entry of the MPCA into the mix via ballast water permitting, however, could further complicate authority and coordination. To date, the agencies have generally worked closely together to coordinate efforts.

To further aid coordination in the state, the Minnesota Invasive Species Advisory Council (MISAC) was established in 2001. MISAC is a consortium of state, federal, and local agencies cochaired by the MDA and DNR. It serves as an advisory and coordinating body that is able to foster possible complementarities among the mandates of the 30 federal, state, tribal, and local agencies and organizations represented. The purpose of the council is to review and share information concerning the current status and management of terrestrial and aquatic invasive species, including animals, insects, plants, and diseases, in Minnesota. Although MISAC's leadership is shared by the two state agencies with the most authority and resources to manage invasive species in the state, its function is to facilitate coordination, cooperation, and information sharing; it does not allocate or control resources and can only recommend policy changes.

MANAGEMENT OF INVASIVES

For years, the DNR's three main goals for management of invasive aquatic plants and animals were to prevent new introductions into the state, prevent the spread of invasives already present, and reduce the impacts caused by invasives. Of the 2008 budget for the DNR Invasive Species Program, 30% was for inspections and enforcement, 10% for education, and 34% for management and control. The remainder of the budget was allocated to administration, state and regional coordination, and research. Continued funding from the watercraft surcharge ($5 every three years) and a $2 fee on nonresident fishing licenses currently generates about $1.6 million per year. In addition, the program began receiving funding from the state's general fund in 2007; this accounted for 37% of its budget in 2008 (ISP 2009).

A major focus of the original Minnesota invasive species legislation and the DNR's ongoing program was preventing the introduction of new invasive species and further spread of invasive species already in the state. Although the legislation and DNR rules provided legal restrictions to prevent transport and introduction, most emphasis was placed on education and voluntary compliance rather than enforcement. The primary role of watercraft inspections is public education and demonstration of proper self-inspection processes (ESP 1994; ISP 2009). Watercraft inspectors are posted at public lake access points, primarily on infested waters; boaters entering or leaving the landing are interviewed, and boats are inspected. Boat operators are informed about the issues and laws, and if aquatic vegetation or invasive species are found, operators are asked to remove them. Inspectors also instruct boaters to drain water from their boats and equipment prior to leaving the water access. Inspectors are not authorized to give citations or warnings but can alert conservation officers in egregious cases, as when a boater refuses to cooperate. Legislation mandates a minimum number of hours be spent on watercraft inspections. More than 20,000 hours of inspections have been performed each year since 1994, with 40,000 to 50,000 boats inspected (ISP 2005, 2009). Effort increased to over 35,000 hours in 2008, partly because of cooperative efforts with local organizations that added inspectors, often on noninfested waters. In addition to watercraft inspections, public service announcements, radio, TV, and print advertisements, billboards and other informational efforts are used to implement the national "Stop Aquatic Hitchhikers!" campaign in the state (ISP 2009). All boat accesses on infested waters are marked with signs that provide information about the invasive species present and the relevant laws.

Although Minnesota was one of the first states to enact laws prohibiting transport of invasive species, the state issued relatively few citations for violations until 2009. In the early years (1991–2002), road checks conducted by conservation officers were used to educate the public; although a number of violations (800 out of 7,400 inspections) were detected during this period, few fines (44) were given (ISP 2005). As a result of several court decisions dealing with reasonable search and privacy, road checks were stopped in 2002, and few citations were issued in subsequent years. The recent increase in funding, in part due to increased concern by the DNR, legislators, and lake groups, resulted in an expansion of effort and training of enforcement officers in 2007 and 2008, when nine enforcement officers were hired to work 50% of their time on invasive species (ISP 2009). The number of citations increased from fewer than 10 to more than 20 per year, and continued increases in enforcement are likely, particularly with increasing pressure from lake associations (e.g., Minnesota Waters 2009).

Several assessments indicate that the watercraft inspection, education, and information strategy was effective at reducing the spread and introduction of new species. A three-state boater survey in 1993 indicated that 70% of Minnesota boaters took precautions to prevent the spread of invasives, but only 39% of Wisconsin and 33% of Ohio boaters did; Minnesotans also were better aware of appropriate actions to prevent the spread of invasives. In 1993, Minnesota laws were more extensive and aggressive, as was boater education, compared with those of other Upper Midwest states. Minnesota boaters were more knowledgeable about invasive species and had already changed their behavior more than the boaters in Wisconsin and Ohio (Gunderson 1994). A similar survey of boaters from five states in 2001 indicated that 90% of Minnesotans took precautions to prevent spread (compared with 70% in 1993), whereas only 82% in Vermont, 46%

in Ohio, 40% in California, and 30% in Kansas took precautions. The differences were attributed to the level of boater public awareness efforts and the variety of methods used in these states (ISP 2005). In addition, responses to questions asked during watercraft inspection indicated that 90% to 95% of Minnesota boaters were aware of Minnesota laws, and 95% took precautions to clean their boats and reduce the likelihood of transporting invasives (ISP 2009). These results suggest that Minnesota regulations and education efforts have altered boater awareness and behavior.

Patterns of spread of aquatic invasives in Minnesota suggest that these attitudes and behaviors have been effective, at least to date. For example, ruffe have been abundant in the Duluth–Superior Harbor since 1990 but have yet to be found in an inland lake in Minnesota. Eurasian watermilfoil introductions initially increased exponentially, but they leveled off and declined during the 1990s after several years of regulation, education, and inspection (ESP 2003). Spiny water fleas were abundant in Lake Superior and a nearby inland lake for a dozen years before spreading to another inland lake in 2003 (Branstrator et al. 2006). It took 10 years before zebra mussels first showed up in an inland lake (ESP 2001), despite being broadly distributed in the Lower Mississippi, St. Croix, and St. Louis River systems, and another three years before they reached a second inland lake in central Minnesota (Lake Ossiwinnimakee in 2003). During this same time, Wisconsin had 40 and Michigan had more than 200 inland lakes with zebra mussels (ISP 2005).

Once these invasive species get spread beyond their original location, the new water bodies become loci for additional spread. Distance to and duration of the nearest infestation was one of the best predictors of spread of Eurasian watermilfoil in Minnesota. Although spread is declining in the Twin Cities metropolitan area, it is accelerating in outstate Minnesota (Roley and Newman 2008). After long periods of little or no new invasion, the spiny water flea and zebra mussel recently expanded rapidly from their outstate locations. Spiny water fleas had spread to a dozen lakes, primarily along the Canadian border, by 2008 (ISP 2009) and to Lake Mille Lacs in 2009. By 2009, Minnesota had 16 inland lakes with zebra mussels (ISP 2010). Prevention efforts appear to delay the spread, but once new loci are created, spread to nearby waters can be rapid. In addition, introduction to connected waters such as the Mississippi River or the Rainy River–Lake of the Woods system can result in rapid spread. Zebra mussels are now throughout the Mississippi River below Brainerd, and spiny water fleas throughout the connected waters of the Rainy River and Lake of the Woods Basin (ISP 2009). Discovery of zebra mussels in the chains of lakes in the Alexandria and Detroit Lakes area in 2009 (ISP 2010) is an example of introduction into one lake leading to infestations in multiple connected lakes.

Early detection, rapid response, and containment are key elements of the new statewide plan (MISAC 2009); prevention of new introductions and prevention of spread within Minnesota are two primary goals of the DNR's Invasive Species Program. However, the likelihood of successful containment or eradication is limited by expense, public acceptance, and the DNR's desire to minimize nontarget impacts. The few successful eradication efforts in Minnesota, such as a small salt cedar population with 40 mature trees near Hibbing (ISP 2005) and

Brazilian waterweed (*Egeria densa*) in Powderhorn Lake, were isolated populations in relatively easy-to-control circumstances. Powderhorn Lake is a small, shallow, isolated (no surface outflow or inflow) water body that regularly has winterkills; aggressive chemical treatment and curtailment of aeration appear to have eliminated the plant here (ISP 2009).

Experience with Eurasian watermilfoil, however, suggests that eradication is not feasible in most lakes, at least with the level of nontarget impacts tolerated by the DNR and the public. Repeated 2,4-D treatments of new infestations failed to eliminate this plant (Crowell 1999). New infestations of zebra mussels in Lake Zumbro were treated with a winter water drawdown that proved, as expected, to be ineffective (ESP 2002). Treatment of the infestation at Lake Osawinamakee in the Upper Mississippi River Basin was limited to copper sulfate treatments at the outlet in an attempt to prevent downstream spread of juveniles, called veligers (ISP 2005). These treatments were not successful, as evidenced by established zebra mussels 30 miles downstream two years later and 50 miles downstream two years after that, in 2007 (ISP 2009).

Although eradication is difficult and not feasible once invaders are widespread, concerted efforts have been effective in some locations. In addition to early detection and rapid response, eradication requires sufficient resources and enforced cooperation to ensure that the eradication effort is carried out and monitored (Simberloff 2009). A more aggressive approach to new invasions would require more advance discussion of options than has occurred to date and a conscious decision to accept substantial nontarget effects. Such an approach has been taken in Australia and California (Anderson 2005). Perhaps inclusion of early detection, rapid response, and containment as key elements of the new statewide plan (MISAC 2009) will spur discussion of a more aggressive approach. However, managers as well as the public will need to be convinced that significant nontarget effects and public inconvenience are acceptable costs to achieve some possibility of eradication or long-term containment. Efforts to contain and eradicate hydrilla in Indiana included quarantining the 700-acre Lake Manitou and extensive whole-lake and targeted chemical treatments costing $500,000 per year, treatments that will be required for four to six years (Alix et al. 2009; Keller 2007).

Once invasives are well established in Minnesota, the state focuses on reducing impacts on the local ecology, society, and economy. The DNR first looks to reduce ecological impacts and minimize harm to natural systems and native species. In fact, the DNR has an extensive aquatic plant control permitting process and Aquatic Plant Management Program to "protect the beneficial functions of aquatic vegetation while allowing riparian property owners to obtain access to public waters" (Enger and Hanson 2009). Minnesota is particularly protective of natural vegetation; bottom barriers and dredging are not allowed, and property owners must have a permit in order to use weed rollers. The state does not permit mechanical harvesting of more than 50% of the littoral area (defined in Minnesota as ≤ 15 feet) or chemical control in more than 15% of the littoral area of a water body, unless the DNR provides a variance. These restrictions are aimed at protecting water quality, native plants, and fisheries habitats. These rules have limited whole-lake herbicide treatments to a few experimental studies and caused tension among lakeshore associations (e.g., Minnesota Waters 2009). The DNR's position on whole-lake herbicide treatments is supported by several experimental studies in Minnesota that have shown reductions of native species and decreases in water clarity with whole-lake herbicide treatments, particularly in eutrophic lakes (Valley et al. 2006).

The DNR will provide assistance and funding only to protected waters that have public water accesses (ISP 2009). As indicated below, the state spends substantially less on control than local governments and private citizens, and it expects private citizens to cover expenses for their access to the resource. However, Minnesota has a fragmented set of government structures with authority to collect funding and make management decisions (discussed in Chapter 6). Two lakes have effective conservation districts, and a number of others have watershed districts that invest in lake issues. But most are unorganized and at best have lake associations without taxing, regulatory, or coordinating authority. Most watershed districts are focused on water quality issues that may or may not include management of invasive species. A few (30) lakes have organized lake improvement districts. Lake improvement districts are a subset of local units of government that focus on lake issues such as water quality, water quantity, and aquatic vegetation. Their funding and authority are delegated to them by the parent government, typically the county. It is unclear why so few lake improvement districts have been formed since they were authorized in 1974, but half (15) of the districts were formed since 2004, primarily to deal with aquatic vegetation (DNR 2009). We expect that more lakes will follow suit to avoid the funding and lack of authority issues faced by lake associations and to be able to take advantage of matching state funds. Strong, lake-based governmental authority (lake boards with taxing and regulatory authority) resulted in aggressive nuisance plant control in Michigan (Nygren 1999), and invasive species may facilitate organization of lake improvement districts that can also address water quality issues.

As a result of public pressure from interest groups such as Minnesota Waters, a statewide organization that includes most lake associations, additional funds have been allocated to the control of invasives. The Invasive Species Program is currently provided with $350,000 of cost-share funding to assist with the treatment of curlyleaf pondweed and Eurasian watermilfoil (ISP 2009). The largest of these efforts, with $230,000 of funding, are pilot projects to assess whole-bay or whole-lake herbicide treatments for curlyleaf pondweed and Eurasian water-milfoil, which require variances and special monitoring to assess effects of the treatments. However, lack of local matching funding has limited the use of these programs and may be spurring designation of lake improvement districts.

The DNR's primary allegiance to public resources and fisheries creates tension with riparian owners. Lake associations and shoreline owners prefer fewer restrictions on their ability to control nuisance vegetation and would like to restrict access and use of lakes, in part to reduce the likelihood of new introductions (Minnesota Waters 2009). However, they also believe that management costs should be largely shared by the state, particularly if the public has access.

Minnesota has invested considerable resources in research on controls, in particular, biological control of purple loosestrife and Eurasian watermilfoil and improved methods of chemical control of Eurasian watermilfoil and curlyleaf pondweed. From 1992 through 2003, the state invested about $150,000 per year in purple loosestrife and Eurasian watermilfoil biological control research (ESP 2002). This research resulted in successful purple loosestrife biological control agents established throughout the state and a large reduction in the use of herbicidal control (ISP 2009). More recently, the state has invested in the development of approaches for early season control of curlyleaf pondweed by the U.S. Army Corps of Engineers (e.g., Skogerboe et al. 2008), and application of these approaches is being assessed with the above-mentioned multiyear, whole-bay or whole-lake pilot program on 14 lakes (ISP 2009). The state has also invested in research to improve ballast water treatment and assess the feasibility of fish barriers to prevent invasion of Asian carp.

ECONOMIC IMPACTS

Attempts to establish the economic costs of aquatic invasive species fall into two categories: large-scale back-of-the-envelope calculations based on control costs (e.g., OTA 1993) and localized estimates of welfare impacts of nonmarket values (e.g., Horsch and Lewis 2009). Lovell and Stone (2005), in their review of the literature on impacts of species that arrived via ballast water, find a wide range of estimates and methodologies used to establish economic damages from invasive aquatic species. The best information about economic impacts of aquatic invasives in Minnesota comes from DNR reports on control decisions made by private landowners. When made by private individuals, control costs provide at least a lower bound on estimates of the benefits of removing invasive species: benefits have to be at least as large as costs of the action to be taken.

In the context of aquatic plants such as Eurasian watermilfoil and curlyleaf pondweed, removal provides temporary relief. Although most existing removal methods allow plants to return, removal does enable lakeshore property owners to use the lake for swimming and boating during summer months. Most removal of aquatic plants in Minnesota lakes requires a permit from the DNR, however, because it is in the public interest to maintain aquatic vegetation for aquatic wildlife. In some cases, particularly in experimental programs for lakewide treatment of curlyleaf pondweed and Eurasian watermilfoil, the DNR has contributed public resources in the form of cost sharing. In most other cases, individuals bear the full cost of aquatic plant removal along their lakeshore. The DNR's annual report on invasive species says, "Lake residents and lake associations who do routine management of curlyleaf pondweed to reduce nuisance areas using both herbicides and mechanical harvesting undertake the majority of curlyleaf pondweed management done in Minnesota" (ISP 2009). Without the presence of these invasive plants, lakeshore property owners would not bear these costs, so the costs represent a true welfare loss to this subset of the population.

Recent information for control costs comes from 2008, when 759 lakes were infested with curlyleaf pondweed and 215 were infested with Eurasian watermilfoil (ISP 2009). From 2005 through 2008, between 6,000 and 7,000 acres of the two plants in offshore areas were treated each year with herbicide, and another 1,600 to 2,500 acres were harvested (Enger and Hanson 2009). Approximately 4,600 acres of the chemical treatments were in a pilot project where the DNR covered 35% ($230,000) of the total cost of $645,623 (ISP 2009). Based on typical costs of herbicide treatment ($300/acre) and harvesting ($250/acre) (Getsinger et al. 2002; similar results were obtained from analysis of DNR records and input from applicators), between $420,000 and $720,000 was spent on the remaining chemical control acres, and between $400,000 and $625,000 was spent on harvesting. Total offshore invasive plant control thus cost approximately $1.5 million to $2 million per year. In 2008, an additional 1,168 acres of shoreline on 6,183 properties were treated with herbicides, and 74.2 acres on 406 properties were treated with mechanical methods (Enger and Hanson 2009). These treatments included native plants, however, so it was not possible to determine the amount of treatment focused on exotic invasives. An estimated 56,000 pounds of 2,4-D was applied for Eurasian watermilfoil control. In addition to the pilot DNR program, the state spent $17,000 on control at launch sites in 2008 and provided an additional $105,000 in matching grants for milfoil control. Although good records were kept of state expenditures for research, control, and matching grants reported by the Invasive Species Program, there is not an effective method of recordkeeping on private and commercial expenditures. A comprehensive reporting system and economic analysis of control costs for all invasives in the state would be useful.

CONCLUSIONS

One way to think about the problem of aquatic invasive species is as an issue of biological pollution. As with other forms of pollution, many aquatic invasive species have been introduced into the environment as by-products of either production or consumption activities. Activities related to consumer behavior, such as the disposal of aquarium products or enjoyment of recreational boating, have also been responsible for the introduction of invasive species into waterways. For example, the initial introduction of Eurasian watermilfoil into the United States may have occurred when the contents of an aquarium were discarded in a water body (Couch and Nelson 1985). Many of the subsequent introductions of this weed to new locations were caused when recreational boaters unwittingly transferred the plant during the course of visiting multiple waterways.

Considering invasive species akin to other forms of pollution has been the idea behind using the Clean Water Act and its associated NPDES permitting system to control discharge of ballast water. Some economists have argued for the use of tools typically applied to other pollution problems—emission taxes, cap-and-trade systems, liability rules—in invasive species control. In several important ways, however, invasive species are unlike other forms of pollution. First, although small amounts of conventional pollutants can be considered harmless because their concentrations are so diluted, living organisms reproduce, grow in size, move, and evolve over time. When species find a new environment hospitable, their numbers grow to be far higher than their numbers at the time of introduction. Invasive species are worse than stock pollutants, because they not only accumulate in the environment, but also grow rather than decay over time. Second, whereas conventional pollutants are by-products of other activities, species introductions may occur intentionally. Species that become problematic to society at large may be thought by some individuals to be beneficial. For example, the common carp (*Cyprinus carpio*) was intentionally distributed far and wide by the federal government in the late 1800s. Reed canary grass (*Phalaris arundinacea*) has been grown as a forage crop and used to stabilize streambanks, yet it is now known to crowd out beneficial native wetland plants (Galatowitsch et al. 1999). Water hyacinth is a known invasive aquatic plant in southern states, yet Minnesota lakeshore owners can purchase it for use in water gardens, from whence it could be illegally introduced to lakes or allowed to escape (ISP 2009).

In some sense, although invasive species introductions share common causes with conventional pollutants, as both are by-products of production and consumption, once an invader has become established, the analogy to disease control becomes apt. Quarantines and other restrictions on the movement of infections are used to contain the spread of a biological contaminant, including aquatic invasives. Because of these differences, new approaches beyond the standard solutions for pollution control must be created or adapted for the invasive species problem. Although a nonzero standard for many pollutants may be acceptable, because the damage caused at low concentrations may be negligible, a zero-tolerance approach for particularly harmful species makes sense; prevention is potentially the best weapon in the fight against invasive species damage.

Minnesota's experience shows, however, that prevention can be an elusive goal when authority is limited by state boundaries. When federal action has been inadequate to prevent the introduction of damaging invasives, Minnesota has been left with a limited set of options: to push for federal action, take matters into its own hands by enacting state legislation, or react and try to contain the invasions that occur. Since 2003, the DNR has advocated for a barrier on the Mississippi River to prevent upstream spread of Asian carp and has funded assessments of appropriate barriers (ISP 2004, 2009). However, federal inaction has allowed bighead and silver carp to reach Minnesota.

In some cases, action on the state level has yielded tremendous benefits. The early loosestrife program showed that a combination of enabling legislation, management action, assessment, and research could tackle a seemingly insurmountable problem. Coordination of programs with other states and the federal government resulted in a protocol of control that has worked to dampen loosestrife's effects and restore damaged wetlands. The species-by-species approach led to a comprehensive program that addressed multiple species in a coordinated fashion. The emphasis on research and assessment holds the possibility of learning about better ways to suppress and manage existing populations.

Minnesota was among the first to restrict and regulate transport of invasive species via recreational boating activities. Although these restrictions were enforced by mandated watercraft inspections and monitoring, the state has emphasized an educational over a punitive approach. This may reflect a belief in the efficacy of education and recognition that relying on enforcement would be costly and difficult from a public relations perspective. Still, a tension exists between the efficiency of a more restrictive regulatory approach and the public attractiveness of a voluntary approach that relies on individual responsibility.

Minnesota's experience with aquatic invasive species raises questions about the degree to which invasive species prevention and control is a public or private good. Who should bear the costs of prevention or introduction, and are these economically limiting or socially important activities? In the context of aquatic plants in inland lakes ringed by private property, decisions about limiting species introductions, deciding on appropriate treatment methods, and paying for invasive species control are often based on who benefits, and the extent to which they benefit, from the quality of the aquatic environment. Management agencies must balance the interests of various nonresident users and the public with those of riparian residents, who do not always agree among themselves. In addition, the importance of keeping the state's resources free of invasive species has a great deal to do with whether the public is willing to bear the nontarget effects, use restrictions, and expense of aggressive control programs. Public discussion on this issue should be held in advance of the arrival of an invader so that agencies responsible for eradication programs can focus on implementation of the appropriate plan. Minnesota citizens share a common interest in limiting invasive species but will have to continue to grapple with what promises to be an ever-changing but persistent concern.

ACKNOWLEDGMENTS

The authors thank Jay Rendall and Chip Welling from the Minnesota Department of Natural Resources for their comments and insights and gratefully acknowledge the NSF Integrative Graduate Education and Research Traineeship (IGERT) program Risk Analysis for Introduced Species and Genotypes (NSF DGE-0653827) for the impetus to collaborate on this project.

REFERENCES

Alix, M.S., R.W. Scribailo, and J.D. Price. 2009. *Hydrilla verticillata* (Hydrocharitaceae): An Undesirable Addition to Indiana's Aquatic Flora. *Rhodora* 111:131–136.

Anderson, L.W.J. 2005. California's Reaction to *Caulerpa taxifolia*: A Model for Invasive Species Rapid Response. *Biological Invasions* 7:1003–1016.

Branstrator, D.K., M.E. Brown, L.J. Shannon, M. Thabes, and K. Heimgartner. 2006. Range Expansion of *Bythotrephes longimanus* in North America: Evaluating Habitat Characteristics in the Spread of an Exotic Zooplankter. *Biological Invasions* 8:1367–1379.

Buck, E.H. 2007. CRS Report for Congress: Ballast Water Management to Combat Invasive Species. www.ncseonline.org/NLE/CRSreports/07Jul/RL32344.pdf (accessed September 28, 2009).

Connelly, N.A., C.R. O'Neill, Jr., B.A. Brown, and T.L. Knuth. 2007. Economic Impacts of Zebra Mussels on Drinking Water Treatment and Electric Power Generation Facilities. *Environmental Managment* 40:105–112.

Couch, R., and E. Nelson. 1985. Myriophyllum spicatum in North America. Proceedings of the 1st Symposium on Watermilfoil *(Myriophyllum spicatum)* and Related Haloragaceae Species. Vicksburg, MS: Aquatic Plant Management Society.

Crowell, W.J. 1999. Minnesota DNR Tests the Use of 2,4-D in Managing Eurasian Watermilfoil. *Aquatic Nuisance Species Digest* 3 (4):42–46.

DHS (Department of Homeland Security, Coast Guard). 2009. Standards for Living Organisms in Ships' Ballast Water Discharged in U.S. Waters; Draft Programmatic Environmental Impact Statement: Proposed Rule and Notice. *Federal Register:*44632–44672.

DNR (Minnesota Department of Natural Resources). 2009. Lake Improvement Districts in Minnesota. http://files.dnr.state.mn.us/waters/watermgmt_section/shoreland/lake_improvement_districts_in_minnesota.pdf (accessed December 13, 2010).

Enger, S., and S. Hanson. 2009. *A Summary of Permitted Management Work for Aquatic Vegetation, Algae, Leeches, Swimmers Itch 2008*. Staff Report 45. St. Paul: Minnesota Department of Natural Resources, Division of Ecological Resources.

EPA (U.S. Environmental Protection Agency). 2008a. National Pollutant Discharge Elimination System (NPDES), Industrial and Commercial Facilities, State 401 Certification Letters. http://cfpub.epa.gov/npdes/docs.cfm?program_id=14&view=allprog&sort=name#certification (accessed October 16, 2009).

———. 2008b. Final Issuance of National Pollutant Discharge Elimination System (NPDES) Vessel General Permit (VGP) for Discharges Incidental to the Normal Operation of Vessels Fact Sheet. http://cfpub.epa.gov/npdes/home.cfm?program_id=350 (accessed September 28, 2009).

ESP (Exotic Species Program). 1993. *Ecologically Harmful Exotic Aquatic Plant and Wild Animal Species in Minnesota: Annual Report for 1993*. St. Paul: Minnesota Department of Natural Resources.

———. 1994. *Ecologically Harmful Exotic Aquatic Plant and Wild Animal Species in Minnesota: Annual Report for 1994*. St. Paul: Minnesota Department of Natural Resources.

———. 2000. *Harmful Exotic Species of Aquatic Plants and Wild Animals in Minnesota: Annual Report for 1999*. St. Paul: Minnesota Department of Natural Resources.

———. 2001. *Harmful Exotic Species of Aquatic Plants and Wild Animals in Minnesota: Annual Report for 2000*. St. Paul: Minnesota Department of Natural Resources.

———. 2002. *Harmful Exotic Species of Aquatic Plants and Wild Animals in Minnesota: Annual Report for 2001*. St. Paul: Minnesota Department of Natural Resources.

———. 2003. *Harmful Exotic Species of Aquatic Plants and Wild Animals in Minnesota: Annual Report for 2002*. St. Paul: Minnesota Department of Natural Resources.

Fetterolf, C.M., Jr. 1980. Why a Great Lakes Fishery Commission and Why a Sea Lamprey International Symposium. *Canadian Journal of Fisheries and Aquatic Sciences* 37:1588–1593.

FWS (U.S. Fish and Wildlife Service). 2007. Injurious Wildlife: A Summary of the Injurious Provisions of the Lacey Act (18 U.S.C. 42; 50 CFR 16). www.anstaskforce.gov/Documents/Injurious_Wildlife_Fact_Sheet_2007.pdf (accessed October 13, 2009).

Galatowitsch, S.M., N.O. Anderson, and P.D. Ascher. 1999. Invasiveness in Wetland Plants in Temperate North America. *Wetlands* 19:733–755.

Getsinger, K.D., A.G. Poovey, W.F. James, R.M. Stewart, M.J. Grodowitz, J.J. Maceina, and R.M. Newman. 2002. *Management of Eurasian Watermilfoil in Houghton Lake, Michigan: Workshop Summary*. Aquatic Plant Control Research Program Final Report. Washington, DC: U.S. Army Corps of Engineers' Engineer Research and Development Center.

Glassner-Shwayder, K. 1999. *Legislation, Regulation and Policy for the Prevention and Control of Nonindigenous Aquatic Nuisance Species: Model Guidance for Great Lakes Jurisdictions*. Ann Arbor, MI: Great Lakes Commission. www.glc.org/ans/pdf/Model-Guide-Fin-June-99.pdf (accessed December 13, 2010).

Grigorovich, I.A., R. Colautti, E.L. Mills, K. Holeck, A. Ballert, and H.J. MacIsaac. 2003. Ballast-Mediated Animal Introductions in the Laurentian Great Lakes: Retrospective and Prospective Analyses. *Canadian Journal of Fisheries and Aquatic Sciences* 60:740–756.

Gunderson, J. 1994. Three-State Exotic Species Boater Survey: What Do Boaters Know and Do They Care? Minnesota Sea Grant. www.seagrant.umn.edu/exotics/boat.html (accessed December 13, 2010).

Horsch, E.J., and D.J. Lewis. 2009. The Effects of Aquatic Invasive Species on Property Values: Evidence from a Quasi-Experiment. *Land Economics* 85:391–409.

IMO (International Maritime Organization). 2009. GloBallast Partnerships. http://globallast.imo.org/index.asp (accessed October 6, 2009).

ISP (Invasive Species Program). 2004. *Invasive Species of Aquatic Plants and Wild Animals in Minnesota: Annual Report for 2003.* St. Paul: Minnesota Department of Natural Resources.

———. 2005. *Invasive Species of Aquatic Plants and Wild Animals in Minnesota: Annual Report for 2004.* St. Paul: Minnesota Department of Natural Resources.

———. 2009. *Invasive Species of Aquatic Plants and Wild Animals in Minnesota: Annual Report for 2008.* St. Paul: Minnesota Department of Natural Resources.

———. 2010. *Invasive Species of Aquatic Plants and Wild Animals in Minnesota: Annual Report for 2009.* St. Paul: Minnesota Department of Natural Resources.

Keller, D. 2007. Hydrilla Invades the Midwest. *Lakeline* 27 (3):23–24.

Krantzberg, G., and C. Boer. 2008. A Valuation of Ecological Services in the Laurentian Great Lakes Basin with an Emphasis on Canada. *Journal of the American Water Works Association* 100:100–111.

Lovell, S.J., and S.F. Stone. 2005. *The Economic Impacts of Aquatic Invasive Species: A Review of the Literature.* Working Paper #05-02. Washington, DC: U.S. Environmental Protection Agency, National Center for Environmental Economics.

Miller, M.L. 2004. The Paradox of U.S. Alien Species Laws. In *Harmful Invasive Species: Legal Responses.* Edited by M.L. Miller and R.N. Fabian. Washington, DC: Environmental Law Institute, 125–184.

Minnesota Interagency Exotic Species Task Force. 1991. *Report and Recommendations of the Minnesota Interagency Exotic Species Task Force.* St. Paul: Minnesota Department of Natural Resources.

Minnesota Waters. 2009. Aquatic Invasive Species in Minnesota's Waters—An Aquademic. www.minnesotawaters.org/sites/default/files/publications/Aquatic%20Invasive%20Species%20in%20Minnesota%E2%80%99s%20Waters%20%E2%80%93%20An%20Aquademic.pdf (accessed December 13, 2010).

MISAC (Minnesota Invasive Species Advisory Council). 2009. A Minnesota State Plan for Invasive Species. Review Draft. http://files.dnr.state.mn.us/natural_resources/invasives/state_invasive_species_plan.pdf (accessed December 13, 2010).

MPCA (Minnesota Pollution Control Agency). 2008. Fact Sheet for State Disposal System (SDS) Permit MNG300000, Ballast Water Discharge General Permit. www.pca.state.mn.us/publications/ballast-permitfactsheet.pdf (accessed October 2, 2009).

NRC (National Research Council). 2008. Great Lakes Shipping, Trade, and Aquatic Invasive Species. Committee on the St. Lawrence Seaway: Options to Eliminate Introduction of Nonindigenous Species into the Great Lakes, Phase 2. Great Lakes Shipping, Trade, and Aquatic Invasive Species. Transportation Research Board Special Report 291. http://onlinepubs.trb.org/Onlinepubs/sr/sr291.pdf (accessed December 13, 2010).

Nygren, J. 1999. Eurasian Watermilfoil in Michigan, Minnesota and Wisconsin: A Study of State Policies. Senior honors thesis, University of Minnesota.

OTA (U.S. Congress Office of Technology Assessment). 1993. *Harmful Non-indigenous Species in the United States.* Washington, DC: U.S. Government Printing Office.

Reid, D. 2008. Ballast Water and Saltwater Flushing: Closing a Gap in the Protection Framework for the Great Lakes. *ANS Update* 14 (1):1. www.glerl.noaa.gov/pubs/fulltext/2008/20080035.pdf (accessed October 2, 2009).

Ricciardi, A. 2006. Patterns of Invasion in the Laurentian Great Lakes in Relation to Changes in Vector Activity. *Diversity and Distributions* 12:425–433.

Rockwell, H.W. Jr. 2003. Summary of a Survey of the Literature on the Economic Impact of Aquatic Weeds. www.aquatics.org/pubs/economic_impact.pdf (accessed October 7, 2009).

Roley, S.S., and R.M. Newman. 2008. Predicting Eurasian Watermilfoil Invasions in Minnesota. *Lake and Reservoir Management* 24:361–369.

Ruiz, G.M., and D.F. Reid. 2007. *Current State of Understanding about the Effectiveness of Ballast Water Exchange (BWE) in Reducing Aquatic Nonindigenous Species (ANS) Introductions to the Great Lakes Basin and Chesapeake Bay, USA: Synthesis and Analysis of Existing Information.* Ann Arbor, MI: U.S. Department of Commerce, National Oceanic and Atmospheric Administration, Great Lakes Environmental Research Lab.

Simberloff, D. 2009. We Can Eliminate Invasions or Live with Them. Successful Management Projects. *Biological Invasions* 11:149–157.

Skinner, L.C., W.J. Rendall, and E.L. Fuge. 1994. *Minnesota's Purple Loosestrife Program: History, Findings, and Management Recommendations.* Special Publication 145. St. Paul: Minnesota Department of Natural Resources.

Skogerboe, J.G., A. Poovey, K.D. Getsinger, W. Crowell, and E. Macbeth. 2008. *Early Season, Low Dose Applications of Endothall to Selectively Control Curlyleaf Pondweed in Minnesota Lakes.* Aquatic Plant Control Research Program Technical Notes Collection (ERDC/TN APCRP-CC-08). Vicksburg, MS: U.S. Army Engineer Research and Development Center.

Valley, R.D., W. Crowell, C.H. Welling, and N. Proulx. 2006. Effects of a Low Dose Fluridone Treatment on Submersed Aquatic Vegetation in a Eutrophic Minnesota Lake Dominated by Eurasian Watermilfoil and Coontail. *Journal of Aquatic Plant Management* 44:19–25.

CHAPTER 14

Flooding and Flood Management

Kenneth N. Brooks, James D. Fallon, David L. Lorenz, James R. Stark, and Jason Menard

F loods result in great human disasters globally and nationally, causing an average of $4 billion of damages each year in the United States (Leitch 2003). Minnesota has its share of floods and flood damages, and the state has awarded nearly $278 million to local units of government for flood mitigation projects through its Flood Hazard Mitigation Grant Program. Since 1995, flood mitigation in the Red River Valley has exceeded $146 million (DNR 2009b). Considerable local and state funding has been provided to manage and mitigate problems of excess stormwater in urban areas, flooding of farmlands, and flood damages at road crossings. The cumulative costs involved with floods and flood mitigation in Minnesota are not known precisely, but it is safe to conclude that flood mitigation is a costly business.

This chapter begins with a description of floods in Minnesota to provide examples and contrasts across the state. Background material is presented to provide a basic understanding of floods and flood processes, prediction, and management and mitigation. Methods of analyzing and characterizing floods are presented because they affect how we respond to flooding and can influence relevant policies.

The understanding and perceptions of floods and flooding commonly differ among those who work in flood forecasting, flood protection, or water resource management and citizens and businesses affected by floods. These differences can become magnified following a major flood, pointing to the need for better understanding of flooding as well as common language to describe flood risks and the uncertainty associated with determining such risks. Expectations of accurate and timely flood forecasts and our ability to control floods do not always match reality. Striving for clarity is important in formulating policies that can help avoid recurring flood damages and costs.

FLOODING IN MINNESOTA

A flood is a hydrologic event that occurs whenever a stream flows over and out of its banks. Small floods are not rare; they can occur in response to snowmelt or excess rainfall. Large floods, produced by heavy rainfall, rapid snowmelt, or both, are rarer than small floods and can cause major property damage.

Minnesota has experienced many large floods since European settlement. The state generally experiences two types of flooding. One type is caused by convectional or frontal rains that result in substantial rainfall and runoff. These can be of local or regional extent but generally affect smaller watersheds more than large ones. Small rivers are susceptible to flooding when intense rainfall occurs over relatively small areas. The risks associated with this type of flood vary depending on soils, land cover, and topographic settings. Areas of high topographic relief, such as the North Shore of Lake Superior or southeastern Minnesota, can experience streamflows that destroy structures built near rivers. Areas of low topographic relief, such as western Minnesota, can be quickly inundated and remain flooded for weeks because of poor drainage.

The second prevalent type of flooding in Minnesota involves spring snowmelt and associated precipitation that often occurs on frozen ground. Large rivers, such as the Red River of the North, the Minnesota River, and the Upper Mississippi, are particularly susceptible to this type of flooding. Large floods result in flows that greatly exceed bankfull (Figure 14.1). "Bankfull" is the elevation of the water level, called stage height, in the stream at which water is at the top of its banks; any flow above bankfull elevation will overflow onto the floodplain. This definition applies to rivers that have a conspicuous floodplain. Bankfull flow is the streamflow corresponding to the bankfull stage; for most streams, it corresponds to an annual

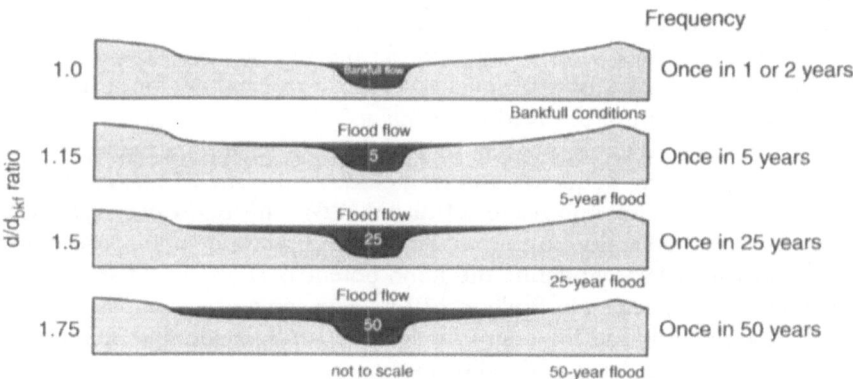

Figure 14.1 *Bankfull flow and flood flow relationships of streams for Rosgen natural "C" channel types*

Notes: Not to scale. Ratios of mean depth (d) to bankfull mean depth (d $_{bkf}$) and associated flood flow frequencies are empirical relationships for natural "C" channel types based on the Rosgen stream classification.

Source: Adapted from Verry (2000).

recurrence interval between 1 and 2 years, most commonly in the range of 1.5 to 1.7 years (Dunne and Leopold 1978; Rosgen 1996).

The larger the flow, the higher the elevation (stage) of the water surface above the bank, and the greater the magnitude of the flood and potential for flood damages; however, the more extreme the flood, the less frequently it occurs. Thus a pattern of increasingly large floods occurs on average at a frequency of once every 50, 100, or 200 years, or a probability of 2%, 1%, or 0.5% per year, respectively. As seen in Figure 14.1, for Rosgen natural "C" channels that are meandering, slightly entrenched streams with well-developed floodplains, general relationships exist between the depth of water at bankfull and the depth at different recurrence intervals (Rosgen 1996; Verry 2000).

It is reasonable to suggest that climate change and human alterations to the landscape have affected, and will affect, the magnitude and frequency of flooding in Minnesota. Urban and agricultural development can increase runoff from the land and likely exacerbate flooding, at least for smaller-magnitude floods that occur more frequently. However, the lack of long-term streamflow data for many unregulated watersheds makes it difficult to quantify or relate "natural flooding," land use change, and climate change. Most hydrologists agree that the large flood events that have occurred during Minnesota's recorded history, such as the 2009 flood on the Red River, would have been substantial regardless of human alterations to the landscape.

Peak flow during a flood is used to determine the probability, commonly expressed as the annual recurrence interval, that a given flow will be exceeded. For example, a flood that has a 1% chance of exceedance in any given year would, on the long-term average, be expected to occur only about once a century; therefore, such a flood would be termed a "100-year flood." However, the chance of such a flood occurring in any given year is 1%. Thus a 100-year flood actually can occur in successive years at the same location. In some cases, annual exceedance probabilities can be based on periods of regulated flow or adjusted when historic data are available. Annual exceedance probabilities for peak flows in 1997 on the mainstem of the Red River ranged between 2% and 0.2%. Federal agencies are requested to use a particular probability distribution, the log–Pearson type III, for estimating frequency and magnitude of the annual maximum flood (IACWD 1982, hereafter referred to as Bulletin 17B). The log–Pearson type III is a flexible distribution that fits most annual flood data fairly well, but no procedure or distribution will accurately define the flood potential of any given stream. The general processes described in Bulletin 17B involve computing summary statistics of peak flows, accounting for unusually large or small values, and adjusting for regional trends in the shape of the distribution.

Results of the Bulletin 17B procedure for the 1997 flood on the Red River at Grand Forks, North Dakota (Figure 14.2), show that this flood was a rare event, but that floods equal to or greater than it in magnitude could occur in the future. An important part of the Bulletin 17B procedure is to describe the uncertainty in the magnitude estimates. For example, the 1% annual exceedance probability for the Red River at Grand Forks determined by the Bulletin 17B procedure is 100,000 cubic feet per second (cfs). Confidence limits are described in terms of

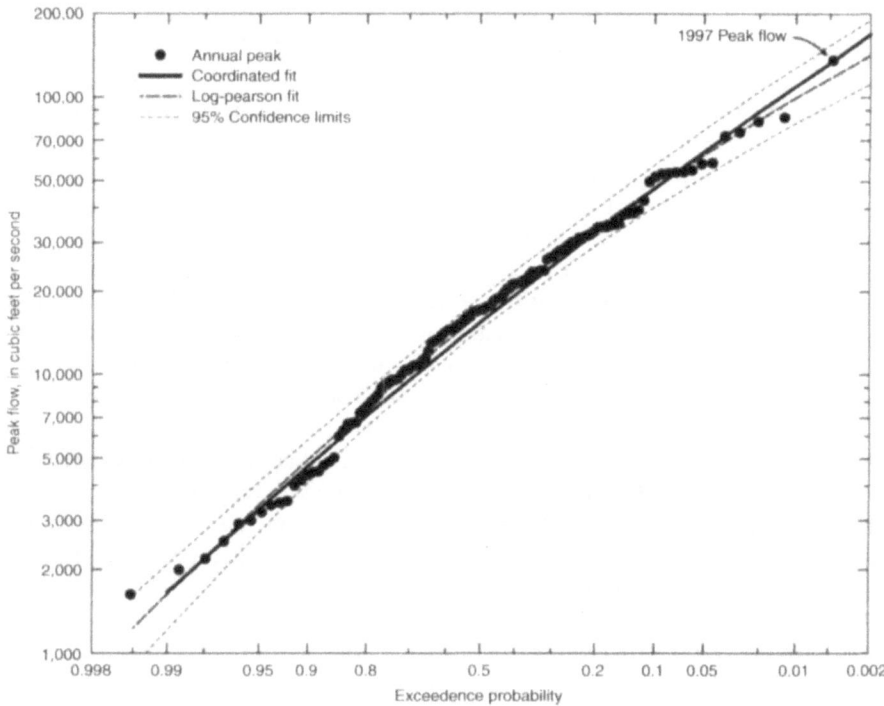

Figure 14.2 *Annual maximum flood frequency curve and confidence limits at Grand Forks, North Dakota*

Notes: Based on data from 1882–2008 using the log-Pearson (Bulletin 17B) procedure. Also shown are the 1997 flood peak and the coordinated frequency curve that stakeholder groups and agencies have agreed on. Source: USACE (2001).

percentage and describe the probability that the true value is within the stated range. For the Red River at Grand Forks, there is a 95% chance that the true 1% flood is between 80,900 and 129,000 cfs. Cities on large rivers generally adopt a frequency curve called the "coordinated value," a curve based on Bulletin 17B but modified with input from stakeholders. The coordinated curve value for the 1% flood on the Red River at Grand Forks is 110,000 cfs (see Figure 14.2).

Rainfall-Driven Floods

Rainfall-driven floods can occur from either summer convective thunderstorms or persistent, long-duration rainfall derived from frontal storms. Convective storms result in more localized flash flooding caused by short-duration, high-intensity rainfall that exceeds the infiltration capacity of soils. Frontal storms with excessive rainfall that persist over several days or weeks can saturate the soils over large areas, resulting in widespread flooding. Following are some examples of rainfall-driven floods in Minnesota.

2007 Flood in Southeastern Minnesota. Record rainfall occurred during August 2007, causing severe floods in parts of the Upper Mississippi River Valley and killing 14 people. Widespread, slow-moving thunderstorms developed along a stationary front stretching from northern Iowa through northern Illinois, while the low-level jet stream transported warm, moist air from the remnants of Tropical Storm Erin into southern Minnesota and southern Wisconsin. The greatest rainfall occurred in southeastern Minnesota and southwestern Wisconsin, where record amounts were recorded (Fallon et al. 2008).

Flooding was severe in four states. In southeastern Minnesota, the most severe flooding occurred in and adjacent to the Whitewater and Root River Basins. Two U.S. Geological Survey (USGS) streamflow-gauging stations in the Root River Basin—at Root River near Houston and Rush Creek near Rushford—recorded peak flows of record (Fallon et al. 2008). Peak flow at the gauging station on the Root River had an annual exceedance probability of less than 1% (100-year flood), while peak flow of Rush Creek near Rushford had an annual exceedance probability of less than 0.5% (200-year flood). Federal disaster areas were declared for 14 counties in Iowa, 6 in Illinois, 8 in Minnesota, and 14 in Wisconsin. Damages were estimated to be greater than $157 million in Minnesota (Minnesota Office of the Governor 2007).

2002 Flood in the Red River of the North. During June 2002, major flooding occurred in the Red River Basin (Wiche et al. 2002). Extensive damage to roads, bridges, and crops occurred throughout the flooded area in northwestern Minnesota and northeastern North Dakota. Thirteen counties in Minnesota were designated as disaster areas.

Several streamflow-gauging stations on the Wild Rice, Marsh, and Roseau Rivers recorded peak stages and flows during the floods. For example, a peak flow of 12,300 cfs occurred on the Wild Rice River at Twin Valley, Minnesota (USGS 2009–2010). This exceeded the previous peak flow, which occurred in 1997, by more than 20% and had an annual exceedance probability between 1% and 0.5%. A peak flow of 16,000 cfs occurred on the Roseau River near Malung, Minnesota, that was 20% greater than the previous peak flow. The annual exceedance probabilities for peak flows on the Roseau River ranged from less than 0.2% below South Fork near Malung and at Ross, Minnesota, to between 0.5% and 0.2% below State Ditch 51 near Caribou. Peak flows on most tributaries to the Red Lake River and on the river itself at Crookston had annual exceedance probabilities between 5% and 10% (USGS 2009–2010).

Snowmelt-Driven Spring Floods

Springtime flooding is common in Minnesota as a result of excessive snowmelt runoff, which often is augmented by rain on snow. Following are some examples of snowmelt-driven floods in Minnesota.

1965 Flood in the Upper Mississippi River Basin. Flooding in the Upper Mississippi River Basin caused more than $100 million in damage to public and private property in 1965 (USACE 1965). The Mississippi River from Aitkin, Minnesota, to Guttenburg, Iowa, as well as major tributaries, exceeded flood stage. Conditions leading up to the flooding included deep ground frost levels, snow moisture content as high as 11 inches, a late spring thaw, and rain occurring with the snowmelt runoff and river breakup.

This flood affected urban areas of Minneapolis, St. Paul, and Winona, Minnesota. The Minnesota River near Jordan crested at 117,000 cfs, an annual exceedance probability of less than 1% and a flow that remains the peak of record for this location (USGS 2009–2010). With other tributaries to the Mississippi just upstream from Minneapolis cresting by mid-April, the combined flows of the Mississippi and Minnesota Rivers resulted in record peak flows and stages on the Mississippi downstream into Iowa. The Mississippi River at St. Paul crested at 26.01 feet, with a corresponding flow of 171,000 cfs, an annual exceedance probability of less than 0.5%, and the highest flow recorded at this location in nearly 150 years (USGS 2009–2010).

1997 Flood in the Red River of the North. Record floods devastated many communities along rivers and streams in the Red River Basin in North Dakota and western Minnesota during spring 1997. The flood peak at Grand Forks corresponded to a 0.5% annual exceedance probability based on the coordinated frequency curve (Figure 14.2). Thousands of people were forced to flee their homes, some permanently, as floodwaters and severe weather caused nearly $2 billion in damages to the region (Macek-Rowland 1997; Macek-Rowland et al. 2001). The flat topography of the basin and shallow channel, characteristics due to the basin's location in the bed of ancient Lake Agassiz, make an extensive floodplain. As a result, widespread flooding occurs with relatively small increases above bankfull flow.

Many of the conditions necessary for large floods were present in the Red River Basin before and during the floods of 1997 (Macek-Rowland 1997; Macek-Rowland et al. 2001). The winter was one of the most severe on record for the upper Great Plains. Record snowfalls were recorded throughout North Dakota and western Minnesota, creating conditions conducive to major flooding. Melting of the snowpack and thawing of the ice began in late March but were inhibited by a late-spring blizzard that occurred during early April. The additional moisture brought by the blizzard added to the increasing streamflows in the Red River Basin.

The peak flow of the Red River at Wahpeton, North Dakota, crested at 12,800 cfs; the annual exceedance probability for the peak flow was between 1.0% and 0.5%. Peak flow on the Red River at Fargo, North Dakota, and Moorhead, Minnesota, was 28,000 cfs. Annual exceedance probability for this peak flow was between 1.0% and 2.0%, and the stage exceeded the record set in 1897. Flow in the Red River at Grand Forks, North Dakota, peaked at 136,900 cfs, and the stage was more than 2 feet higher than the record set in 1897. The annual exceedance probability for this peak flow was between 0.5% and 0.2%. The stage continued to rise to 54.35 feet. In addition to peak flow on the mainstem, peak flows on many tributaries to the Red River had about 10% or greater annual exceedance probabilities during the 1997 floods (Macek-Rowland 1997; Macek-Rowland et al. 2001; USGS 2009–2010).

2009 Flood in the Red River of the North. Record flooding also occurred in the Red River Basin during spring 2009. Several factors contributed to record flooding and related to weather conditions the previous fall and winter. The foundation for flooding began in fall 2008 when heavy rains saturated the ground (MCWG 2009). When the ground froze in late fall, it trapped the moisture and left little opportunity for snowmelt to infiltrate. Record snowfalls occurred during the winter season, and the area had the wettest March on record. Other factors that contributed to the floods included the moisture content of snowfall, rapid snowmelt, and additional precipitation during snowmelt. The winter was also one of the coldest on record, and ice jams became a threat on streams and rivers. The cold temperatures added additional ice thickness to the rivers, which increased ice jams. However, spring flooding was a product of more than just precipitation, runoff, and the presence of ice jams. The low gradient of the Red River and the northerly direction of the flow of the river contributed to the severity of flooding as well, as did human factors, including channel obstructions such as bridges, levees, roadways, and floodplain development.

THE UNCERTAINTY OF PREDICTING FUTURE FLOODS

Although ranking floods helps illustrate the relative magnitude of a flood, it has limited use for evaluating future risk. Flood probability analysis can be used to determine annual exceedance probability, the probability that a future flood of a magnitude equal to or greater than a specific flow will occur in a given year. The uncertainty associated with predicting the probabilities of future floods has two aspects. The first pertains to the length of streamflow records from which frequency relations are derived. As discussed above, the derived flow value for a given probability (e.g., a 1% annual exceedance probability of 100,000 cfs) is associated with a confidence interval, a range of values that include the true value. For example, as illustrated in Figure 14.2, there is a 95% probability that a true 1% flood is between 80,900 and 129,000 cfs. As streamflow records cover an increasing number of years, confidence intervals around the 1% flood become narrower; that is, we have more confidence in our estimate of the 1% flood with a longer historical record. This confidence, however, is based on the assumption that past streamflow records are representative of future streamflows. The second aspect of uncertainty is that other sources can influence our prediction of future floods, particularly changes in flood frequency and magnitude caused by shifts in climatic patterns and human-induced changes in watershed and channel conditions.

Technological advances provide tools that allow us to better forecast floods, delineate and map flood-prone areas, and simulate and evaluate climatic and land use changes and their effects on flood forecasting. However, present technology does not offset the uncertainty associated with predicting the future. Acquiring an understanding of the causes of floods and determining the risks associated with different landscapes and river basins are necessary to develop response plans. This entails knowledge of the probabilities associated with floods of different magnitudes. Traditionally, we have analyzed historical streamflow records and based our estimates of flood frequency on that information. With the ever-changing climatic conditions (coupled with periodicity of temperature and precipitation patterns) and land use conditions, this approach is problematic. At present, no better alternatives are available for characterizing flood frequency and developing response plans.

Trends of peak flows also are important, because they may indicate changing levels of risk to emergency and infrastructure managers. Historical annual peak-flow time-series data for selected USGS streamflow-gauging stations in the Midwest were analyzed to determine the presence and subsequent magnitude of any trend through time at each station. Examination of the trend in peak-flow magnitude for the period 1958 to 2007 did not indicate a systematic basinwide trend in either direction. Of the 147 gauging stations analyzed, 83 indicated an upward trend, 60 a downward trend, and 4 no trend (Ryberg 2009).

RESPONSES TO FLOODS: MANAGING FLOODS AND FLOOD DAMAGES

Many options exist for controlling or managing floods and mitigating flood damages, including flood-control measures such as dams, levees, and diversions; flood-proofing of buildings and structures; flood forecasting and warning; responses to forecasts, such as evacuation and construction of temporary dikes; and flood damage avoidance through floodplain management and zoning (Brooks et al. 2003; Dunne and Leopold 1978). Flood forecasting is essential for warning people of impending flood hazards but does not provide long-term mitigation measures to cope with floods. Long-term measures include programs such as the National Flood Insurance Program (NFIP), which provides incentives for avoiding human occupation of the active floodplain (FEMA 2010). In contrast to many other states, Minnesota has few flood-control dams and reservoirs and does not have the extensive levee system that exists along much of the Mississippi River in downstream states.

Structural measures to manage or control floods have proved to be successful in many cases; however, they can have serious drawbacks. In particular, extending levees along a river can create unwanted downstream effects, as discussed below. The report by the International Joint Commission (IJC 2000) following the 1997 flood on the Red River considered options of reducing flood flows by adding 0.7 million acre-feet of flood storage to the existing 1 million acre-feet in the upper parts of the basin. Economic, social, and environmental issues associated with increased flood storage, however, precluded further consideration. Lessons learned from Leopold's analysis were considered by an Interagency Levee Task Force (ILTF) composed of federal agencies and those of five states: Iowa, Illinois, Indiana, Wisconsin, and Missouri. This task force identified nonstructural alternatives that included the purchase of permanent easements on floodplain lands, flood-proofing of buildings, relocation of buildings outside of floodplains, and other alternatives (USACE 2009). Although Minnesota was not part of this effort, the results of this task force complement ongoing efforts in the state to develop future flood management options and policies.

Levees and Flooding: Lessons from the 1993 Flood on the Mississippi River

When floods occur, overbank flow spreads onto the floodplain and is slowed by the resistance of the vegetative surfaces of the floodplain. As the flood peak recedes, floodwater is stored on the floodplains of most rivers. Consequently, the magnitude of the flood peak in downstream areas is attenuated in contrast to rivers whose channels have either been deepened (channelized) or become constricted within levees or flood walls that disconnect the river and its floodplain.

The 1993 flood on the Mississippi River caused $10 billion in damage and was one of the most devastating in the 20th century. Postflood evaluations by Leopold (1994) indicated that record flood stages were measured at 44 locations in the Upper Mississippi River and 49 in the Missouri. Levees caused these stages to be higher than they would have been if the levees were not in place; furthermore, these higher stages were accompanied by higher flow velocities because the channel was constrained. In places where the levees were not breached, the higher stages did not cause an increase in flood damages, and their flood-control benefit was realized by adjacent landowners. However, where the levees failed, the higher flood stages caused substantially greater devastation than would have occurred without them.

Following his evaluation, Leopold pointed out that by preventing rivers from overflowing onto their floodplains, levees increase river stages by restricting the width of the rivers and eliminating temporary storage of floodwaters on the floodplains. He concluded that flood damages could have been reduced by allowing some floodplain areas to flood, and that flood-control policy should consider the costs and benefits of using easements to restrict development on floodplains in lieu of levees. Such an approach would be particularly attractive in areas where levees have been breached during a flood. If natural floodplain storage along rivers is not utilized to some extent, and if levees continue to be constructed, it appears that downstream areas will become more vulnerable to increased flood damages.

Floodplain Management

A floodplain, in geologic terms, is the depositional feature adjacent to a stream that is formed by the stream when it exceeds its banks and deposits sediments. As defined by planners, engineers, and agencies, the term refers to any area susceptible to surface water flooding (DNR 2009a). Floodplain management is a zoning approach that controls human activities and development on the 1% chance (100-year) floodplain (Figure 14.3). Local units of government identify permitted land uses on the floodplain, and floodplain regulations are administered

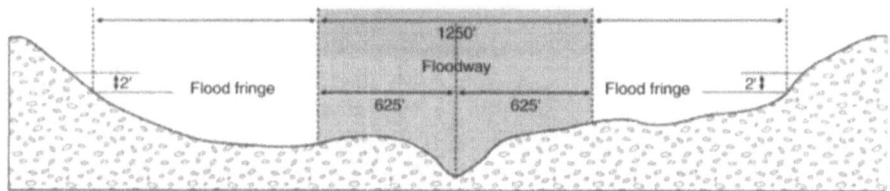

Figure 14.3 *Floodway regulation areas in the lower Minnesota River Valley*
Note: Not to scale.
Source: Garvey et al. (1986).

by local zoning authorities (DNR 2006). The channel and the portion of the floodplain area required to pass the 100-year flood without increasing the water-surface elevation more than a designated height (6 inches in Minnesota) are referred to as the "floodway." The floodway portion of a floodplain is zoned to prohibit most types of development that involve structures or filling. Portions of the 100-year floodplain outside of the floodway but still susceptible to flooding are known as the "flood fringe" (DNR 2009a).

Local zoning authorities must adopt the minimum federal and state standards but can have more restrictive regulations. For example, they could regulate certain activities (e.g., hospitals, fire stations, or hazardous-waste disposal sites) in a 500-year floodplain rather than just in a 100-year floodplain, or they could prohibit any structural development in a floodplain. Minnesota requires that the regulatory flood protection elevation be 1 foot higher than the federal standard of the 100-year flood event.

Any community can enroll in the National Flood Insurance Program (NFIP). The Federal Emergency Management Agency (FEMA) produces Flood Insurance Rate Maps (FIRM) that designate high, medium, and low flood risk areas. People with structures in the 1% (100-year) floodplain must have flood insurance before they can acquire mortgage loans from federally regulated lenders. People living outside of the floodplain also can purchase flood insurance if their community is enrolled in NFIP; their insurance rates are less expensive than for structures in the floodplain. FEMA mapping focuses on watercourses, but residents near lakes and basins or in lower areas are at risk from flooding, especially from intense rains.

Minnesota's recovery following the 1997 flood on the Red River indicates a progressive approach that entails considerable cooperation and coordination among agencies and communities. Following the 1997 flood, which had levee failures and damages in excess of $1 billion, the immediate flood recovery needs were met by state and federal emergency management and humanitarian help from the Red Cross and others (Lokkesmoe 2007). The Disaster Recovery Act Task Force, which serves as a clearinghouse for local requests, was initiated by the state immediately following the flood. Local requests for disaster assistance included housing acquisition; removal of structures; and construction of pump stations, floodwalls, and levees. In terms of long-term recovery, the U.S. and Canadian governments directed the IJC to conduct a study and make recommendations.

Minnesota's Flood Mitigation Program was established to provide cost-share funds to local governments for purposes of reducing the risk of future flood damages. This program has four priorities, in this order: (1) acquiring and removing homes from the floodplain; (2) constructing flood protection projects such as pump stations, small levees, and floodwalls; (3) obtaining funding for federal flood-control projects; and (4) after local communities are protected, working toward the development of impoundments and other measures to reduce flood magnitude. According to Lokkesmoe, "The potential for repetitive damages from floods has been significantly reduced in Minnesota," and "the cities' commitment to floodplain management will reduce the risk of future flood damages" (2007, 13).

Recurring floods in the Red River of the North have led to changes in flood policies, new programs, and changes in existing programs and institutions to facilitate coordination and cooperation among local, state, and federal organizations and agencies (Halliday and Associates 2009; IJC 2000). In particular, the following changes have occurred:

- new institutional arrangements: the IJC, Red River Basin Commission, and International Red River Board, consisting of government representatives from Canada and the United States, with the objective of addressing issues following the 1997 flood on the Red River;
- a focus on small floods along tributary streams in the Red River Basin as well as major floods in the Red River;
- emphasis on cooperation and innovation, with an integrated watershed management approach;
- increased emphasis on farmstead ring dikes;
- Canadian policy changes, including a new regulation for designated flood areas, new emergency management legislation, and floodway expansion;
- changes by the U.S. Army Corps of Engineers that support integrated basinwide consideration of projects working with the Minnesota Red River Watershed Management Board and its North Dakota counterpart;
- new state building codes that include flood-proofing measures;
- substantial improvements in flood forecasting, including improved data networks and data collection that support forecasting, model development, and improved communication of forecasts; and
- mitigation that emphasizes structural measures and removal of high-risk structures from the floodplain.

Although the responses to the 1997 Red River flood have been largely positive, Leitch (2003) provides examples of local development occurring in flood-prone areas at elevations below the 100-year regulatory floodplain as a result of real estate transactions. Through subdividing properties, selling lots via "developers' agreements," and obtaining variances that are approved by local government, housing development has continued to occur in floodplains, leaving property owners vulnerable to repetitive flood losses that can lead to increasing costs for flood insurance. Since the inception of the NFIP, more than $3.5 billion in payments have been made nationally for repetitive-loss properties (Pollnow and Garren 2003). This example points to the importance of establishing binding policies at the local level regarding development and variances on floodplains.

Floods, Land Use, and Integrated Watershed Management

Focusing attention on watersheds and integrated watershed management (Gregersen et al. 2007) is a viable approach to complement flood reduction and flood-management efforts, as indicated in the IJC (2000) report and the Natural Resources Framework Plan by the Red River Basin Commission (RRBC 2005). This approach is based on the realization that the actions of people on the watershed, floodplain, and channel can affect the magnitude of the more "moderate" floods and the frequency and extent of flood damages that occur. People's use of the land can contribute directly to the magnitude and frequency of certain levels of flooding through increasing urbanization and the associated expansion of impervious areas on the landscape, loss of natural floodplain storage, fragmentation of forestland, mining, artificial drainage of agricultural lands, and associated loss of wetland storage. Consequently, an integrated watershed management (IWM) approach that complements floodplain management warrants consideration in managing floods and flood damages (Halliday and Associates 2009; IJC 2000).

An IWM approach to mitigate flood damages entails several considerations. Human activities that change the watershed or river channel can influence the amount and timing of runoff and streamflow (Brooks et al. 2003; Calder 2005; Ice and Stednick 2004; Verry 2000). The extent to which human activities affect the magnitude and frequency of floods is determined largely by the loss of storage in the watershed. Storage in a watershed is limited, so when rainfall or snowmelt occurs at sufficiently high rates or over a sufficiently long time, all available storage on the landscape becomes filled; from that point on, runoff to streams becomes equal to the input of precipitation. Furthermore, when meandering streams are straightened or wetlands drained and connected to stream channels, the speed at which water flows downstream can increase, resulting in higher peak flows.

In Minnesota, losses of watershed storage can be attributed to changes in vegetative cover and soils, drainage of wetlands and lakes, urbanization, channel modifications, and disconnecting rivers from their floodplains. What has not been clear is the extent to which such losses affect floods of different magnitudes. Studies by Magner et al. (1993), Lu (1994), Miller (1999), and Verry (2004) indicate that wetland drainage and changes in vegetation by agricultural and forestry activities can increase the magnitude of floods ranging from 4% to 50% annual exceedence probability, but not that of floods with a 1% probability. Perhaps the most drastic human-induced increases in runoff are caused by urbanization, where vegetative cover and permeable soils become replaced with impervious surfaces, such as roads, parking lots, and rooftops. Although urbanization can increase the magnitude of flood peaks in small watersheds, such increases are greater for frequent small floods than for large floods (Konrad 2005). Furthermore, because urbanization is not prevalent in a large percentage of Minnesota's river basins, the effects are more pronounced on a local level than basinwide.

In flood mitigation, the role of IWM needs to entail actions at the local and state levels that include measures such as enhancing microstorage, proposed by the IJC (2000), which suggested using depressions in the landscape and placing gates on culverts to hold back floodwaters in upper reaches of the Red River. Wetland restoration, riparian management, stream channel restoration, and encouraging the planting of native perennial vegetation and crops are components of an IWM approach. More central to IWM is the concept that natural resources be managed on the basis of natural landscape boundaries (e.g., ecoregions) rather than political boundaries. In fact, many water resource professionals would argue that watersheds or catchments, which are finer-scale than ecoregions, are the correct scale for management. Watershed boundaries are specified in the Red River Basin Natural Resources Framework Plan (RRBC 2005).

Improving Communication

Communication during flood forecasting and floods is key to managing floods and flood damages. When floods like the 2009 spring floods in the Red River occur, a tremendous amount of coordination is needed among local, state, and federal government officials to orchestrate timely and appropriate responses. It is generally agreed that communication and coordination among agencies, cities, and the public improved greatly between the 1997 and 2009 floods in the Red River Valley (Buan 2009; Lejcher 2009; Wiche 2009). Response efforts begin with local emergency managers because they are close to the event and first on the scene. Every county in Minnesota has a professional emergency manager, but in some cases, local response efforts begin with and continue to be coordinated by local officials.

With record river crests occurring in most major drainage basins in the Red River during spring 2009, National Weather Service (NWS) river forecasts were critical for the communities and individuals during their flood fights. Data and real-time information needed for river forecast models depend largely on federal, state, and local agencies. The backbone of the river gauge network, needed to provide data for river forecast models, is streamflow data from the USGS. Streamflow data and flood-related information are shared among government agencies through several means, including Internet, email, and daily coordinating conference calls. Improvements in interagency communication and coordination that occurred between the 1997 and 2009 floods helped support the flood fight. To some extent, this can be attributed to taking a watershed approach in which local groups within watersheds communicated more effectively with government agencies. Strong leadership was provided by the NWS, which facilitated communication and coordination among agencies and citizens. Nevertheless, issues of communicating flood forecasts during the 2009 Red River flood emerged at a Rainfall-River Forecasting Joint Summit held in St. Paul in October 2009 and attended by federal agencies, local officials, and stakeholders (USACE 2008). One issue centered on a lack of understanding by local communities of how to interpret long-range flood probability forecasts provided by the NWS. At the heart of the issue was how local communities should respond to long-range flood forecasts (those given 30 to 90 days out) that have different probabilities of occurrence, recognizing that as flooding becomes more imminent, appropriate responses to mitigate damage become better known.

The use of probability terminology remains a lingering and long-term impediment to communicating and understanding risks and uncertainty of large floods. Considerable confusion arises with the terminology commonly used in describing floods and developing mitigation strategies. Terms such as the 100- or 500-year recurrence interval often send the wrong message to the public, or the message is misunderstood. The concepts of flood frequency and annual exceedance probabilities are not clear to many; for example, many incorrectly assume that once a 100-year flood occurs, such a flood will not occur again for 99 years. Improved and clear communication among technical flood analysts, the public, and policymakers is essential so that better choices are made concerning economic development within and adjacent to flood-prone areas. For one thing, rather than using the terminology of recurrence intervals, it may be better to emphasize the idea that large floods with a low probability of occurring in any given year actually have a much higher risk when considered over a period of several years. For example, the probability of a 100-year flood, which has a 1% chance of occurring in any given year (annual exceedance probability), has a 26% probability of occurring over a period of 30 years, which is a concept highly relevant to homeowners, who typically have 30-year mortgages (an example emphasized by DNR 2006).

STATUS OF MINNESOTA'S FLOOD MANAGEMENT POLICIES

Given the issues and examples related to floods and flood mitigation discussed earlier in this chapter, a question that might be asked is whether Minnesota has the policies in place to allow for and promote viable flood mitigation. To address this question, consider the Association of State Floodplain Managers' 10 guiding principles for effective floodplain management programs (ASFPM 2003):

1. State floodplain management programs need strong, clear authority.
2. State floodplain management programs should be comprehensive and integrated with other state functions.
3. Flood hazards within the state must be identified and the flood risks assessed.
4. Natural floodplain functions and resources throughout the state need to be respected.
5. Development within the state must be guided away from flood-prone areas; adverse impacts of development both inside and outside the floodplain must be minimized.
6. Flood mitigation and recovery strategies should be in place throughout the state.
7. The state's people need to be informed about flood hazards and mitigation options.

8. Training and technical assistance in floodplain management need to be available to the state's communities.
9. Levels of funding and staffing for floodplain management should meet the demand within each state.
10. Evaluation of the effectiveness of states' floodplain management programs is essential and successes should be documented.

In response to the 10 principles above, the DNR website provides details concerning floods, floodplains, and mitigation through its series of Floodplain Information Sheets (DNR 2010), as well as information on flood insurance and FEMA. The information is in place to support an effective floodplain management program by the state; however, coordination and cooperation with local watershed management districts and federal agencies is the key to addressing issues of recurring floods such as those in the Red River Basin.

Policymakers have many resources at their disposal in formulating flood policy in Minnesota, from local to regional to federal levels. The state has had a solid record of local community involvement through its 45 watershed management districts, which are governed by a Board of Commissioners from the counties that have land within the watershed district boundaries. Educational, training, and related support are provided to local communities by the Minnesota Association of Watershed Management Districts (MAWD 2010).

Numerous organizations exist to support floodplain management and flood policy in Minnesota and the region. There is considerable local watershed involvement with state and federal agencies and interagency organizations, including the Red River Basin Commission (RRBC 2010); Regional Flood Risk Management Team (RFRMT n.d.); Minnesota Association of Floodplain Managers (ASFPM n.d.); and Silver Jackets (2010), an interagency team composed of the USACE, FEMA, and other state and federal agencies. The purpose of such organizations is to leverage information and resources and improve public risk communication through a united effort. As a result, the organizations appear to be in place, and coordination efforts among citizen groups and local, state, and federal agencies have been improved, particularly in the aftermath of the 1997 Red River flood. Improving flood management and mitigation seems to center around enhanced communication and understanding of risk management.

CONCLUSIONS

Minnesota experiences flooding from both extreme rainfall events in summer months and snowmelt runoff that commonly is compounded by spring rain on the snow. All river basins in Minnesota are susceptible to intense rainfall floods that can be localized from convective storms or more widespread from frontal rainfall. Responses to floods in the state have benefited from active participation of local citizens, cities, and watershed management organizations, coupled with improved communication and coordination by local, state, and federal agencies. Minnesota has a good combination of local, state, and federal cooperation in fighting floods and mitigating their effects. Thus the state has the appropriate ingredients for good flood policy.

As discussed earlier, Minnesota has not stressed the development of major flood-control dams and reservoirs or the construction of extensive levee systems to control or manage floods. Instead, the state has relied largely on a combination of flood forecasting and construction of temporary dikes and levees before a flood occurs. In 2010, Minnesota developed a comprehensive flood hazard mitigation program (Minnesota Statutes Section 103F.161) to make the state more flood-resilient by constructing permanent levees, floodwalls, and impoundments and removing structures from the floodplain.

Communication and coordination during flood forecasting and the flood events are stressed as keys to managing floods and flood damage. National Weather Service river forecasts will remain critical, requiring data and real-time information from local, state, and other federal agencies.

Minnesota is likely to keep its emphasis on nonstructural flood alternatives, such as the purchase of permanent easements on floodplains and flood-proofing buildings or relocating them outside of floodplains. Local zoning authorities are required to adopt federal and state standards, or even more restrictive regulations, in order to participate in the NFIP. State maximum standards in floodways limit or prohibit many types of development.

Focused attention on watersheds and integrated watershed management (IWM) needs to complement flood reduction and flood management efforts. The approach is based on the realization that the actions of people can affect flooding and flood damages. Land use change contributes directly to the magnitude and frequency of flooding. Consequently, IWM that complements floodplain management should be pursued with the purpose of reducing flood damages. This can include using IWM to enhance storage by using depressions in the landscape as well as other means to hold back floodwaters. IWM should include wetland restoration, riparian management, stream channel restoration, and encouragement of planting native perennial vegetation and crops. An important concept to stress is that natural resources need to be managed on the basis of watersheds, rather than political boundaries. Minnesota has achieved local community involvement through watershed management districts and other local organizations that are essential to supporting floodplain management and flood policy.

Uncertainty associated with predicting the magnitude and likelihood of future floods and the implications for floodplain management and economic development adjacent to rivers need to be better understood by citizens, developers, and policymakers. Determining flood risks poses serious challenges to flood analysts and emergency management personnel, particularly in the provision of information that could be used by policymakers for flood mitigation policies or work. One of the challenges is in understanding the level of uncertainty or unpredictability that comes with trying to accurately predict flood frequencies and magnitudes. When not fully understood or communicated properly, this uncertainty can seriously affect floodplain and floodway delineation and can create considerable concern among local communities and government agencies.

REFERENCES

ASFPM (Association of State Floodplain Managers). 2003. Effective State Floodplain Management Programs. www.floods.org/PDF/Effective_State_Programs_Final.pdf (accessed December 1, 2009).

———. No date. Home page. www.mnafpm.org (accessed December 14, 2010).

Brooks, K.N., P.F. Folliott, H.M. Gregersen, and L.F. DeBano. 2003. *Hydrology and the Management of Watersheds.* 3rd ed. Ames, IA: Iowa State Press.

Buan, S. 2009. Personal communication between Steve Buan, National Weather Service, and the authors.

Calder, I.R. 2005. *Blue Revolution: Integrated Land and Water Resource Management.* 2nd ed. London: Earthscan.

DNR (Minnesota Department of Natural Resources). 2006. Floodplains and Floodplain Management. http://files.dnr.state.mn.us/publications/waters/floodplain_basics.pdf (accessed October 9, 2009).

———. 2009a. Division of Waters. www.dnr.state.mn.us/waters/watermgmt_section/floodplsin/rfpe.html (accessed August 13, 2009).

———. 2009b. *State of Minnesota Flood Hazard Mitigation Grant Expenditures in the Red River Valley, Minnesota as of August 2009.* St. Paul: DNR Waters.

———. 2010. Division of Waters: Publications, Information Sheets, Presentations. www.dnr.state.mn.us/publications/waters (accessed December 14, 2010).

Dunne, T., and L.B. Leopold. 1978. *Water in Environmental Planning.* San Francisco: W.H. Freeman and Company.

Fallon, J.D., H.S. Garn, M.L. Harris, and K.D. Lund. 2008. Floods from Record Rains in Illinois, Iowa, Minnesota, and Wisconsin, August 17–30, 2007. Paper presented at Upper Mississippi River Basin States Cooperators' Roundtable for the USGS Cooperative Water Program. November 1–2, 2007, Dubuque, IA.

FEMA (Federal Emergency Management Agency). 2010. The National Flood Insurance Program. www.fema.gov/business/nfip (accessed December 14, 2010).

Garvey, E., P. Gersmehl, and D. Brown. 1986. *Minnesota Water Rights and Regulations.* WRRC Public Report No. 5. St. Paul: University of Minnesota, Water Resources Research Center.

Gregersen, H.M., P.F. Folliott, and K.N. Brooks. 2007. *Integrated Watershed Management: Connecting People to Their Land and Water.* Cambridge, UK: CABI, Cambridge University Press.

Halliday, R., and Associates. 2009. How Are We Living with the Red? A Report by R. Halliday and Associates to the International Red River Board. www.ijc.org/php/publications/pdf/ID1633.pdf (accessed December 1, 2009).

IACWD (Interagency Advisory Committee on Water Data). 1982. *Guidelines for Determining Flood-Flow Frequency.* Bulletin 17B. Washington, DC: Hydrology Subcommittee, Office of Water Data Coordination, U.S. Geological Survey.

Ice, G.G., and J.D. Stednick, eds. 2004. *A Century of Forest and Wildland Watershed Lessons.* Bethesda, MD: Society of American Foresters.

IJC (International Joint Commission). 2000. Living with the Red: A Report to the Governments of Canada and the United States on Reducing Flood Impacts in the Red River Basin. www.ijc.org/rel/pdf/001590part1e.pdf (accessed December 14, 2010).

Konrad, C.P. 2005. Effects of Urban Development on Floods. U.S. Geological Survey Fact Sheet 076-03. http://pubs.usgs.gov/fs/fs07603 (accessed December 14, 2010).

Leitch, J.A. 2003. Floodplains and the Tyranny of Small Decisions. *ASFPM News and Views* 15 (5):1, 10–11. www.floods.org/Newsletters/News_Views/NV_October03.pdf (accessed December 1, 2009).

Lejcher, T. 2009. Personal communication between Terry Lejcher, Minnesota DNR, Division of Waters, and the authors.

Leopold, L.B. 1994. Flood Hydrology and the Floodplain. In *Coping with the Flood: The Next Phase*, edited by G.F. White and M.F. Myers. *Water Resources Update* 95 (Spring): 11–15.

Lokkesmoe, K. 2007. Disaster Recovery following the 1997 Spring Floods in Minnesota. *Water Resources IMPACT* 9 (5):12–13.

Lu, S-Y. 1994. Forest Harvesting Effects on Streamflow and Flood Frequency in the Northern Lake States. PhD diss., University of Minnesota.

Macek-Rowland, K.M. 1997. *1997 Floods of the Red River of the North and Missouri River Basins in North Dakota and Western Minnesota.* Open-File Report 97-575. Washington, DC: U.S. Geological Survey.

Macek-Rowland, K.M., M.J. Burr, and G.B. Mitton. 2001. Peak Discharges and Flow Volumes for Streams in the Northern Plains, 1996–97. U.S. Geological Survey Circular 1185-B. http://pubs.er.usgs.gov/pubs/cir/cir1185B (accessed December 14, 2010).

Magner, J.A., G.D. Johnson, and T.J. Larson. 1993. The Minnesota River Basin: Environmental Impacts of Basin-wide Drainage. In *Industrial and Agricultural Impacts of the Hydrologic Environment*, vol. 5, pp. 147–162. Edited by Y. Eckstein and A. Zaporozec. Alexandria, VA: Water Environment Federation.

MAWD (Minnesota Association of Watershed Management Districts). 2010. Home page. www.mnwatershed.org (accessed December 14, 2010).

MCWG (Minnesota Climatology Working Group). 2009. Climate Journal. http://climate.umn.edu/doc/whatsnew.htm (accessed December 11, 2009).

Miller, R.C. 1999. Hydrologic Effects of Wetland Drainage and Land Use Change in a Tributary Watershed of the Minnesota River Basin: A Modeling Approach. MS thesis, University of Minnesota.

Minnesota Office of the Governor. 2007. Governor Pawlenty Signs Flood Relief Appropriation Bill Following One-Day Special Session. www.governor.state.mn.us/mediacenter/pressreleases/2007/PROD008309.html (accessed October 11, 2007).

Pollnow, C., and E. Garren. 2003. Repetitive Flood Losses: Benefitting All NFIP Policyholders by Addressing an Unacceptable Burden. *ASFPM News and Views* 15 (6):1, 4. www.floods.org/Newsletters/News_Views/NV_December03.pdf (accessed December 1, 2009).

RFRMT (Regional Flood Risk Management Team). No date. Home page. www.mvd.usace.army.mil/rfrmt (accessed December 14, 2010).

Rosgen, D. 1996. *Applied River Morphology.* Pagosa Springs, CO: Wildland Hydrology.

RRBC (Red River Basin Commission). 2005. Red River Basin Natural Resources Framework Plan. www.redriverbasincommission.org/Services/NRFPnonsdlstchFINAL.pdf (accessed December 1, 2009).

———. 2010. Home page. www.redriverbasincommission.org (accessed December 14, 2010).

Ryberg, K. 2009. Personal communication between Karen Ryberg, U.S. Geological Survey, and the authors.

Silver Jackets. 2010. Silver Jackets: Many Agencies One Solution. Home page. www.iwr.usace.army.mil/nfrmp/state/index.cfm (accessed December 14, 2010).

USACE (U.S. Army Corps of Engineers). 1965. *1965 Spring Floods, Post Flood Report.* St. Paul: USACE.

———. 2001. *Final Hydrology Report: Hydrologic Analyses—The Red River of the North Main Stem, Wahpeton/Breckenridge to Emerson, Manitoba.* St. Paul: USACE.

———. 2008. Tri-Agency Fusion Cell, Rainfall-River Forecasting Joint Summit. Saint Paul, October 19, 2009. http://mvs-wc.mvs.usace.army.mil/fusion/fusion.htm (accessed June 7, 2010).

———. 2009. Raising the Standard: Interagency Flood Risk Management in the Midwest. www.iwr.usace.army.mil/ILTF (accessed December 1, 2009).

USGS (U.S. Geological Survey). 2009–2010. National Water Information System: USGS Water Data for Minnesota. http://waterdata.usgs.gov/mn/nwis (accessed November 30, 2009; March 3, 2010).

Verry, E.S. 2000. Water Flow in Soils and Streams: Sustaining Hydrologic Function. In *Riparian Management of Forests of the Continental Eastern United States.* pp. 99–124. Edited by E.S. Verry, J.W. Hornbeck, and C.A. Dolloff. Boca Raton, FL: Lewis Publishers.

———. 2004. Land Fragmentation and Impacts to Streams and Fish in the Central and Upper Midwest. In *Lessons for Watershed Research in the Future: A Century of Forest and Wildland Watershed Lessons.* pp. 129–154. Edited by G.G. Ice and J.D. Stednick. Bethesda, MD: Society of American Foresters.

Wiche, G. 2009. Personal communication between Gregg Wiche, U.S. Geological Survey, and the authors.

Wiche, G.J., K.G. Guttormson, S.M. Robinson, G.B. Mitton, and B.J. Bramer. 2002. *June 2002 Floods in the Red River of the North Basin in Northeastern North Dakota and Northwestern Minnesota.* Open-File Report 2002-278. Washington, DC: U.S. Geological Survey.

CHAPTER 15

Policy Complexity and Competing Uses in a "Water Rich" State

John K. Helland

*B*eing "water rich," as many people describe Minnesota, also has led to water policy complexity in the politics of water disputes and law through the many organizations—both public and private—that govern and assist the management of water bodies in the state. This chapter discusses Minnesota's water policy setting, the continuing issue of agricultural drainage versus wetland protection, the evolution of state public waters authority, the layers of management and competing "advocacy agencies" on water policy needs, and some examples of cooperation and coordination among agencies and groups on state water issues of common concern.

THE STATE'S WATER POLICY SETTING

Because Minnesota's early reliance on agriculture caused it to be the driving sector of economic prosperity, it can be said that the history and politics of state water policy has somewhat followed suit. When Minnesota achieved statehood in 1858, one of the first laws enacted allowed private corporations to be formed "for the purpose of draining lands and creating privileges."[1]

Early settlers thought the state's abundant water was the enemy, thus the extensive practice of drainage; in the modern era, Minnesota now has a statutory policy to conserve rainwater where it falls on the land. Ever since statehood, policymakers and water managers have tussled over the local rights of agricultural drainage versus the protection and restoration of wetlands.

Known as the "Land of 10,000 Lakes," Minnesota is rich in freshwater resources. It is second among the states in area of surface water, with over 12,000 lakes 10 acres or more in size, 92,000 stream miles (including drainage ditches), and an abundance of groundwater. Three huge drainage basins originate from

Minnesota: the Mississippi River, which drains to the Gulf of Mexico; the Red River of the North, which drains to Hudson Bay; and the Great Lakes, which drain to the St. Lawrence River and the Atlantic Ocean. Because of these major headwater areas, Minnesotans feel a stewardship obligation to care for them, not only for the state's citizens, but for citizens living in other states where they flow.

Although Minnesota's water resources are abundant, they are not unlimited, and agriculture, industry, domestic uses, recreation, and ecosystem services all compete for water. Local control is a paramount principle of the state's governance system. And with all its water, local units of government, along with special-purpose units of the state and federal governments, have been given various degrees of authority to manage the resource. Water policy can be found in many places—in state statutes, programs, administrative rules, and agency guidelines; the accumulation of state and federal court decisions; the decisions of local governments; and many applicable federal laws—and it changes over time. Water policy in Minnesota evolved away from wetland drainage toward protection and restoration of wetlands; away from prior destructive farming practices toward the statewide adoption of soil and water conservation practices; and away from discharge of raw sewage and pollutants into water bodies toward the adoption of water quality standards and their enforcement. Minnesota's water policy also reflects decisions and actions made in other policy areas, including energy and land use, transportation and agriculture, public health and education, state budgeting and economic development, and even electoral politics and campaigns. Water policy can ignore complexity but not avoid it.

Although Minnesota has been recognized nationally as a leader in the management of water resources, it does not have a comprehensive state water policy. The water policy is more patchwork in nature, reacting to crisis situations; it is strong in many areas, weak in some, and silent in a few. Areas where the state's water policy is strong include meeting the basic human needs for water in Minnesota; ensuring that water remains an important economic asset; slowing and sometimes reversing many destructive water use trends; and maintaining a high level of technical expertise. The state's abundance of water resources provides it with a somewhat broad margin of error. Minnesota's people and elected officials historically have been fast to respond to immediate and long-term threats to the quality of its surface water, and increasingly to its groundwater as well. This public commitment is reflected in the water policy that has been formulated over the decades.

WETLANDS VERSUS DRAINAGE: THE CONTINUING STRUGGLE

Wetlands have been important points of contention in Minnesota landscape management since European settlement. Early in the settlement process, wetlands were regarded as wastelands that had to be drained to make the land agriculturally productive. Governor Ramsey, speaking to the Minnesota legislature in 1861, commented that wetlands were exceptionally valuable because they were easily drained, highly productive, and would become some of the most valuable lands in the state (BWSR n.d.).

Wetland drainage was strongly encouraged early in Minnesota's history but also was a regulated practice. Since 1887, Minnesota statutes have provided a structure for governance of drainage practices. Early regulations encouraged drainage, allowed landowners to petition the county for permission to drain and to request support for drainage, provided a structure for monitoring drainage practices, and served as a basis for evaluating whether drainage of a specific wetland represented a societal benefit. One could argue that this law, passed nearly 125 years ago, sets up the concept of evaluating practices on the basis of ecosystem services, although the language of ecosystem services did not evolve until more than 100 years later. The state's early laws permitted each water problem to be solved locally, with little thought given to the impact on other areas or future generations. All kinds of local drainage units were created in a willy-nilly fashion until 1897, when a three-member state commission was formed (Helland 1999).

PUBLIC WATERS AUTHORITY

The Great Depression of the 1930s started to change how Minnesota managed water. The economic turmoil and severe drought greatly slowed down drainage activity. Counties began to place ditch liens against property owners on foreclosed farmland to protect the counties against ditch maintenance costs, and many ditching enterprises were ill advised, with the land serving no consequential use after drainage, because the farms were abandoned. The Minnesota Department of Conservation was created in 1931, replacing the state drainage commission, and the concept of "public waters" was strengthened. Basically, Minnesota waters that served a beneficial public purpose were subject to control by the state.[2] The state would protect not only the amount of public waters, but also the "containers," the lakes, streams, and wetlands where the water was confined (DNR 2010).

In 1937, a law was passed that established the water appropriation permit program to appropriate water, or to build dams.[3] The Conservation Department also was given authority to establish permanent and legal lake levels. At the same time, following new federal law, soil conservation districts were established in the state to curtail soil erosion. A Legislative Interim Commission on Water Conservation, Drainage, and Flood Control was convened in 1953 to review past drainage laws and their ability to meet modern needs. Their deliberations led to the creation of the Water Resources Board (later becoming part of the present Board of Water and Soil Resources), which could authorize the creation of geographic watershed districts along watershed boundaries. Once created, watershed districts could take over drainage systems inside their boundaries. County boards were required to evaluate the effects on the environment and natural resources when considering a drainage project, and the number of petitioners required to initiate a project was increased. The commissioner of conservation (now natural resources) also was required to evaluate environmental and conservation impacts before a drainage project could be established (MWPB 1981).

That review and evaluation process, as well as the influence of societal valuation of wetlands, has resulted in increases and decreases in the rate of wetland drainage over the last 75 years. After the Second World War, there were high demands for agricultural productivity and high rates of wetland drainage. By the late 1960s and early 1970s, U.S. society was actively engaged in an environmental movement that increased society's perceived value for wetlands and decreased the rate of wetland drainage. As society perceived greater benefits from undrained wetlands for a range of services, state and federal agencies began a practice of purchasing wetlands for conservation purposes (Helland 1996).

Because of widespread concern over what constituted public waters, which required a state permit if they were altered, the legislature established the Public Waters Inventory Program in the late 1970s to map all public/protected waters and wetlands. This was done on a county-by-county basis and, for the first time, gave specific statutory protection to types 3, 4, and 5 wetlands of a certain size. Today Minnesota is conducting drainage system inventories in an attempt to modernize drainage records throughout the state and monitor the protection of required buffer strips along the ditches. Under the Board of Water and Soil Resources (BWSR), a Drainage Work Group composed of affected stakeholders meets regularly to make recommendations on clarifying antiquated provisions in the state's long-standing drainage law.

When Minnesota's landmark wetland protection law passed in 1991, after three years of protracted legislative debate between agricultural and environmental interests, it contained 24 specific exemptions from the replacement ratio in draining or filling wetlands. Over the next five years, the law was further amended three times by agricultural interests, to reduce the replacement ratio in creating wetlands in certain areas; to broaden the definition of agricultural land, thereby increasing acreage that would be exempt; and to allow local governments to devise alternative wetland plans if they felt that their landscape was unique enough to deserve it. Chapter 7 in this book details the "no net loss" policy created by the Wetlands Conservation Act of 1991 and wetland protection under the federal Farm Bill provisions of recent vintage.

STATE WATER MANAGEMENT AGENCIES

Because of Minnesota's water rich nature and local control principle, government water resource managers are found in various capacities throughout the state. There are six state agencies that deal with major water policy programs: the Department of Natural Resources, Pollution Control Agency, Department of Health, Board of Water and Soil Resources, Department of Agriculture, and Environmental Quality Board.

The Department of Natural Resources (DNR) is responsible for the water appropriation permit system, groundwater monitoring and conservation principles, water recreation and invasive species concerns, flood mitigation management, shoreland and scenic river protection, water supply and planning, dam safety and water levels, streamflow monitoring and water mapping, and the Great Lakes Compact. The DNR has the most extensive water management responsibilities among the state agencies.

The Pollution Control Agency (PCA) protects, improves, and restores water quality and is responsible for the feedlot and impaired waters programs, acid precipitation, clean-water monitoring, stormwater and septic system standards, and leaking landfills and underground storage tanks. It is also responsible for water quality regulation, monitoring, and enforcement under the Federal Clean Water Act.

The Department of Health administers the Safe Drinking Water Act in public water supplies, regulates wells and borings, protects source water and wellheads, establishes groundwater health risk limits, and is responsible for public health safety in administering the federal Safe Drinking Water Act.

The Board of Water and Soil Resources provides technical and financial assistance to local units and landowners on water resource concerns, for comprehensive local water planning, and for wetland management and protection. This is the "helping agency," providing direction and assistance to watershed districts and soil and water conservation districts.

The Department of Agriculture assists farmers with best management practices for water quality and integrated pest management and regulates pesticide control. Because of the Groundwater Protection Act and the 1998 animal feedlot laws (described below), their agricultural water management responsibilities have greatly increased over the past 25 years.

The Environmental Quality Board, composed of representatives from the state agencies with environmental responsibilities along with five citizens, provides for environmental review oversight, is responsible for overall state water-planning coordination, makes water policy recommendations, and develops a state water plan (EQB 2008).

Besides these six agencies with major water duties, the Minnesota Public Facilities Authority, Department of Transportation, and Department of Public Safety also have small programs relating to water. In addition, other agencies concerned with water management include the Metropolitan Council and watershed management organizations in the Twin Cities area, watershed districts covering about one-third of the state, soil and water conservation districts, lake improvement districts, and lake conservation districts. A virtual plethora of water managers abounds.

In 1969, the House of Representatives Committee on Land and Water Resources created two subcommittees to study and analyze water governance. One conclusion was that the administrative system for water was so large and cumbersome that few, if any, governmental officials and citizens could have a clear understanding of the system. Similarly, at about the same time, a different legislative committee concluded that significant changes in administrative structure were required if societal goals were to be met. They suggested that those goals might be met through greater involvement by municipalities and counties or through assigning greater powers to special-purpose districts.

The committee at the time felt that special-purpose districts were the less preferred option, arguing that these functions are the proper purview of counties and municipalities (Waelti 1974).

Further interim hearings and action during the 1980 legislative session made legislators realize that it is impossible to divorce state-level management from local water management, because each affects the other. This began a 30-plus-year debate and discussion over coordination and consolidation of the state agencies dealing with water, and who should primarily be responsible on the local level. A major drought cycle in the late 1970s led the legislature to create the Water Planning Board, whose responsibilities included creating a water framework and related land resources plan, coordinating water programs, and involving citizens. A 1980 law asked the board to conduct a local water management study for possible clarifications and improvements in water authorities. This kicked off a number of studies on water management coordination or consolidation, which have probably averaged one every two years since Earth Day 1970 (MWPB 1982).

Some changes have been made since then: the Water Planning Board eventually was merged into the Environmental Quality Board, which later formed the Water Resources Coordinating Committee; and BWSR was created by the merger of the Water Resources Board, Soil and Water Conservation Board, and Southern Minnesota Rivers Basin Council. Some characterize Minnesota's water-governing system as a "collection of advocacy agencies." Each water resource management agency has a distinct perspective. The system meets the needs of various interest groups and gives them a voice in state government decisionmaking that they might not have had with only one agency.

Major decisions are made with full public scrutiny. Water resource issues are complex and far-reaching, and agencies dealing with agriculture, health, public safety, natural resource management, and pollution control all have legitimate interests in these issues. The external checks and balances of the system can foster creative tension and diversity in dealing with the complex issues. In 1986, a Minnesota house legislative report summarized all the attempts up to that point on water coordination and consolidation (Helland 1986) and concluded: "Despite administrative complexity and the fragmentation and overlap that *may* occur among state water management agencies, Minnesota traditionally has supported a system of strong, competing agencies, each concerned with its own duties and specific goals. In political terms, an advocacy system promotes competition and increases the public representation of each goal or interest, while highlighting political choices. Conflicts and tradeoffs in such a system are meant to be solved through the political rather than the administrative process."

So the Minnesota water governance system is built on existing institutions and the result of more than 50 years of modern advances in the understanding of water-related problems. It reflects the state's strong political and cultural tradition of direct citizen participation in government decisionmaking and involving constituent groups in policy formation.

COOPERATION AND COORDINATION IN WATER MANAGEMENT

Several times in recent decades, mainly in reaction to newfound environmental concerns, Minnesota agencies with water programs have banded together willingly to help solve a particular concern. Two examples discussed here are the 1989 Groundwater Protection Act and a 1998 law that addressed animal feedlot problems. During the mid-1980s, several concerns received public attention because of their threats to groundwater. Wells in southeastern Minnesota were found to be contaminated with nitrates from fertilizers and animal waste. In addition, volatile organic chemicals from leaking petroleum storage tanks were found to be entering groundwater aquifers. Pesticide products causing contamination also became an increasing concern, as did the fact that many of the state's solid waste landfills were found to be leaking and polluting groundwater and surface water (MSPA 1989).

Based on two key water-related strategies, the Ground Water Protection Strategy by PCA and the Strategy for the Wise Use of Pesticides and Nutrients by the Environmental Quality Board (EQB), the governor and the legislature each sprang into action on the issue. The governor through EQB convened all the major state water agencies and the Minnesota Geological Survey and gathered public input to draft a major groundwater protection bill.

The senate formed an ad hoc water committee to concentrate on groundwater issues during the interim before the 1989 session. In the state house, legendary environmentalist Willard Munger and his Environment and Natural Resources Committee held a number of hearings on protecting groundwater. When it was time to introduce the house bill in the 1989 session, Munger decided that explaining and lobbying the 10 separate articles of the bill would be equally shared by himself and four other authors. This was a rare occurrence at the legislature, where usually only a bill's main author leads the push to passage.

A wide coalition of interest groups both worked on and helped educate others on the need for a new groundwater protection law. They included departments of the University of Minnesota, League of Cities, Association of Counties, Association of Townships, Soil and Water Conservation District Association, Minnesota Association of Watershed Districts, many statewide environmental groups, and even some business organizations. All were instrumental in moving the bill toward eventual passage.

Although the two largest farm organizations, the Minnesota Farm Bureau and Minnesota Farmers Union, generally supported the principles of the bill when introduced, they, along with certain agribusiness concerns and individual farmers, were nervous about how strongly the protective measures might affect them. At one hearing in the House Agriculture Committee, after the bill had passed the House Environment Committee, Representative Munger sensed that a great number of less restrictive amendments were going to be offered, and also that the bill might be defeated in the committee. He sent an aide out to find the speaker of the house, who was supportive of strong groundwater protection, and after some messages to the chair of the Agriculture Committee, the bill sailed on to the next committee in line to hear it with only a few minor amendments.

Former governor Elmer Anderson, long a supporter of conservation measures, wanted an exemption from using groundwater in a "once-through cooling system" for air-conditioning his business company offices. It was reluctantly granted after tailoring an amendment to fit only his company's situation. A powerful agricultural legislator who represented what was largely an agricultural district told an aide to then-governor Rudy Perpich before the final vote on the legislative floors that the law had better not affect farmers for two years. Therefore, the effective date on some agricultural provisions and restrictions was delayed.

The main goal of the Groundwater Protection Act was to ensure that groundwater was to be maintained in its natural condition, free from any degradation caused by human activities. The legislature acknowledged that this may not be always practicable but encouraged the development of methods and technology that would make preventing degradation possible. All the major state water agencies had duties and goals to make the act work. Several actions since 1989 to achieve the legislative goal have been successfully implemented. Stronger water conservation efforts were achieved and promoted by state and local agencies. New or increased water fees were added to reflect the cost of the resource use, and local water-planning grants were funded to assist local planning efforts. Also, greater groundwater monitoring and testing for pollutants were required.

In 1999, a new House Subcommittee on Groundwater was formed to review the 10-year history of the Groundwater Protection Act and identify evolving groundwater issues. It was apparent from the review that the act had advanced broad societal goals, although it was difficult to attribute specific changes to the act itself. The act was a common civic good that brought diverse interests together in a bipartisan manner. It led to a collective vested interest that left a legacy for future generations by setting goals to protect the public's drinking water. By the act's investment in and new knowledge of the groundwater resource, it created the potential to save future public dollars in identifying groundwater quantities and preventing its future contamination. Groundwater, as an important natural resource, became better understood for future needs and management (Helland 2001).

Feedlot Legislation

A similar, but somewhat smaller, effort in state agency cooperation and coordination occurred during the 1998 legislative session, when a series of laws was enacted to address the growing concern over large-scale animal feedlots and their pollution problems. The Department of Agriculture (MDA) had been given several new environment-related responsibilities in the 1989 Groundwater Protection Act, including strengthening pesticide and fertilizer regulation, waste pesticide collection, nonpoint-source pollution control, groundwater and surface water monitoring, and best management practice adoption. This effort continued in 1998, when the legislature decided to concentrate on animal feedlots. The MDA was to adopt new rules for the construction of manure storage facilities. The legislature also assessed the need for a noncommercial manure applicator training and certification program, required licensing for commercial animal waste technicians and certification for manure testing laboratories, and made funding available for farm manure digester technology.

The Minnesota Pollution Control Agency (MPCA), after originally setting standards for animal feedlots shortly after it was created in 1969, was told by the legislature to adopt more modern rules over the issuance and denial of feedlot permits, use a set schedule for issuing National Pollutant Discharge Elimination System (NPDES) permits for feedlots with 1,000 or more animal units, and place a two-year moratorium on construction of certain swine waste lagoons. Also, feedlot program grants to counties were increased; and the MPCA was directed to work with Minnesota Extension Service, MDA, BWSR, local units of government, producer groups, and appropriate federal agencies to notify and educate agricultural producers regarding any changes in animal feedlot rules.

The State Planning Agency, with the assistance of the EQB, was given the tasks of compiling a four-year generic environmental impact statement (GEIS) until 2002, on issues relating to livestock production, and developing new environmental review rules on phased actions for animal feedlots. The BWSR received new grant moneys for soil and water conservation districts to cost-share on feedlots. The Department of Labor and Industry, in consultation with the MPCA, MDA, and Health Department, was tasked with making changes to standards for hydrogen sulfide exposure levels within livestock confinement facilities. The Board of Animal Health was to issue permits to rendering-plant owners in order to prevent the spread of disease from transporting carcasses of domestic animals. Counties, with MPCA approval, could assume responsibility for processing feedlot permit applications under certain standards, but they could adopt more stringent standards if they thought it was necessary. They also had public notice and reporting requirements on new animal feedlots. In addition, the legislative auditor was asked to report to the legislature on the adequacy of enforcement of the animal feedlot rules and the effectiveness of counties in bringing feedlots into compliance with the rules (Helland 1998).

Although the $3 million GEIS helped inform the MPCA's rulemaking for feedlots and helped curb the severe odor problems from large facilities, it did not address the controversial issues between the agricultural and environmental communities. The State Planning Agency decided to obtain a consensus strategy among all stakeholders before the GEIS process was begun. In doing so, issues of contention between the disparate groups were left on the table.

Both the Groundwater Protection Act of 1989 and the 1998 laws that focused on feedlots show how state water agencies can cooperate and coordinate with local units of government and special-purpose districts to handle major problems. Other enactments, such as the Reinvest in Minnesota Resources Act and Wetland Conservation Act, also required the agencies to work together. These all led up to the Clean Water, Land, and Legacy Amendment of 2006, with all the major water agencies being represented on the Clean Water Council, sharing funding for specific duties to clean up impaired surface waters. However, the agencies had to wait until 2009 to receive new money for groundwater and drinking-water issues.

CONCLUSIONS

From the days of early statehood into the early part of the 20th century, the growth of local drainage councils and commissions has evolved into major state water agencies, local units of government as part of their duties, and some special-purpose districts that have stayed the course for the better part of 50 years. It all makes sense to those who work with or govern the water management system, but for those newly initiated through either a political office or a project requiring multiple water permits, it is confusing, Byzantine, and seemingly redundant. However, each governmental unit has a specific water task to monitor or accomplish, little overlap occurs, and interested citizens can have public input if they so desire.

The studies and reports on consolidating water programs and agencies into one big bureaucracy probably will continue to appear, but when they do, the agency spokespersons and the constituent groups that support them always seem to convince legislators and others that no system is perfect and Minnesota's does work in the end.

There has been some interest recently in re-creating a legislative water commission or possibly a standing water committee under an existing commission, such as the Legislative Citizen-Commission on Minnesota Resources. The well-liked Legislative Water Commission, established in the 1989 Groundwater Protection Act to provide a forum for legislators to learn about water issues and gain expertise, served nobly until it was disbanded after six years because of budget cuts. During its existence, it provided feedback on water issues to members of the executive branch and legislators who did not serve on the commission.

An idea that seems to be gathering support lately is that all water monitoring, which is now spread among at least five state agencies, be consolidated under one agency for better focus and ease of data collection. Another idea taking root is that Minnesota needs an integrated water information and management system that is accessible to public and private users in order to understand the whole water system. Real concern about water resources seems to arise only during crisis situations, such as drought or flood. There is little serious public education about the limits of water resources.

The Minnesota Environmental Quality Board has suggested that the state's reputation as "water rich" is not as well deserved as it once was. The growing and significant demands on its renewable water resources make water supply and water quality management special concerns, especially with a projected population growth of 1 million anticipated in the next 25 years.

It seems clear that the politics of Minnesota water policy will always have some element of agricultural interests versus environmental protection as part of the debate. Although drainage and the wetland protection issue may diminish over time, water quality problems from nonpoint-source pollution, in both surface water and groundwater, will continue in the coming decades.

Whatever Minnesota faces in the foreseeable future concerning water policy needs or a streamlined management structure, the state is fortunate in this dire economy to have stable funding for new initiatives. In 2008, Minnesota voters passed a constitutional amendment to increase the sales tax by three-eighths of 1%, with the money divided among environmental and arts programs until June 30, 2034. Approximately $70 million annually of this legacy money will be available for protecting, restoring, and enhancing surface water and groundwater quality and protecting drinking-water sources. The University of Minnesota Water Resources Center recently led development of a 25-year detailed framework plan, working cooperatively with state agencies, local governmental units, special-purpose water districts, and nonprofit environmental groups, to identify policy, research, monitoring, and evaluation issues in order to achieve sustainable groundwater and surface water use in the years ahead. This should prove immensely valuable for policymakers and citizens alike in targeting legacy dollars to the right water resource concerns.

We may be able to echo a famous Minnesotan on National Public Radio, Garrison Keillor, by paraphrasing: "Minnesota—where the rivers are strong, the lakes are good-looking, and the fishing is above average" (EQB 2000, 3).

NOTES

1. Laws of Minn., 1858, Ch. 73.
2. Minn. Stat., § 103A.201.
3. Laws of Minn., 1937, Ch. 468.

REFERENCES

BWSR (Board of Water and Soil Resources). No date. History of Wetland Regulation and Conservation in Minnesota. www.bwsr.state.mn.us/wetlands/wca/history.html (accessed January 18, 2011).

DNR (Department of Natural Resources). 2010. History of Water Protection. www.dnr.state. mn.us/waters/watermgmt_section/pwpermits/history.html (accessed December 17, 2010).

EQB (Environmental Quality Board). 2000. Minnesota Watermarks: Gauging the Flow of Progress, 2000–2010. www.gda.state.mn.us/pdf/2000/eqb/wtr_mrk.pdf (accessed December 17, 2010).

———. 2008. Managing for Water Sustainability: Report of the EQB Water Availability Project. Availability Project. www.eqb.state.mn.us/documents/Managing_for_Water_Sustainability_12-08.pdf (accessed December 17, 2010).

Helland, John K. 1986. *State Water Management: Reorganization and Consolidation*. St. Paul: Minnesota House of Representatives, Research Department.

———. 1996. *Wetlands: A Brief Historical Background on Wetland Issues and a Chronology of State Laws since 1858*. Information Brief. St. Paul: Minnesota House Research Department.

———. 1998. *1998 Law Changes Affecting Minnesota Animal Feedlots*. Information Brief. St. Paul: Minnesota House Research Department.

———. 1999. The Drainage Issue. Information Brief. St. Paul: Minnesota House Research Department. http://news.minnesota.publicradio.org/features/2005/10/10_steilm_impairedditches/drainage.pdf (accessed January 18, 2011).

———. 2001. A Survey of the Groundwater Act of 1989. St. Paul: Minnesota House of Representatives Research Department. www.house.leg.state.mn.us/hrd/pubs/gdwtract.pdf (accessed December 17, 2010).

MSPA (Minnesota State Planning Agency). 1989. *The Minnesota Ground Water Protection Act of 1989: A Summary.* St. Paul: MSPA.

MWPB (Minnesota Water Planning Board). 1981. *Toward Efficient Allocation and Management: Special Study on Local Water Management.* St. Paul: MWPB.

———. 1982. *Special Study on Local Water Management.* St. Paul: MWPB.

Waelti, John J. 1974. *Minnesota Water and Related Land Resource Policies with Emphasis on the 68th Legislative Session.* St. Paul: Water Resources Research Center, University of Minnesota.

PART VI

MOVING TOWARD
A PROGRESSIVE FUTURE

CHAPTER 16

Finding a Path to Water Sustainability

John R. Wells

M innesota will be home to well over a million new residents in the next 25 to 30 years. Where they live, work, and play, and how they engage in these activities, will have a profound effect on the state's water resources. Forces outside the state's direct control, such as those emanating from the global economy or global climate change, also will have a marked influence on Minnesota waters. Under any foreseeable scenario, Minnesotans can expect the water-related problems they face to increase in severity and complexity as the state's people, economy, and communities expand.

The experience of the state's Impaired Waters Program provides a case in point. The state list for 2010 includes 3,000 surface water quality impairments—those water bodies and river segments that do not meet a water quality standard—but this statistic is based on monitoring of only 17% of the state's rivers and 28% of its lakes.[1] Unfortunately, the rule of thumb is that about 40% of assessed waters make the impaired waters list, and Minnesota has 12,000 lakes over 10 acres and 69,000 miles of natural streams and rivers (DNR 2010).

With the state's problems being both local and global in nature, the management challenge facing Minnesotans is daunting. How we respond to these threats will determine whether we can sustain those assets that we most value and whether our children and our children's children will experience the quality of water resources that we enjoy today.

FIVE PRINCIPLES FOR SUSTAINABLE WATER MANAGEMENT

Given this complex set of forces, the challenge for policymakers is to help Minnesotans develop a sustainable relationship with their land and water resources. The concept of *integrated water resources management* suggests that a sustainable relationship requires the coordination of water, land, and related resources in a way that benefits people and economies without compromising ecosystems (GWP 2000). The Minnesota legislature has reached similar conclusions in recent years, passing legislation requiring planning for "sustainable water use" defined as use that "does not harm ecosystems, degrade water quality, or compromise the ability of future generations to meet their own needs."[2]

This chapter suggests five principles to guide future management that build on this law and the concepts of sustainability and integrated water resources management that underlie it: *interconnectedness, informed decisionmaking, precaution, transparency,* and *accountability*. These principles provide an operational checkpoint and guide to decisions in working toward sustainable water management.

Principle #1: Interconnectedness

The first principle of sustainable water management is interconnectedness. Whether the subject is polluted runoff that impairs a lake's water quality or the pumping of groundwater that reduces flows or changes water quality in a stream or wetland, the principle of interconnectedness recognizes that land use, water quality, and water quantity are all interconnected. The legislature recognized this principle in an explicit way with a new law restricting the appropriation of groundwater in order to protect surface water supplies, water quality, and ecosystems.[3] The idea of integrated water resources management was devised to help water managers recognize these interconnections in a systematic way.

The principle also incorporates the notion, more broadly, that water's social, economic, and ecological characteristics are interconnected. Figure 16.1 provides a conceptual framework for understanding the basic relationships among social, economic, and ecological systems. It acknowledges that the economy is not some abstract force independent of communities or the environment, but rather a natural subset of those systems that is dependent on and responsible for their

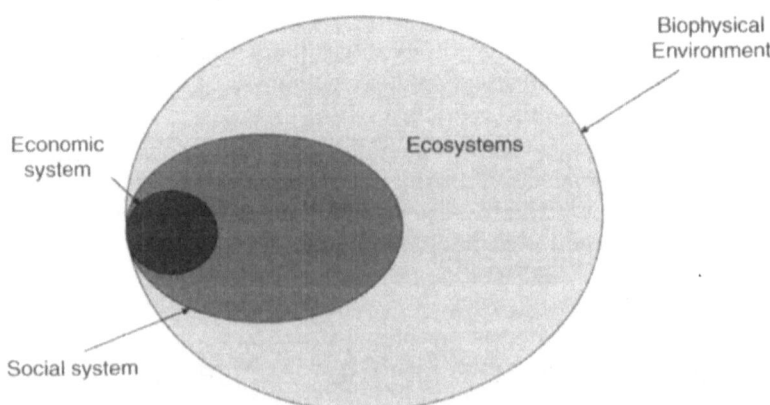

Figure 16.1 *Essential relationships of sustainability*

Source: Kranz et al. (2004).

well-being. It illustrates that humans are an integral part of nature and the water resources it contains, not apart from them.

One of water's most profound connections is to land. What we do on the land, whether for economic development, energy production, food and fiber production, housing, or transportation, affects the demands we place on water, the quality of water, the amount of water that is stored and the amount that runs off the land, and the timing of all the above. What we do on the land affects whether we have too much water or too little, and whether it is as clean as we want and need it to be. Further, because individual actions on the land may add up to large effects over space and time, the idea we call "cumulative impacts"—that many small actions can have collectively large short- or long-term consequences, or both—is an important element of the interconnectedness principle. Management that ignores these relationships and this interconnectedness is at risk of having to revisit false or incomplete solutions.

Principle #2: Informed Decisionmaking

The second principle of sustainable water management is informed decision-making. Nothing else can substitute for a clear understanding of water resources and their connection to human activities.

In making water allocation decisions, for example, an understanding of normal- and drought-period water budgets, and how these budgets might be altered by climate change (as discussed in Chapter 12), is important. (A water budget details the quality, quantity, and timing of water coming into a water unit, such as an aquifer or watershed; what happens to it as it moves within and through the unit; and its condition, timing, and quantity when and where it leaves the unit.) Without such knowledge, a planner could not fathom the implications of a community plan for water and of water for that community plan, and a manager could only guess about the consequences of a permit decision. The same may be said for understanding the implications of a variety of other state actions on water resources, such as the subsidy for an ethanol production facility or the expansion of a wastewater treatment plant. Ignorance of these relationships may hamper understanding of the choices people make and blunt efforts to safeguard water resources for the future.

Achieving such understanding requires collection and interpretation of the data needed to describe the water system. It requires use of this information to determine relationships and system responses to natural and human forces. Figure 16.2 depicts the natural ecosystem processes and social and economic drivers that informed decisionmaking must illuminate. It lists the ways in which ecosystems respond to the demands people place on them. Over time, this interaction produces the goods and services we seek as a society, as well as changes, both good and bad, to the natural and social capital that provide them.

Informed decisionmaking also requires concerted efforts to understand the views of affected interests, including those of local governments and citizens. If we understand that everything is connected, it follows that we must engage the people who have something to do with those connected parts and use information and

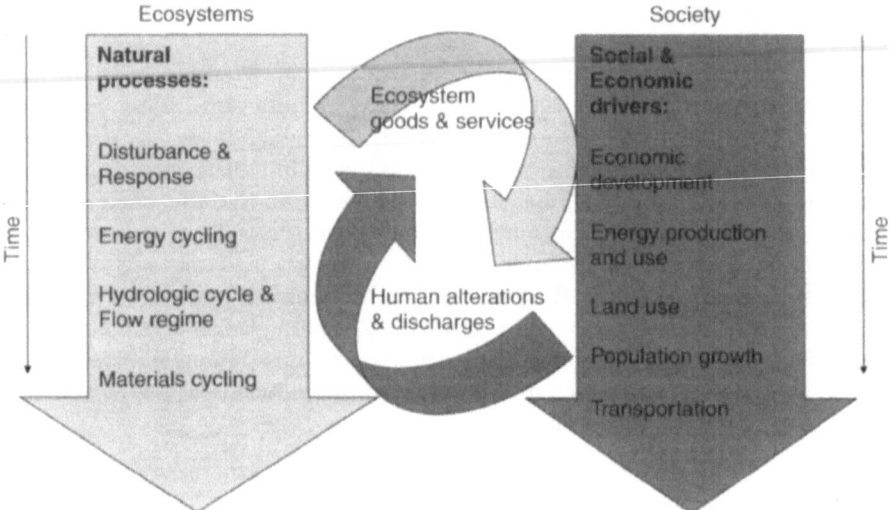

Figure 16.2 *Ecosystem processes and drivers*

Source: WICP ACWI (2005).

science to understand just how they interact, as well as how their interest might affect the resource.

Given the complexity of the natural environment and the ways people may affect it, scientists, planners, managers, and citizens need ready and reliable access to information. This may be information in its raw form or in a form that helps them understand the implications of the choices people make. Information systems must store, maintain, and manage data, but they also must help people understand the resource and the likely future consequences of decisions that affect it.

Four steps of information delivery—the collection, management, analysis, and dissemination of information—each require attention to ensure that the information informs decisionmaking in support of sustainable water management. When these efforts fall short, such as when a budget cut necessitates that a stream gauge or a monitoring well be dropped from the state's network, our understanding of the resource and ability to assess long-term trends or determine aquifer capacity suffers.

Informed decisionmaking is an important principle because the pursuit of sustainable water management cannot succeed without it, or without the understanding that water managers, water users, and water interests gain from it. However, society moves on in the face of incomplete information, which is why people and managers also need to take precautions.

Principle #3: Precaution

The third principle of sustainable water management is the precautionary principle. This principle argues that we should not take steps today that may put our resources, and therefore our future, at risk tomorrow. It argues that we must take precautions in the face of scientific uncertainty, or lack of information and understanding, to ensure that we do not make choices that may have significant, irreversible consequences. The legislature adopted this concept in the context of the Great Lakes Compact: "The current lack of full scientific certainty should not be used as a reason for postponing measures to protect the basin ecosystem."[4]

Society's widespread reliance on synthetic chemicals illustrates the importance and challenge of the precautionary principle. It is hard to make easy-to-reverse choices when considering the use of 90,000-plus chemicals society has already introduced into the environment and the 2,000 to 3,000 additional chemicals it introduces each year. We know enough to regulate something on the order of 100 to 200 chemicals today, and although most of the others may be benign and serve good purposes, we do far too little to understand unintended consequences of the potentially problematic ones or their metabolites. Whether or not emerging "green" chemistry approaches may enable us to change this equation in the future, it is clear that society will never get ahead of the curve by a reactive regulatory approach alone.

In addition to caution in the face of uncertainty, the precautionary principle suggests a commitment to respecting limits to growth, preventing the emergence of problems, and addressing existing problems in a sustainable way. The legislature made these commitments in the Ground Water Protection Act of 1989, establishing a series of protection strategies and declaring in its purpose statement the precautionary state goal that "ground water be maintained in its natural condition, free from any degradation caused by human activities."[5]

People also can prevent problems without fully understanding system limits simply by avoiding wasteful or inefficient uses of the water resource. Ultimately, such uses waste money and opportunity, as well as the resource. The legislature recognized the wisdom of this with the adoption of a conservation rate structure requirement for public water suppliers in 2008.[6]

Solutions that merely shift problems downstream or defer them for future resolution again may waste money or opportunity and risk problem escalation. A management framework that only reacts to issues and problems is unlikely ever to be sustainable. The legislature has recognized that preventing future water quality impairments is at least as important as correcting existing ones.[7] So, apparently, have Minnesotans. In a recent informal public survey conducted by the University of Minnesota (CSAC 2011), respondents opined that Clean Water Legacy moneys should be spent in equal portions on preventing waters from becoming impaired and cleaning them up once impaired.

Taking precautions to make certain that Minnesotans understand their impact on water resources will help them live within their means and avoid those choices that put the state's environment and communities at risk. To do this, policymakers need to do a better job of making information about water resources and their importance available to the school systems and the general public.

Principle #4: Transparency

The fourth principle of sustainable water management is transparency. The transparency principle asks that managers make decisions in full public view and in a way that is easily understood. This is a guiding precept for water because, as the legislature recently put it, "the quality of life of every Minnesotan depends on water."[8] Given the importance of water in their lives, people need to see what decisionmakers are doing and why, in order to understand where their values and priorities fit with those of the decisionmakers, and if they do not fit, why. When people are well informed, they are in a better position to do their part even if it means changing course. They also are more likely to support expenditures of money and resources necessary to make informed decisions.

The legislature has recognized the importance of this principle in any number of mandates, but its 1991 adoption and 2009 amendment of the state's overall strategic and long-range planning law illustrates the importance state government places on public engagement. Not only is the state's planning officer directed to seek public advice about the future, but he or she must also "stimulate public interest and participation in the future of the state."[9]

Transparency is especially important in today's environment because so many of the individual choices that each citizen makes every day contribute to Minnesota's water-related problems. Whether it is a detergent used to wash dishes, grass clippings that wash into the storm sewer, or the thermostat being set too high or low, Minnesotans become part of a complex problem—but potentially the best hope for a realistic solution. A transparent management framework helps people see how they may contribute to a problem and how they may help.

Principle #5: Accountability

The fifth principle of sustainable water management is accountability. The accountability principle requires that people and institutions take responsibility for making sustainable water management decisions. This means taking responsibility for ensuring that each of the principles is addressed in a unified and economical manner, because ultimately, neither accountability nor sustainability can occur without unified, informed, cautious, open, and efficient management. It also means being held responsible for achieving sustainable outcomes. Thus, the accountability principle requires routine mechanisms for checking whether people and institutions adhere to these provisions and routine evaluation of how well they do so. By taking such steps, people and institutions will be in a position to understand when and how they may need to make adjustments to ensure sustainable outcomes.

Although follow-through by decisionmakers may sometimes be called into question, recognition of the need to hold agencies and programs accountable to broad state goals has been embraced by both the legislative and executive branches of government in Minnesota for some time. In 2009, the legislature established a number of provisions calling for the evaluation of programs and projects associated with expenditures of funds under the new Clean Water, Wildlife and Legacy

Amendment. The legislature established major planning processes for each fund to ensure that moneys were spent in a deliberate, goal-oriented manner.[10]

Finally, for accountability to make a difference, it must lead to regular course changes where inspection shows that other approaches can be taken to better effect. The legislature set up just such a process when it required the evaluation of wetland restorations completed under the Clean Water Fund, with reports back to key legislative committees.[11] When embraced in this manner, the principle of accountability ensures progress toward sustainability and allows its verification.

UNDERSTANDING CONTEXT

The principles establish a series of tests to guide practitioners in the pursuit of sustainable water management. How well they are applied today will strongly influence the challenges future Minnesotans face. Having consistent ground rules for how people think about, and use in responding to, water-related issues is important. Yet it is also important to understand why decisions often wander from what policymakers might choose to do in the abstract. To illustrate this, three case studies in the management of water quantity, water quality, and land use are examined.

Water Quantity Management

In 2008, the Environmental Quality Board (EQB) and Freshwater Society issued independent reports calling attention to the lack of forward-looking approaches to the allocation of Minnesota's water resources. The EQB found that "the state does not collect or process sufficient water-related information to know with certainty overall whether it is managing water resources sustainably." Although the board found that Minnesota law governing the allocation of water resources is "comprehensive and thorough," it also pointed out that for a variety of reasons, the state applies this body of law more effectively *in response to* applications for water use (i.e., reactively) than it does in a proactive way. The board noted that "the state has only recently begun to consider whether its water supplies are sufficient to meet the long range seasonal requirements of communities, businesses and ecosystems" (EQB 2008).

The Freshwater Society recognized the department's tendency to focus on individual appropriation requests as they come in when it concluded that "the Minnesota Department of Natural Resources issues pumping permits for wells on a case-by-case basis. The agency does not deny permits based on the anticipated cumulative impact of each new well it approves, and the agency lacks authority to restrict development where groundwater is scarce" (Freshwater Society 2008).

Still, the need to shift the focus to what may be sustainable is becoming widely accepted. In a 2005 report to the legislature, the Minnesota Department of Natural Resources (DNR) noted: "Consideration of the whole hydrologic system requires analysis and evaluation of each proposed water withdrawal in the context of that whole. To minimize impacts and plan for sustainability, it is not sufficient to treat each well or proposed withdrawal as separate from other past, existing, and proposed uses of the regional supply" (DNR 2005).

The Granite Falls Energy case study (see Box 16.1) shows the importance of following the principles of sustainable water management in a water quantity context. It also provides a good reminder that water issues do not occur in a vacuum, but interconnect with other sectors, in this case the energy, transportation, and agricultural sectors. State economic development policy played a role through Minnesota's Job Opportunity Building Zone program, which provided the company tax breaks to locate in the area. Additionally, the case study shows an innovative state effort to grapple with the allocation of water from a confined, uncertain, and limited aquifer supply source.

Box 16.1. *Granite Falls Energy*

Jim Sehl, Minnesota Department of Natural Resources.

Granite Falls Energy appropriates groundwater for the production of ethanol. Because of concerns regarding the ability of the aquifer to support the withdrawal, in June 2005 the Minnesota Department of Natural Resources (DNR) issued a limited water appropriation permit for the project until the plant could find an alternative source of water. The water appropriation permit set thresholds on the aquifer in order to protect the system from overpumping. The threshold language included two action levels. The first action occurs when the amount of water above the top of the confined aquifer is at or below 50% of the prepumping static water level. The permittee is notified that water levels are declining at a rate that may not be sustainable, that actions must be taken to reduce water consumption at the plant, and that exploration for alternative sources of water must commence. The second action occurs when the available head above the confined aquifer is at or below 25% of the prepumping water level. If the water level drops to or below the 25% threshold, the permittee is notified, and the permit is suspended until water levels have recovered.

Production at the Granite Falls Energy plant began in late 2005. As anticipated, water levels in the source aquifer declined at a steady rate (see Figure 16.3); however, water levels in the aquifer did not reach the 50% threshold, because leakage from an overlying aquifer provided sufficient water to the source aquifer to halt continued declines. Unfortunately, declining water levels in the overlying formation and well interference problems with surrounding domestic wells began to occur. Because of the significant drop in water levels in the source aquifer and a prediction that water levels would eventually slip below the 50% and potentially the 25%

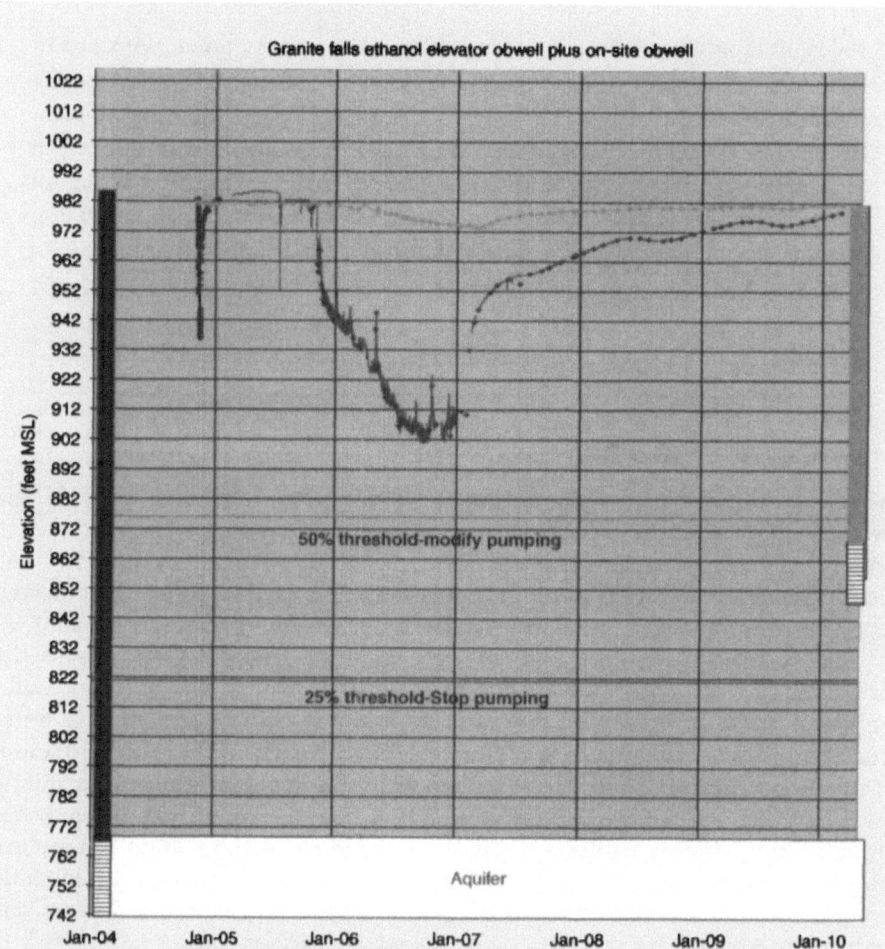

Figure 16.3 *Multiple aquifer response to pumping by granite falls energy*

threshold, the DNR recommended that Granite Falls Energy accelerate the process of installing a water intake system on the Minnesota River. The plant stopped appropriating groundwater in early 2007 and is currently using surface water from the river to supply water needs for the facility. Future expansion of the Granite Falls Energy plant is an ongoing topic of discussion. The DNR is concerned about the use of surface water as the sole source of water for the facility and is working with Granite Falls Energy to develop a long-term plan for use of the aquifer to supplement the use of river water.

Figure 16.3 is a graphic depiction of water levels in the aquifer system at Granite Falls Energy. The black lines are water level readings for the source aquifer used by the plant. The black dots are hand readings. The column on the left side depicts the well used to monitor the source aquifer, and the column on the right represents the well used to monitor the upper aquifer, where most domestic wells obtain water. The light gray line on the graph represents water levels in the upper aquifer. As water levels in the aquifer used by the plant decline, the water level in the upper aquifer also begins to decline, resulting in well interference problems. Note that water levels in the source aquifer did not reach the 50% threshold, which is represented by the horizontal line. Water levels began to rebound immediately after the plant stopped pumping, as did water levels in the upper aquifer system. However, they have not completely recovered, and there is concern that use of this aquifer as a long-term source of water, even at a lower rate and volume, may not be sustainable.

Water Quality Management

After a yearlong study, the Citizens League's Water Policy Task Force reached some eye-opening conclusions about the state's efforts in managing water quality: "In Minnesota's current water governance system, government entities bear the lion's share of the responsibility to assure the public has access to clean water. This system is not effectively protecting and improving the state's waters" (Citizens League 2009).

The perfluorinated chemicals (PFCs) case study (Box 16.2) highlights the principles of sustainable water management in the context of a complex water pollution problem. The study illustrates the intimate relationships among water quality and water quantity, surface water and groundwater, and land use and water. It also shows interconnections that result in the solution of one problem expanding and complicating others. It is a story about how people have treated the land over time, making choices without a basic understanding of the natural system and potential consequences. Although the day-to-day "business" of society frequently compels decisions to be made whether or not sufficient information exists, the results can be costly.

Box 16.2. *Perfluorinated Chemicals as Tracers of Groundwater–Surface Water Connectivity*

Virginia Yingling, Minnesota Department of Health.

A recent investigation of perfluorinated chemicals in the cities of Oakdale and Lake Elmo, Minnesota, illustrates the intimate connection between groundwater and surface water (MDH 2008). PFCs are a family of chemicals used, among other things, to create nonstick or stain-resistant coatings, such as Teflon® and Gore-Tex®. They are extremely persistent in the environment and highly mobile. One of the simpler PFCs, perfluorobutanoic acid, even has been used as a tracer in fractured bedrock studies (Flury and Wai 2003).

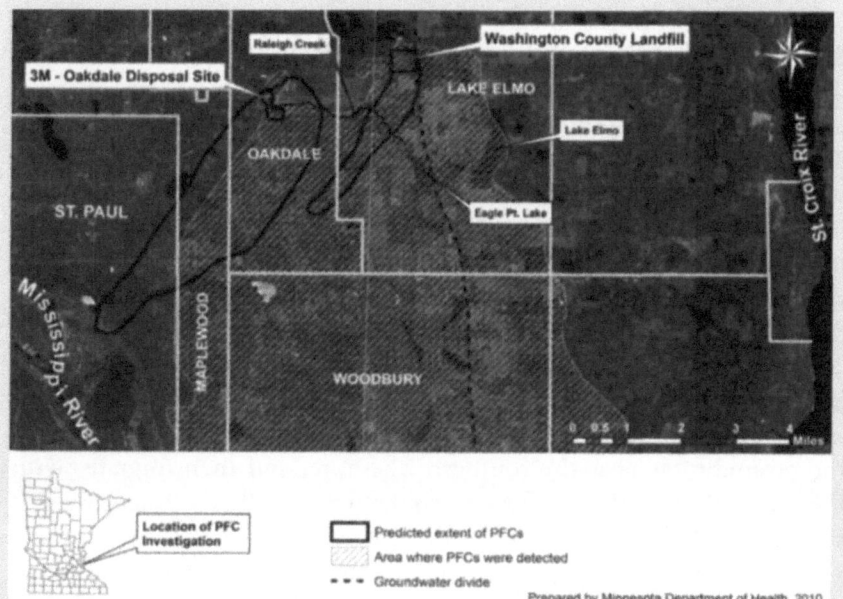

Figure 16.4 *Locations of PFC disposal sites and detection and nearby major features*

Source: Based on MDH (2010).

The 3M Corporation disposed of PFCs and other manufacturing wastes at two sites in Lake Elmo and Oakdale between the mid-1950s and mid-1970s (Figure 16.4). In 2004, perfluorooctanoic acid and perfluorooctane sulfonate were detected in the Oakdale city wells. Preliminary hydrogeologic modeling of potential contaminant movement predicted large, but otherwise typical, contaminant plumes emerging from the sites and following the regional groundwater flow to the southwest toward the Mississippi River. However, sampling of more than 450 wells indicated a much more widespread and complicated pattern of PFC transport, including the crossing of a groundwater divide.

For example, perfluorooctane sulfonate (PFOS) was detected in only a few monitoring wells at the Lake Elmo disposal site at concentrations generally less than 1 part per billion (ppb) and was not detected in any of the private wells immediately down-gradient of the site. However, PFOS was detected in many wells farther down-gradient at concentrations higher than those found in the landfill. It turned out that PFOS was present only in wells southwest of Raleigh Creek, which originates in wetlands at the Oakdale disposal site where PFOS was present at very high concentrations (over 3,000 ppb). Sampling of Raleigh Creek confirmed that PFCs, including PFOS, were discharging from the Oakdale disposal site into the creek. Near where Raleigh Creek enters the city of Lake Elmo, it becomes an intermittent or "losing" stream. This means that during most of the year, the water flowing in the creek infiltrates through the streambed into the groundwater and migrates with the regional groundwater flow to the southwest, contaminating hundreds of private wells in western Lake Elmo. PFC concentrations in wells immediately down-gradient of the creek are nearly identical to the concentrations measured in the creek, even though these wells were completed in bedrock at depths up to 180 feet.

During periods of the year when Raleigh Creek maintains some flow through its entire length, a portion of the PFCs it carries are discharged into Eagle Point Lake. From the lake, some of the PFC load reenters the groundwater near the southern lakeshore, and then migrates with the regional groundwater flow to contaminate wells south and west of the lake. Additionally, some of the PFC load migrates through the wetlands along the eastern shore of the lake, entering Lake Elmo and then reentering the groundwater near the eastern lakeshore, resulting in trace-level contamination of wells near the lake. It is this last leg of the PFC journey, from Eagle Point Lake to Lake Elmo, that allows the PFCs to migrate across the groundwater divide.

The PFC investigations in the Lake Elmo–Oakdale area demonstrate the ability of contaminants to migrate into areas entirely unanticipated if groundwater and surface water were independent systems. In fact, groundwater and surface water in the study area are simply different expressions of a single, complex hydrologic system.

Land Use Management

The state has long recognized the importance of land use to water management. The legislature adopted shoreland, floodplain, wild and scenic rivers, and critical area protection laws in the early 1970s. The Environmental Quality Board and its water policy predecessor, the Water Planning Board, identified land use as a key concern and the need to integrate land use and water management as a key principle in each of the state's three decadal water plans (MWPB 1979; EQB 1991, 2000).

The Mississippi Headwaters case study (Box 16.3) highlights the connections between land use and water management. The Mississippi Headwaters Board experience is a study of politics, local responsibility, and the use of land along a prized natural resource. Citizens with an independent streak warded off federal management of the Mississippi headwaters with the argument that they could do the job better at the local level. In this case, "local" came to mean a regional board composed of representatives from eight county governments. However, the land

Box 16.3. *Local Protection for the Headwaters of the Mississippi River: A Case Study in the Evolution of River Protection*

Molly McGregor, former administrator, Mississippi Headwaters Board.

It made sense to add the headwaters of the Mississippi River to the National Wild and Scenic Rivers System in 1975. The act was intended to protect the nation's outstanding natural, cultural, scenic, scientific, and recreational river values, and this stretch of the Mississippi was picturesque, relatively undeveloped, and highlighted by a popular state park.

Following a recommendation to include the Mississippi headwaters in the national rivers system, the Department of Interior launched its study of the river. The study coincided with a political battle within Minnesota about a federal environmental initiative to expand the Boundary Waters Wilderness Area. Local opposition to that proposal had deep political repercussions in the state, undermining efforts to establish federal protection for the Mississippi and spawning the idea of a local protection alternative.

Within two years, leaders from northern Minnesota and the Department of Natural Resources agreed on an organization for managing the river based on the state's Wild and Scenic River Act, which uses local zoning to achieve a common management approach. The proposal called for each member county to appoint a commissioner to a board responsible for certifying zoning actions of the member counties, planning for the lands under its jurisdiction, and protecting the natural, cultural, scenic, scientific, and recreational values of the river's first 400 miles. Finalized in 1980, the proposal was endorsed by the National Park Service, enacted into Minnesota law in 1982, and supported by matching state and county funds. The eight counties signed on, and the Mississippi Headwaters Board was established.

Using zoning to protect the river was a new and untested approach, and within 10 years of the board's founding, most of the undeveloped parcels within the river's protected corridor had been platted into the minimum lot size allowed by the headwaters law. At 5 acres each, the lots were larger than typical in Minnesota shoreland zones. More troubling, however, was the lack of rationale for the selection of 5 acres as an optimum lot size for development that would not impair the river or diminish its natural, cultural, scenic, scientific, and recreational values. Indeed, although words to this effect are found in each of the major laws guiding river protection—the National Wild and Scenic Rivers Act, the Minnesota Wild and Scenic Rivers Act, and the law establishing the

Mississippi Headwaters Board—an operational definition of these values was never developed or applied. Defining these values could establish an operational basis for river protection by creating measures to evaluate activities in the corridor.

In 1988, the Mississippi Headwaters Board initiated a volunteer water quality monitoring program to measure the health of the river as a basis for demonstrating progress in river protection. The program was recognized for its educational benefits but also represented an expansion of the concept of river protection into water quality, which typically had not been considered by the recreation-oriented strategies of the 1960s and 1970s. However, again, no functional tie was made between the results of monitoring and success of the program—and zoning—in protecting river water quality.

Water quality and the complexity of ecological linkages have added a new dimension to river protection since the 1970s. Yet existing programs tend to treat these activities as merely educational or informational. Instead, these concepts should empower local boards to add science to decisionmaking and require developers and others to demonstrate how proposals will affect runoff and transport of nutrients to rivers, as well as how the rivers' ecosystem services can be retained in the face of development.

use management tool of choice of local government, the zoning ordinance, served as a blunt instrument that had less than the desired effect in preserving the river's natural, cultural, scenic, scientific, and recreational values.

CONCLUSIONS

The case studies offer a retrospective, real-world test of the principles we suggest. The principle of interconnectedness is illustrated by each but perhaps most graphically by the perfluorinated chemicals case study, in which land and water, surface water and groundwater, water quality and quantity connections each had a profound effect. The principle of informed decisionmaking was well illustrated by the Granite Falls Energy case, in which managers approved the appropriation of groundwater despite an uncertain supply. They did so, however, with deliberate precaution, illustrating application of the third principle by instituting an early warning system to trigger the use of alternative water sources when availability of groundwater proved insufficient.

The Mississippi Headwaters case illustrates the principles of transparency and accountability, because decisions were made in the open at public meetings of county boards and the Headwaters Board. But one also might ask if the process was sufficiently transparent for people to get the message, see their part, and understand the behaviors they needed to change. Further, it is hard to argue that accountability was sufficient in a process that never examined whether the actions taken had any effect on the program's stated goals.

Because Minnesota will be home to well over a million new residents in the coming decades, and because our relationship with water can only get more complicated and complex, those who must care for the state's water resources should benefit from a set of guiding precepts. And because it is impossible to know exactly what the future may bring, the principles of interconnectedness, informed decisionmaking, precaution, transparency, and accountability should provide the compass Minnesota requires to navigate its waters of the future.

NOTES

1. The percentage of lakes tested includes those monitored by remote sensing for aquatic recreation suitability.
2. Laws of Minn. 2009, Ch. 172, Art. 3.
3. Laws of Minn. 2010, Ch. 361, § 55.
4. Minn. Stat. § 103G.801.
5. Minn. Stat. § 103H.001.
6. Minn. Stat. § 103G.291.
7. See, e.g., Laws of Minn. 2009, Ch. 172, Art. 2, § 4.
8. Laws of Minn. 2010, Ch. 361, Art. 2, § 48.
9. Minn. Stat. § 4A.01.
10. See, e.g., Laws of Minn. 2009, Ch. 172, Art. 3, § 30.
11. Minn. Stat. 2009 Supp. § 114D.50, subd. 6.

REFERENCES

Citizens League. 2009. To the Source: Moving Minnesota's Water Governance Upstream. Report of the Citizens League Water Policy Study Committee. www.citizensleague.org/publications/reports/482.RPT.To%20the%20Source.pdf (accessed December 17, 2010).

CSAC (Citizen Stakeholder Advisory Committee). 2011. Report on Activities: Minnesota Water Sustainability Framework. www.wrc.umn.edu/prod/groups/cfans/@pub/@cfans/@wrc/documents/asset/cfans_asset_290474.pdf (accessed January 14, 2011).

DNR (Minnesota Department of Natural Resources). 2005. *Options for Sustainable Management of Ground Water.* St. Paul: DNR, Division of Waters.

———. 2010. Minnesota Facts & Figures: Lakes, Rivers and Wetland Facts. www.dnr.state.mn.us/faq/mnfacts/water.html (accessed December 17, 2010).

EQB (Environmental Quality Board). 1991. *Minnesota Water Plan* www.eqb.state.mn.us/documents/MinnesotaWaterPlan1991.pdf (accessed December 17, 2010).

———. 2000. Minnesota Watermarks: Gauging the Flow of Progress, 2000–2010. www.gda.state.mn.us/pdf/2000/eqb/wtr_mrk.pdf (accessed December 17, 2010).

———. 2008. Managing for Water Sustainability: Report of the EQB Water Availability Project. www.eqb.state.mn.us/documents/Managing_for_Water_Sustainability_12-08.pdf (accessed December 17, 2010).

Flury, M., and N.N. Wai. 2003. Dyes as Tracers for Vadose Zone Hydrology. *Review of Geophysics* 41:1002–1005.

Freshwater Society Guardianship Council. 2008. *Water Is Life: Protecting a Critical Resource for Future Generations.* Report to the Freshwater Society Board by the Freshwater Society Guardianship Council. http://freshwater.org/images/stories/PDFs/publications/Water-is-Life-Report.pdf (accessed December 17, 2010).

GWP (Global Water Partnership). 2000. *Integrated Water Resources Management.* Technical Advisory Committee (TAC) Background Papers No. 4. Stockholm, Sweden: GWP.

Kranz, R., S. Gasteyer, H.R. Heintz Jr., R. Shafer, and A. Steinman. 2004. Conceptual Foundations for the Sustainable Water Resources Roundtable. *Water Resources Update* 127:11–19.

MDH (Minnesota Department of Health). 2008. *Public Health Assessment on Perfluorochemical Contamination in Lake Elmo and Oakdale, Washington County, Minnesota.* St. Paul: MDH.

———. 2010. Perfluorochemicals in Minnesota. www.health.state.mn.us/divs/eh/hazardous/topics/pfcs/index.html (accessed January 17, 2011).

MWPB (Minnesota Water Planning Board). 1979. *Toward Efficient Allocation and Management: A Strategy to Preserve and Protect Water and Related Land Resources.* St. Paul: MWPB.

WICP ACWI (Water Information Coordination Program, Advisory Committee on Water Information). 2005. Sustainable Water Resources Roundtable: Preliminary Report. http://acwi.gov/swrr/Rpt_Pubs/prelim_rpt/index.html (accessed December 17, 2010).

CHAPTER 17

Lessons and Opportunities for Engagement by Present and Future Policymakers

K. William Easter and Jim Perry

*I*n order to have significant effect, water policy must take a holistic approach that considers the physical, social, and economic conditions within watershed, aquifer, and river basin contexts (Chapter 16). One of the key challenges of such an approach is that political boundaries hardly ever closely match surface water and groundwater boundaries (Chapter 3). When designing water policy, this challenge must be addressed and procedures developed to overcome this mismatch in boundaries. Education can be an important input in trying to bridge this difference. However, education can be effective only if policymakers and the public assign a higher priority to water resource issues and come to recognize the highly interconnected nature of our water resources. What people and firms do in one part of a river basin, watershed, or aquifer will impact other parts. This is clearly illustrated by the PFC and Granite Falls case studies discussed in Chapter 16.

If Minnesota's water policy is to foster sustainable water use, management decisions should be based on five principles (Chapter 16): First, as argued above, management must recognize the *interconnectedness* of our water resources. Water is a nearly universal solvent that flows both above ground and below ground. Second, *informed decisions* are important and must be based on sound information. This means that the state needs to collect the necessary data and fund the research needed to provide information for decisionmakers. Third, the *precautionary principle* applies when making major water resource decisions, because we never have enough information to be completely certain what future impacts will be. Fourth, decisionmaking needs to be as *transparent* as possible. If people fully understand why a decision or policy has been made, it is more likely that they will support the policy or decision. Finally, water managers and policymakers need to be *accountable* so that mistakes or misunderstandings can be corrected. Also, if they are not accountable, they tend to take unnecessary risks.

How well has Minnesota done in addressing its water resources using an integrated approach that is based on sound information? How well does current water policy meet Minnesota's future needs, and to what extent have precaution, transparency, and accountability guided these policies? Positive responses to these questions are critical as Minnesota addresses its growing demands for clean water and energy, and an expanding urban population. This it must do while trying to be a good headwaters steward.

STRENGTHS OF MINNESOTA'S WATER POLICY

Several policy initiatives are helping Minnesotans make more informed decisions. One is the sentinel lakes program called Sustaining Lakes in a Changing Environment (SLICE), which uses key lakes throughout the state to monitor the health of aquatic ecosystems over time (Chapter 3). This will allow the state to detect lake water quality problems at an earlier stage than has been possible in the past. In the case of groundwater, new insights from research and data collection have provided policymakers with a much better understanding of the complexity and heterogeneity of the state's groundwater. This has resulted in improved groundwater policy, such as restrictions on the use of the deep Mount Simon Aquifer under the Twin Cities (Chapter 8).

An example of water policy that has been transparent, precautionary, and accountable has been the Wetlands Conservation Act (WCA, Chapter 7). The WCA is a comprehensive and effective model of grassroots and interagency cooperation. Minnesota policy over the past 100 years has gone from draining its wetlands to protecting and restoring them, representing one of the major water policy changes in the state during the 20th century. The recent agreement and cooperation between the U.S. Army Corps of Engineers (USACE) and the state of Minnesota regarding dredging and placement of dredge material in navigable waters is another positive example (Chapter 10). This has been particularly important for navigation and other uses of the Mississippi River and Lake Superior. The new policy clearly takes into account environmental problems caused by dredging and disposal of the dredged material. Environmental issues are now on a more equal footing with navigation.

Minnesota has been very active in establishing interstate and international water management boards, committees, and councils to help set policies and guidelines for the management of international and interstate waters (Chapter 2). It has also been active in establishing international agreements for the Great Lakes and its international rivers. In its management of these international water bodies, Minnesota has supported active participation by local leaders in the areas most closely tied to these water bodies.

Another strong point of the state's water policy has been its requirement that individuals or firms that want to directly extract water from any groundwater or surface water source must obtain a permit from the Minnesota Department of Natural Resources (DNR). To obtain the permit, applicants must show that their proposed use will not damage existing ones. This has been an important policy for preventing the overuse of water, particularly groundwater. However, this does cause significant enforcement challenges for the DNR and may become a larger issue in the future (Chapter 16).

Minnesota has made important strides in its flood-control efforts. It has considerable local watershed involvement with state and federal agencies. An important aspect of these flood-control efforts is the enhanced communication and understanding of flood risk management. Communication during both the forecasting and flooding stages is key. An active coalition of local groups, and state and federal agencies is working together in Minnesota to more effectively mitigate and adapt to flooding events (Chapter 14).

Finally, the largest and most striking success is that Minnesota recently passed the Clean Water, Land, and Legacy Amendment. The Amendment was passed by Minnesota voters in 2008 and provides funding for at least 30 years to improve water quality throughout the state. Sales taxes raised through the Amendment provide a dedicated source of funding. This clearly shows the strong commitment Minnesota voters have to their water resources. However, this political commitment raises the danger of false expectations, a sense among voters that this will solve all of Minnesota's water problems. As discussed below, funding of projects is just one step in Minnesota's efforts to maintain and improve its water resources. Additional, ongoing funding will be needed to support basic management of the state's water resources.

GAPS AND INCONSISTENCIES IN THE STATE'S WATER POLICY

Although Minnesota's water policymakers have shown innovation in response to the changing demands placed on the state's water resources, have these changes been enough? New demands mean that existing water policy will need to change to fill gaps and reduce inconsistencies, unnecessary overlaps, and complexities. Yet several of Minnesota's water problems cannot be solved by changes in state water policy alone. For example, probably the key water quality problem facing Minnesota is nonpoint-source water pollution. Both federal water and agricultural policies will need to undergo significant change before the state can achieve the needed reductions in nonpoint-source pollution. Current federal and state water quality efforts have had little impact on agricultural practices, as shown by the relatively constant level of nonpoint-source pollution over the past 20 years (Chapter 11).

Another issue that goes beyond policy is funding. Funding at both state and federal levels is inadequate to meet the growing demands for urban water infrastructure (Chapter 9). This is a growing problem as urban populations increase and old water systems deteriorate. In addition, funding is needed to make Minnesota's invasive species control efforts effective (Chapter 13). Funding for international and interstate entities also is critical for these entities to be effective in helping manage Minnesota's international and interstate waters (Chapter 2).

Adoption of improved methods for economic evaluation of water infrastructure investments should complement any additional funding (Chapters 1 and 9). If the state does obtain additional funds, it needs to make sure they are used effectively.

Probably one of the more difficult policy questions that arise from Minnesota's current water management structure is who is in charge. No single state agency has authority and responsibility to ensure that water use and allocation decisions internalize all the impacts they have on water quality and quantity throughout a river basin. This was not so important in the past, when populations were mostly rural and impacts on water relatively small. Today the situation has changed dramatically, and Minnesota now has to fully consider downstream impacts. It is critical that state lawmakers take action to ensure coordinated leadership in the management of water resources. Many people and experiences in other settings suggest that a single-agency approach is neither practical nor desirable. Having all controls under the discretion of a single agency puts the fox in charge of the henhouse. However, a legislatively mandated coordinating council with specific powers and reporting responsibilities would help agencies better coordinate their activities and ensure that priority issues are addressed effectively.

As part of the effort to improve coordination, transparency, and accountability, changes need to be made in the state's water law. In any such change, it is important to reduce the complexity and overlaps among the water laws and state regulations. Current water law has been put in place without much effort to make sure new additions to the law do not overlap or create inconsistencies with existing laws (Chapter 5).

Another area of concern is the failure of local governments to develop or enforce effective floodplain zoning, as well as zoning of property bordering scenic areas and lakes. Local government may zone the floodplain for no development, but when a developer proposes a variance, it is usually granted with little consideration of the broader picture (Chapter 14). This makes it difficult to prevent the cost of flood control from growing rapidly, as more people move onto the floodplain. A similar problem with development is that local zoning variances are frequently granted allowing builders of large houses to ignore setback requirements and locate their house right on the lakeshore. Yet another issue involves within-lake activities (e.g., dock size), some of which are regulated by local lake associations and others by the DNR, and enforcement problems arise for both regulating authorities. Complicating the problem is the political nature of water management: in 2010, Governor Pawlenty rejected two sets of proposed DNR rules that have been in the making for years (Meersman 2010). Further complicating the issue is that the DNR does not have adequate staff to fully enforce rules on Minnesota's many lakes.

Adjustments or changes in water policy may be necessary to make policies more effective in providing high-quality water resources. For example, the law needs to be changed so that it is much easier for the DNR to deny a water permit request. For denial, the agency should have to show only that there is a reasonable probability that the new permit would compete with or damage existing water uses, including environmental uses. Another example involves disposal of dredge material from Lake Superior. The state law may need to be changed so that the state is allowed more flexibility in working with the USACE on identifying viable disposal sites. Currently, the state and the USACE are in conflict over dumping dredge material in deep holes in Lake Superior (Chapter 10).

If flood information and its communication are to improve, two changes are necessary. First, as climate changes, altering rainfall patterns and intensities, it is important to develop better methods for forecasting flooding events. Policymakers can no longer depend on historical streamflows, because the future is likely to be quite different from the past (Chapters 12 and 14). Second, better ways are needed to describe flood risks to property owners near rivers, streams, and lakes. Talking about the risk of a 100-year flood does not give property owners a good sense of the actual risk. Rather than report that an area may be subject to a 100-year flood, it would be clearer to say that in the next 30 years, there is a 26% chance that the location will be flooded (Chapter 14).

Finally, as Minnesota faces growing water demands and climate changes increase drought periods while reducing groundwater recharge, it is time to consider more effective water pricing policies, as recommended by the new 2010 law on conservation rate pricing. The state needs water-pricing guidelines for local governments. These guidelines should provide examples of pricing strategies that can reduce water use and pollution. This could include increasing block pricing and seasonal or conservation pricing (Chapter 1), as recently begun in the Twin Cities. These pricing methods could be used for water supply and wastewater discharges, with significant charges for both. Firms and households need economic incentives to make better use of their water resources.

FUTURE CONCERNS AND FUTURE ACTIONS

Three factors will require Minnesota to give water management and policy an even higher priority. First is the increasing population and demand for high-quality water, particularly by the growing urban areas. Second, climate change will make water management and policy even more difficult. The state is likely to have more frequent and longer droughts and floods but less chance for groundwater recharge, and therefore less water for drinking, industry, and agriculture (Chapter 12). Third, as the demand for energy grows, so will the demand for water. Minnesota may well have to completely revamp its view as a water-abundant state. Can the state continue to have cooling for electrical power production as its major water use? Can it continue to promote the production of biofuels, which uses up to 19 gallons of water for every gallon of fuel produced?

With expected climate change, cold-water trout streams will be at risk as the temperatures rise. This could be a big issue for sportfishing, tourism, and ecological services. Minnesota may also have to strengthen its wetlands policy as the potential for flooding increases and more areas in the landscape are needed for water storage. To meet these challenges, water policy must be adaptive, flexible, and supported by extensive research.

A number of future actions have been suggested in the book. Some of these recommendations, such as the following, will have a comprehensive impact on water policy:

- coalesce and coordinate Minnesota water policy by forming a legislatively mandated, powerful coordinating committee that has specific reporting requirements, ensuring transparency in leadership for all state water resource programs;
- develop more integrated state water policy and management that clearly take into account the interconnectedness of water resources, both surface water and groundwater;
- to facilitate this change, revise, streamline, and rationalize state water law to eliminate redundancies and unnecessary complexities, and fill gaps to meet the water demands of the 21st century;
- develop active programs to foster watershed districts and their cooperation with lake associations where feasible;
- establish a state-level task force charged with developing guidelines for water pricing designed to help local governments conserve water; and
- invest in improving Minnesota's system of water resource monitoring and assessment, as the public cannot be expected to support a very expensive decisionmaking process using total maximum daily load (TMDL), which is based on knowledge of only about 20% of the resource and whose performance is inadequately assessed and focused on surface water.

Other recommendations are designed to deal with specific issues and include the following:

- provide funding for research and data collection to improve methods for predicting floods as climate changes;
- develop an adaptive strategy to deal with climate change and the pressure it places on water resources;
- identify and protect groundwater recharge areas to help offset the impact of large rainfall events, which will increase with climate change;
- develop buffer zones around these recharge areas, as well as around rivers and lakes where possible;
- develop a comprehensive reporting system for invasive species that includes the cost of their control; and
- develop better and more effective ways to manage and suppress invasive species.

Minnesota water policy is at a crossroads. The state has been on the leading edge of, and a model for, state-level water resource management. This is appropriate given that it is water rich, demographically diverse, and has a strong tradition of state government innovation. Its water resource leadership has resulted in high-quality waters that have a high value for Minnesotans and draw millions of visitors annually.

However, all of this may be at risk. Agency dilution has led to addressing surface water, groundwater, urban stormwater, and agricultural drainage as if they had discrete identities. Rapid changes that are currently occurring, such as in climate, urbanization, agricultural policy and practice, and the ways people value wetlands, are influencing the quality, quantity, and timing of water moving across and through the landscape. Yet Minnesotans have shown very strong support for transparent, accountable water decisionmaking, most notably in passage of the Clean Water, Land, and Legacy Amendment.

The clarion call is sounding for Minnesota's water. To honor the values Minnesotans express for the state's water resources, it is imperative to take a coordinated approach to state water management, and the coordinating unit must ensure development and implementation of integrated approaches to surface water and groundwater management. It must strive to look ahead using scenarios and other tools that allow it proactively to establish clear priorities, act on these priorities, and assess current actions and adapt future ones based on analyses of performance. Minnesota has the opportunity to again be among the top states in the nation in terms of its water leadership and policy, but achieving this requires strategic and thoughtful action, beginning with its institutional and organizational structure.

REFERENCES

Meersman, Tom. 2010. Governor Deep-Sixes Plan to Restrict Lake Properties. *Minneapolis–St. Paul Star Tribune*, August 14, B1 and B5.

Index

Note: *Page numbers in italics indicate figures and tables.*